# INTEGRATED PLANT PATHOLOGY

*Editors*

***S.M. REDDY***

Professor Emeritus
Kakatiya University, Warangal

***H.N. GOUR***

Maharana Pratap University of Agriculture
and Technology, Udaipur - 313 001 (India)

**SCIENTIFIC PUBLISHERS (INDIA)**

P.O. BOX 91 JODHPUR

*Published by:*
Scientific Publishers (India)
5-A, New Pali Road, P.O. Box 91
JODHPUR - 342 001 (India)
Tel.: +91-291-2433323
Fax: +91-291-2613449
E-mail: info@scientificpub.com
www.scientificpub.com



ISBN: 978-81-7233-560-1

**Integrated Plant Pathology**
*Eds. S.M. Reddy & H.N. Gour*
Reprint from Review of Plant Pathology,
ISSN: 0972-9712; Vol. 4 (2008) pp. 1-466

Cover Design: Tanay Sharma

Laser typeset : Rajesh Ojha

Printed in India

**Ann Rev** *Review of Plant Pathology*
*Volume 4 (2008)*

Indian Society of Mycology and Plant Pathology, Maharana Pratap University of Agriculture and Technology, Department of Plant Pathology, Udaipur - 313 001, Rajasthan, India

---

## EXECUTIVE COUNCIL FOR THE YEAR 2007-08



---

Ann Rev

*Review of Plant Pathology*
*Volume 4 (2008)*

# PREFACE

We had launched the Annual Review of Plant Pathology with great hopes and aspirations, aiming to provide perspectives and an opportunity to sift, analyse and synthesize the enormous information available as an integrated whole. That is what had been inked in the opening lines of preface of the 1st volume (2002). Since then, we have been striving to our best of efforts all through these years to synthesise, analyse and churn the subject in plant pathology discipline stocking all these in holistic manner to put as an integrated whole in the form of subsequent volumes of Annual Reviews of Plant Pathology. All previous editors have toiled hard for quality material to be given every time. It has really served as a green oasis in the desert land of literature in Asian countries particularly for all those teachers and students who were desirous of having an upto date rich subject matter yet at a very substantial quite affordable price. It is indeed sizzling to repeat again from the first preface that those trying "to get to the bottom" can look horizontally and see if there is something more rewarding, more interesting and more beautiful outside their confinement. Breadth of vision may trigger their depth of vision, or else, there is every chance of super specialists losing plant pathology and vice-versa. The review of plant pathology is an ocean, which can embound the rivers of diversified talents emerging from various directions offering an elixir of knowledge for all those who show their stark thirst to drink through this bank of ocean.

Dear readers, it has been quite enjoying times for the editors to receive the excellent coordinated cooperation from all the authors, who have contributed in this volume with their best of efforts. The efforts have been enlightening on the latest emerging topics such as integrated disease management, molecular basis of host-pathogen interaction, quorum sensing and mob attack, bio-control, molecular biology of viral pathogens etc.

The ISMPP executive committee has in its meetings held at Hyderabad resolved to rename the publication as "Review of Plant Pathology". The present volume has been over due since several constraints like sluggish communications and 'some unavoidable' factors coupled with our hunger for presenting quality texts by all angles to the readers resulting this over mature baby to see this twilight on this day. We really thank all the erudite authors for their help and the long forbearance, which they exhibited after submission of their articles, we apologise for this inordinate delay. Dr. Pankaj Sharma, Asstt. Professor, MPUAT, Udaipur deserves our special thanks for editorial assistance in several respects.

We invite your valued suggestions, guidance and critical evaluation of this work helping us to bring better volumes in future proving rather more informative.

Thanking you all.

**Editors**

**Ann Rev**

*Review of Plant Pathology*
*Volume 4 (2008)*

# CONTENTS

*Rev. Plant Pathol. Vol. 4, 2008*
*Indian Society of Mycology and Plant Pathology*
*Scientific Publishers (India), Jodhpur, pp. 1-28.*

# PHYTOPATHOGENIC FUNGI: FROM GENE-FOR-GENE TO GENOMICS

***A. Pain, A. K. Dhar* and C. Chattopadhyay*****

Senior Computer Biologist, The Pathogen Sequencing Unit, The Wellcome Trust Sanger Institute, Genome Campus, Hinxton, Cambridge CB 10 1SA, UK (ap2@sanger.ac.uk)
*Adjunct Associate Professor, San Diego State University, Biology Department, 5500 Campanile Drive, San Diego, California 92182, USA (adhar@sciences.sdsu.edu); Senior Scientist, Advanced Bionutrition Corporation, 6430 Dobbin Road, Columbia, Maryland 21045, USA (adhar@abn-corp.com)
**Senior Scientist (Plant Pathology), National Research Centre on Rapeseed-Mustard (ICAR), Sewar, Bharatpur 321303, India (email: chirantan_cha@yahoo.co.uk)

KEY WORDS: Fungi, plant, diseases, resistance, genetics, genomics

## ABSTRACT

Only a small group of fungal species is successful as plant pathogens through employment of a wide array of molecular mercenary to cause disease. Knowledge of molecular mechanisms governing developmental and physiological adaptations of pathogens that allow plant pathogenic fungi to invade plants, colonize tissues and control host defence reactions is limited. This review provides an overview of the genetics of host-pathogen interactions in plants from gene-for-gene theory to genomics approaches for exploring fungal pathogenesis. The utility of functional genomics approaches such as Expressed Sequence Tags (ESTs), microarray analysis, Serial Analysis of Gene Expression (SAGE), high-throughput transposon-mediated insertional mutagenesis, Green Fluorescent Protein (GFP)-based reporter systems, RNA interference (RNAi) and proteomics for elucidating the molecular mechanisms involved in fungal pathogenesis is discussed.

## INTRODUCTION

The ability to cause plant disease is a complex trait that, thankfully, occurs in only a small group of fungal and bacterial species. This small group of fungal and bacterial pathogens are the ones successful in overcoming the host defence and deploy a wide array of molecular techniques to cause disease. Our understanding of the molecular mechanisms governing the developmental and physiological adaptations of these pathogens that allow them to invade plants, colonize tissues and inhibit host defence reactions is limited. However, this situation is set to change as the genome sequencing of a large number of fungal and bacterial plant pathogens is at different stages of completion (8). Since the genome of *Haemophilus influenzae* was sequenced in 1995 (35), genomes of many prokaryotes have been completely sequenced (www.genomesonline.org). The comparative analysis of genomes of bacterial strains and species provided valuable insights into the evolution of virulence and host adaptations. It appears that there are several forms of co-evolution between hosts and pathogens, some involving reciprocal adaptation and others a combination of adaptations and speciation (118). Two of these forms of co-evolution have had a major effect on the evolution of interactions between pathogens and host plants. One is the evolution of defence and counter-defence in interactions between pathogens and host plant, which may have led to the radiation of species through a process of escape-and-radiation co-evolution (119). The other is the gene-for-gene co-evolution (36, 37, 38, 39).

## GENE-FOR-GENE INTERACTIONS IN THE EVOLUTION OF FUNGAL PATHOGENS

Much of the work on breeding for resistance to fungal pathogens in agricultural crops has assumed that virulence in fungal pathogens and resistance in plants are governed by gene-for-gene interactions (36). As different aspects of ecology of species of fungi and plants were incorporated into recent empirical and mathematical studies of gene-for-gene interactions, the original hypothesis has been expanded to accommodate a family of views.

**Table 1.** Expected compatibility between homozygous genotypes in a single-locus gene-for-gene interaction.

| Pathogen genotype | Host genotype | |
|---|---|---|
| | RR | rr |
| VV | Incompatible | Compatible |
| vv | Compatible | Compatible |

R is a dominant host gene conferring resistance to the pathogen and r is a recessive host gene conferring susceptibility. V is a dominant pathogen gene conferring avirulence and v is a recessive pathogen gene conferring virulence

The gene-for-gene hypothesis states that for each gene that conditions a reaction in the host there is a corresponding gene that conditions pathogenicity in the pathogen (i.e., counters the host defence gene). The linked genes of the host and pathogen are known as corresponding genes. Presently, the gene-for-gene hypothesis is interpreted to mean that for each gene determining resistance in the host there is a corresponding gene for avirulence in the pathogen with which it specifically interacts. According to this hypothesis, the occurrence of a resistant or incompatible reaction depends on both the presence of a gene for resistance (R) in the host and the corresponding dominant gene for avirulence (V) in the pathogen. If a host lacks a specific resistance gene, the corresponding avirulence gene cannot be detected in the pathogen and *vice-versa* (Table 1). In diploid host-pathogen associations, the interactions occurring at any particular corresponding gene pair or locus may be further complicated by the occurrence of the resistance or avirulence locus in a heterozygous state. This generates a further five genetic combinations (VvRR, VvRr, Vvrr, VVRr and vvRr). Since good health is the rule and disease the exception (resistance or R dominant to susceptibility or r, avirulence or V dominant to virulence or v), these combinations are generally phenotypically indistinguishable from other compatible or incompatible interactions. The specificity of these interactions, changing from compatible to incompatible as the same host is challenged by different pathogenic genotypes and *vice-versa* (Table 1), is the basic of gene-for-gene interaction. Often a gene-for-gene interaction is inferred from the specificity of a host-pathogen interaction that is coupled with a genetic basis of resistance. Only a few genetic analyses of specific host-pathogen interactions have been done involving only a handful of resistance and avirulence genes.

One of the practical implications of studying gene-for-gene interactions is development of strategies to counter virulence in some fungal pathogens with single genes for resistance in plants. This theory has served as a guiding principle in breeding crops for resistance to fungal pathogens. Since the recognition of the dominant, single-gene nature of resistance against some fungal pathogens, breeding in many crops has involved sequential release of new varieties carrying different resistance genes as a form of evolutionary arms race. This has resulted in microevolution as pathogens have responded to these selective forces by acquiring the corresponding genes for virulence. The best examples of these reciprocal responses are found among cereal rusts (*Puccinia*) and potato late blight (*Phytophthora infestans*). The sequential deployment of single-gene resistance has led to changes in particular multilocus virulence phenotypes (races) in the pathogen and/or appearance of totally new races. Gene-for-gene interactions involving analysis of a few genes for resistance and avirulence have been determined in some plant-pathogen associations (Table 2). Tight linkage groups occur among both resistance and virulence genes (19, 73).

Although breeding programmes and models of gene-for-gene interactions in agriculture often focus on how natural selection will favour increased virulence in local pathogen, studies suggest that the dynamics and evolution of these interactions is sometimes governed by gene flow over long distances. Use of genetic fingerprinting confirms that some epidemics in crops are caused by spread of one or two clones over large areas. Brown *et al.* (12) found that an epidemic of barley powdery mildew (*Blumeria graminis*) in 1980s in UK was initiated largely by two clones able to overcome barley cultivars carrying resistance allele *Mla13*. These results argue for a broader ecological and geographical view of the evolution of gene-for-gene interactions in some crops rather than a view based on minimising the rate of natural selection for virulence in local pathogen populations.

There are two approaches to genetic basis of host-pathogen interaction in natural populations: (i) comparisons of host family lines grown in natural situations, and (ii) detailed study of individual host x pathogen isolates and formal genetic analysis of resistance. The former method compares the total resistance phenotype of individuals or family lines against a genetically variable pathogen population. Thus, the action of major genes for resistance in the host and avirulence in its corresponding pathogen is likely to be masked by interplay between gross morphology and resistance factors. Several virulence and avirulence alleles were identified in a large-scale Europe-wide survey of *Leptosphaeria maculans* isolates (113). The alternative approach of studying individual host x pathogen isolates has come from increasing interest in the dynamics of individual types of resistance. Many elements of gene-for-gene interactions, such as the resistance conferred by single genes with major phenotypic effects tightly coupled between host lines and specific pathogen genotypes, are found in a wide array of natural host-pathogen associations (14, 15, 54, 59, 98). Studies that concentrate on interactions in single populations tend to give the impression of tightly co-evolved gene-for-gene associations between host and pathogen. Simultaneous study of variation in hosts and pathogens in a number of populations of *Linum marginale* and *Melampsora lini* in the same geographic region provided a new perspective to our understanding of how plant and pathogens continue to co-evolve (17, 18, 60). This association is an endemic Australian version of the interaction used by Flor (36) to first elucidate the gene-for-gene concept. It provides further evidence that gene-for-gene interactions do occur in natural populations. The study indicates the spatial scale over which co-evolving associations are likely to develop (16). In the *Linum-Melampsora* association, migration occurs at a high rate, which ensures interchange between individual populations and rapid re-establishment of those genes lost through extinction. Changes in the plant system are slower than in the pathogen. However, within individual populations all host plants are vulnerable to attack regardless of the resistance genes and the virulence structure of associated pathodeme. Changes in the relative frequency of individual resistant lines appear to show little correlation with the frequency of pathogen races in the associated pathodeme. Thus, at the spatial scale of individual host and pathogen demes, there is no clear pattern of reciprocal response in genetic structure of interacting populations. Alternatively, individual populations could be strongly influenced by genetic drift and gene flow that occurs among demes within the same epidemiological region. The sum of these processes across many plants and pathodemes constitute the metapopulation at which gene-for-gene co-evolution takes place and genetic variation in resistance and virulence genes is maintained.

Table 2. Gene-for-gene interaction between plants and pathogens

| **Fungal pathogens** | **Fungal pathogens** |
|---|---|
| Basidiomycetes Uredinales (rusts) | Ascomycetes Erysiphales (powdery mildews) |
| *Avena-Puccinia graminis* | *Hordeum-Erysiphe graminis* |
| *Avena-Puccinia coronata* | *Secale- Erysiphe graminis* |
| *Coffea-Hemileia vastatrix* | *Senecio- Erysiphe fischeri* |
| *Glycine-Phakospora pachyrhizi* | *Triticum-Erysiphe graminis* |
| *Helianthus-Puccinia helianthi* | |
| *Linum-Melampsora lini* | Oomycetes Peronosporales |
| *Triticum-Puccinia graminis* | *Lactuca-Bremia lactucae* |
| *Triticum-Puccinia recondite* | *Medicago-Peronospora trifoliorum* |
| *Triticum-Puccinia striiformis* | *Sorghum-Peronosclerospora sorghi* |
| | *Glycine-Phytophthora megasperma* |
| Basidiomycetes Ustilaginales (bunts and smuts) | *Solanum-Phytophthora infestans* |
| *Avena-Ustilago avenae* | Other fungi |
| *Hordeum-Ustilago hordei* | *Brassica-Plasmodiophora brassicae* |
| *Triticum-Ustilago tritici* | *Malus-Venturia inaequalis* |
| *Triticum-Tilletia caries* | *Callistephus-Fusarium oxysporum* |
| *Triticum-Tilletia controversa* | *Hordeum-Rhynchosporium sacalis* |

3, 14, 23, 43, 54, 79, 94, 100, 110, 119

Several models of gene-for-gene interactions (table 2) predict the rate at which a pathogen population will overcome a set of major resistance genes in the host plant (72, 103). The metapopulation structure of interacting populations may potentially magnify or mitigate the 'boom-and-bust' cycles expected in some genetic models of plant-pathogen interactions.

Combining results from agriculture, natural populations and mathematical models, it appears that the gene-for-gene view of co-evolution is developing into a family of hypotheses that differ slightly in their assumptions and predictions. Thompson and Burdon (119) suggested a hierarchy in this family based on the complexity of population structure and the evolutionary processes driving genetic changes in plants and their pathogens, *viz.*, (a) local agricultural gene-for-gene coevolution, (b) geographic agricultural gene-for-gene coevolution, (c) panmictic gene-for-gene coevolution and (d) metapopulation gene-for-gene coevolution. Continued progress in breeding and management for durable resistance in plant populations (6) and in the understanding of evolutionary dynamics of the gene-for-gene coevolution in natural populations will require more detailed studies.

## FUNCTIONAL GENOMICS APPROACHES TO EXPLORE FUNGAL PATHOGENESIS IN PLANTS

Since completion of the *Saccharomyces cerevisiae* genome sequence (44), the genome sequencing of a number of parasitic and symbiotic fungi has been initiated and are completed or nearing completion (Fig. 1, http://www-genome.wi.mit.edu/annotation/fungi/). These include the model ascomycetes *Neurospora crassa*, *Ashbya gossypii* and *Aspergillus nidulans*. Analysis of the *N. crassa* genome (~40 megabases) provided insights into aspects of *Neurospora* biology including the identification of genes potentially associated with red light photobiology, genes implicated in secondary metabolism, and possession of the widest array of genome defence mechanisms known for any eukaryotic organism. These defence genes include a process unique to fungi called repeat-induced point (RIP) mutation, that had a profound impact on genome evolution (41). The genome of *Ashbya gossypii* with 9.2 megabases, encoding 4718 protein-coding genes represents the smallest genome of a free-living eukaryote yet characterized. More than 90% of the *A. gossypii* genes show a pattern of synteny with *S. cerevisiae* that contained 300 inversions and translocations that have occurred since divergence of these two species (27).

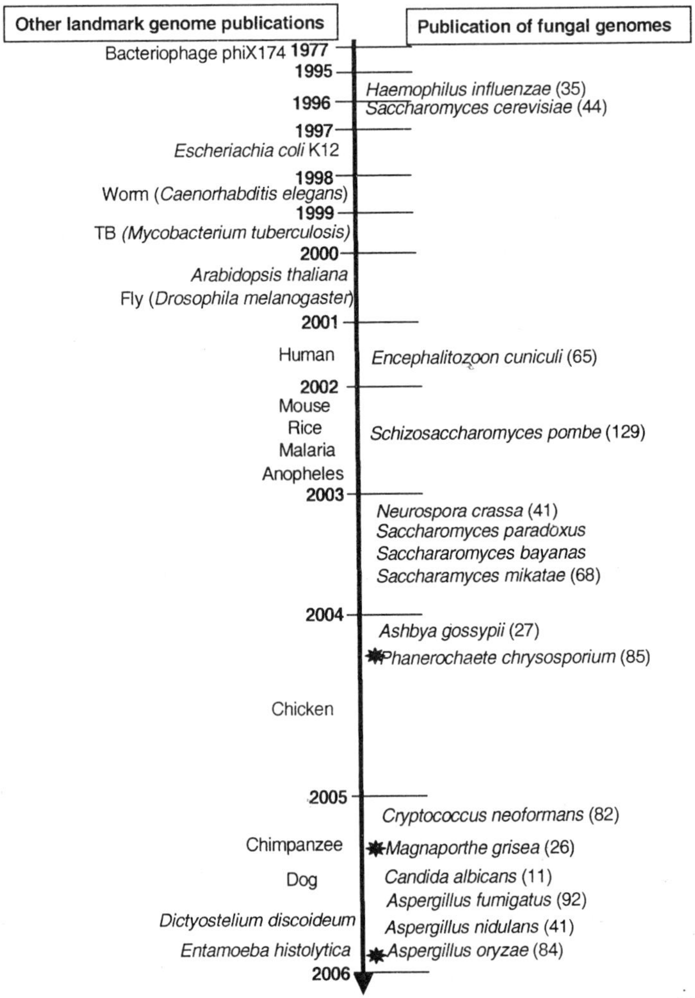


**Figure 1.** An overview of the evolving chronology of genome publications for fungal pathogens and other major organisms. The plant pathogenic fungal genomes have been indicated with an asterisk (*). A detailed list of ongoing and nearly completed fungal genome projects can be found in the websites: http://www.broad.mit.edu/annotation/fgi and http://www.genomesonline.org/

Unlike the bacterial pathogens, our current understanding of the genomic adaptations in pathogenic fungi is rather poor. However, this situation is changing rapidly with the availability of the genomes of a few phytopathogenic fungi (96). Genome analysis of the opportunistic human pathogen *Cryptococcus neoformans* (82), the rice blast pathogen, *Magnaporthe grisea* (26), and the lignocellulose-degrading white rot fungus *Phanerochaete chrysosporium* (85) provided an insight into adaptations required by pathogenic fungi to cause disease. Additionally these data partially answers the query as to whether human and / or plant pathogenic fungi possess a unique set of genes with specific functions dedicated to pathogenic activities. The genome of *M. grisea* encodes a large and diverse set of secreted proteins, including number of proteins containing unusual carbohydrate-binding domains. The rice blast pathogen also possesses an expanded family of G-protein-coupled receptors, several new virulence-associated genes, and large numbers of enzymes involved in secondary metabolism. Consistent with a role in fungal pathogenesis, the expression of several of these genes is up-regulated during early stages of infection (26). More recently, comprehensive analyses of genomes of three *Aspergillus* species (*A. fumigatus*, *A. nidulans* and *A. oryzae*) have been described (41, 84, 92, 97). Genome comparison showed that *A. oryzae* has an additional 7-9 Mb in the genome compared to the other two aspergilli. There are several *A. oryzae*-specific gene blocks (i.e. lacking synteny with *A. fumigatus* and *A. nidulans*) scattered in a mosaic pattern throughout the genome. Analyses of the *A. oryzae* specific gene blocks revealed expansion of genes involved in secondary metabolism, amino acid metabolism and amino acid / sugar uptake transporters (84). In addition, sequence comparison identified over 5,000 non-coding sequences being conserved across all three species, despite about 200 million years of evolution (41).

The availability of genomic sequences for these fungi will provide valuable insight into genome plasticity and the evolution of fungal pathogens. The fungal genome sequence will also serve as a platform to identify gene function and regulatory networks. In addition, genome-wide comparisons coupled with experimental approaches to elucidate the individual gene function will enable us to understand the molecular basis of fungal virulence. Expression profiling efforts identified a number of genes associated with defence signal transduction or antimicrobial action (128). Large-scale investigation of gene expression using functional genomics approaches such as large-scale ESTs, SAGE, microarray, high-throughput transposon-mediated insertional mutagenesis, RNAi, and proteomic analysis, will unravel the inherent complexity of host-pathogen interactions. Such studies will allow further insight on the interplay of a large number of host and pathogen gene products and the consequence of this interplay on pathogen recognition and the evolution of disease resistance (4, 21, 121, 130).

*Expressed sequence tags (ESTs)*

ESTs are short, single-pass, and partial sequences of randomly selected clones from a cDNA library of a given tissue or cells. EST sequences are derived from either the 5' and/ or 3' ends of the transcript. The 5'-ESTs generally include a portion of the encoded protein, and often provide information about the potential function of the expressed gene. On the other hand, the 3'-ESTs generally include the non-coding region of the mRNA and are more variable. However, 3'-ESTs, due to their inherent variability, are more unique than whole genes and thus can be used to distinguish transcripts within a gene family. The sequencing and analysis of a large number of ESTs provides a snapshot of the total gene activity in a cell under a given condition (2). The frequency of occurrence of an EST in a cDNA library is largely determined by the abundance of that transcript, if they are obtained from non-normalised cDNA libraries. The EST sequences, after filtering for contaminants, vector sequence, and quality of the reads, are assembled (clustered) into contigs (containing two or more EST sequence reads) and singletons, then searched against appropriate sequence databases to try to identify gene function based on sequence similarity. Due to low cost and the ease of producing and analysing sequence information, EST mining has greatly facilitated new gene discovery and has proven valuable in genomic studies in phytopathogenic as well as symbiotic fungi (102, 106). Using ESTs, genes involved in the early stages of infection in rice blast fungal pathogen, *M. grisea* were identified (58). As part of the Consortium for the Genomics of the Microbial Eukaryotes (COGEME) project, a relational database containing EST sequences from several plant pathogenic fungi was developed (111, 112) and is publicly available at http://cogeme.exeter.ac.uk. The current version (version 1.5) of the COGEME Phytopathogenic

Fungi and Oomycete EST database contains a total of 59,765 ESTs. In a recent study, comparative clustering of expression patterns of a large set of ESTs from phytopathogenic and saprophytic fungi was used to identify candidate pathogenicity factors (112). In the absence of the fully sequenced genome of a given fungal plant pathogen, generation of large-scale ESTs and their subsequent analysis often give us a fairly good idea about the genetic diversity and gene structure in that organism.

*Microarray analysis*

Microarray analysis has revolutionized our understanding of the dynamics of gene expression during the last decade (108, 109) and is constantly evolving. Gene expression profiling using microarray allows measuring the expression of thousands of genes in parallel in a single experiment and thereby, allows predicting the function of genes, identifying novel transcripts, link genes to biochemical pathways, and developing hypotheses about transcriptional regulation and gene regulatory networks (33). For example, using an oligonucleotide-based microarray, genes that consistently showed altered mRNA expressions in maize during *Cochliobolus carbonum* pathogenesis were identified (7, 90). High-density cDNA microarrays were used to analyse transcript profiles of the plant pathogenic fungus *Blumeria graminis* f. sp. *hordei*, the causal agent of barley powdery mildew (10). The expression profile revealed a group of RNAs whose abundance were correlated with the expression of a gene (*cap20*) known to be required for virulence in *Colletotrichum gloeosporioides*, an important anthracnose causing fungal pathogen of several horticultural crops (avocado, strawberry, papaya, yam, mango, cocoa, olive, lupin, etc.). These candidate genes appeared to be critical for pathogenicity in *B. graminis* (10). Microarray analysis has the potential to address many important issues in plant-fungal pathogen interactions, such as comparing the spectra of genes affected by different pathogen species, pathotypes, or by same pathogen in different plant genotypes, monitoring the dynamics of gene expression changes over time in a plant from initiation through establishment of fungal pathogenesis, and comparing gene expression at infection site versus sites away from the point of infection. Such studies have begun to reveal new features of plant-pathogen interactions (26, 66, 67, 80, 115) and identify candidate targets for developing therapeutics for economically important fungal diseases in plants.

An array based on the *Candida albicans* genome sequence was used to identify components of transcriptional regulators that control the yeast from vegetative to hyphal stage morphogenesis and the complete sexual cycle (89, 123) and to reveal candidate targets for itraconazole treatment in *C. albicans* (24). A similar microarray-based approach was used to study mutualistic (symbiotic) associations in fungi and to identify novel symbiosis-regulated genes in the *Eucalyptus* – *Pisolithus* ectomycorrhizal association (126). Availability of whole genome sequences of *M. grisea* allowed researchers to design oligo-based arrays represen-ting all of the 13,666 predicted genes in the fungus. This array is now commercially available from Agilent Technologies, Inc. (Palo Alto, California). The simultaneous availability of whole genome arrays in rice has been used extensively to identify components of the host's response to microbial colonisation by comparing transcription profiling of rice – *M. grisea*, rice-*Fusarium monilifo-rme* and rice-mycorrhizal associations (48). More recently, tempe-rature-sensitive expression of a distinct set of genes was identified using an array representing the whole genome sequence of *A. fumigatus* (92). The same array was used for Comparative Genomic Hybridization (CGH) with a closely related sexual species, *Neosartorya fischeri,* which revealed ~700 genes of *A. fumigatus* are either absent or highly diverged in *Neosartorya* (92).

*Serial Analysis of Gene Expression (SAGE)*

SAGE is a high-throughput strategy for monitoring global gene expression using gene-specific tags. SAGE is not as comprehensive as a method as microarray technique but it does not require the complete genome sequence to be effective. SAGE has proven itself an excellent method for identifying small coding sequences that have been missed during genome annotation by standard *in silico* methods. Briefly, cDNA derived from cellular mRNA is cleaved into precise 15-bp segments (tags) that are ligated together and then sequenced. The transcript abundance is determined by the frequency of occurrence of gene-specific tags, and thus, allows the differentiation of gene expression patterns among the members of a gene family. SAGE, therefore, allows a qualitative and quanti-tative estimation of the transcriptome (86, 91, 131). Unlike microarray analysis, the genes that are studied by SAGE are not pre-selected, and therefore, SAGE provides a better representation of gene activity in a cell. SAGE has been utilised to investigate the gene expression patterns during the growth of yeast cells (57, 63) and to explore the relationship between transcript abundance and protein accumulation in yeast (49). Using SAGE, mRNA expression during pathogenesis-related development and host colonization has been examined in *B. graminis* (117) and in *M. grisea* (56).

*High-throughput transposon-based insertional mutagenesis*

One of the most powerful techniques for functional genomic analysis is the ability to create a large library of single mutants of the organism in question. With the availability of whole genome sequences, this technique could be extended to genome scale to allow the study of the function of thousands of genes. For model organisms such as *S. cerevisiae*, *Caenorhabditis elegans*, *Drosophila melanogaster*, *Arabidopsis thaliana,* and mice, large mutant collections are already available. In the case of plant pathogenic bacteria, methods such as transposon mutagenesis and signature-tagged mutagenesis have allowed high-throughput forward genetics to be carried out to generate libraries of non-pathogenic mutants for subsequent characterisation (55, 120).

The large-scale gene disruption strategies developed for *S. cerevisiae* are yet to be applicable to most of filamentous fungi. However, such efforts are already underway in a few phytopathogenic fungi, including *M. grisea* (e.g., *Magnaporthe grisea - Oryza sativa* Interaction Database at http://www.mgosdb.org). Filamentous phytopathogenic fungal species have proven more recalcitrant to molecular genetic manipulation, largely due to the inefficiency of introducing DNA in to these organisms for transformation. However, there has been major progress in developing high-throughout mutagenesis technologies that are applicable to filamentous fungi. These include transposon-arrayed gene knockout (TAG-KO) system (52), the use of the *Tc1-mariner* transposable element that was originally derived from *Fusarium oxysporum* (77) and subsequently used to generate insertional mutants in *M. grisea* (125); the *impala* based transposition system (a copy of *impala* inserted in the promoter of *niaD* gene from *A. nidulans*) in generation of mutants in *A. fumigatus* (34); the *Agrobacterium tumefaciens*-mediated transformation (ATMT) in *F. oxysporum*, *M. grisea* and *Cryphonectria parasitica*, using conidia, and thus avoiding the need to generate protoplasts (87, 88, 99,

107) and a PCR-based tool (called Double-Joint PCR) for rapid gene manipulation in filamentous fungi (132). New high-throughput insertional mutagenesis approaches have now been taken to generate large library of mutants to identify gene function in many phytopathogenic fungi including *M. grisea* and *Mycosphaerella graminicola* (1, 52, 81).

*GFP-based reporter systems*

In recent years, Auto Fluorescent Protein (AFP)-based technologies, such as use of Green Fluorescent Protein (GFP) have become important tools for microbiologists and microbial ecologists studying such processes microbe-plant interactions, biosensors, biofilm formation, and horizontal gene transfer (78). The ability to use different AFP markers (e.g., GFP, YFP, CFP and DsRed) with differing fluorescent spectra within a single cell has allowed simultaneous, real-time monitoring of several aspects of microbial physiology and gene expression *in situ*. In practice, a DNA fragment encompassing the promoter region of a candidate gene is fused to a promoterless reporter AFP gene, such as the GFP. This fusion construct is then introduced into the wild-type strain, either on a plasmid or on the chromosome, then the expression of GFP is assayed either in a microplate reader or under fluorescent microscope. This provides a tremendous insight into different aspects of biotic interactions. Furthermore, the integration and use of AFP-based reporter constructs with other markers and technologies is facilitating a systems biology approach to study microbial ecology (78). GFP-based reporter systems have also been used to analyse promoter activities and to localize candidate fungal genes in plants. Technologies are improving to adapt this system to generate GFP-tagged promoter or gene fusions at the genome scale. There is a considerable amount of literature available on use of these techniques in studying plant pathogen interplay including fungus-plant (76, 104) or fungus-fungus interactions (83).

*RNA interference or RNAi*

RNA interference (RNAi) is a phenomenon in which small double-stranded RNA (dsRNA) molecules induce sequence-specific degradation of homologous single-stranded RNA (ssRNA) (28). Initially, the dsRNA is digested into 21-23 nucleotide fragments called small interfering RNA (siRNA). The siRNAs bind to a nuclease complex to form the RNA-induced silencing complex (RISC) generating an immobilized 21-23 nucleotide ssRNA. The active RISC complex bound RNA then targets the complimentary mRNA and cleaves the target mRNA ~12 nucleotides from the 3'-terminus of the siRNA (51). The silencing of endogenous gene expression through what is now known as RNAi was first observed in plants (61), when it was called RNA silencing. Later a similar phenomenon was reported in plant resistance to viruses (29, 50, 105). These phenomena are termed co-suppression, post-transcriptional gene silencing (PTGS), or viral-induced gene silencing (VIGS). Fire *et al.* (32) showed that injection of dsRNA into *C. elegans* was ten-fold more potent in silencing target gene expression than either sense or anti-sense RNA alone and coined the term, RNA interference (RNAi). A genome-wide RNAi project, based on transient expression, was undertaken in *C. elegans* that resulted in description of the phenotype of 10% of the 16,757 genes (5, 40, 64). There is great potential to apply RNAi-mediated gene silencing technology in the study of plant pathogenic fungi, since RNA-mediated gene silencing has been observed in many fungi, including the filamentous fungus *N. crassa* (20, 114). Antisense RNA-based methods have been successfully used to control gene expression in many filamentous fungi including *N. crassa*, *A. oryzae*, *C. albicans*, *Cryptococcus neoformans* and *M. grisea* (25, 45, 46, 62, 71, 133). De Backer *et al.* (25) used a similar antisense-based approach to identify genes critical for growth in the pathogenic yeast *C. albicans*. Recently, using RNAi approach, a cluster of NIF transcriptional activators involved in spore formation in *Phythophthora infestans* has been identified (116).

RNAi-based systems are particularly useful when traditional gene disruption techniques do not work, mainly due to functional redundancy as in cases of paralogous gene families or gene duplications. Since RNAi-mediated gene silencing is a homology-based system, it can be used at the level of gene families or could be used for characterisation at the individual gene level by generating mutants with different gene expression levels as a result of intrinsic variation in the efficacy of interference. In addition, it is possible to target many unrelated genes, allowing epistasis analysis (22). The ability to create targeted loss-of-function phenotypes by RNAi will greatly expedite the identi-fication of gene function in plant–fungus interactions in the years ahead.

*Proteomics*

The proteome is defined as the entire protein complement of a genome of a given organism (47). Analysis of the proteome offers the possibility of identifying protein abundance, localisation, interactions and post-translational modifications. Proteomics includes various approaches that integrate protein separation and characterization. An example is the use of 2D-gel electropheresis of protein with mass spectrometry to generate a catalogue of expressed proteins (69) combined with other analytical techniques, such as mapping protein-protein interactions using yeast two-hybrid (Y2H) screening (127).

The Y2H method aims at identification of physical interactions between proteins by co-expression of gene constructs within a single cell of *S. cerevisiae*. The method was first developed by Fields and Song (31) and used the modular nature of the GAL4 transcriptional activator protein. Two chimeric proteins were produced within the same cell, one was the DNA binding domain of GAL4 fused with Protein A and the other GAL4-activation domain fused with Protein B. The transcription of a given reporter gene is activated only if the Protein A and Protein B physically interact. In a comprehensive study of this kind in *S. cerevisiae*, Uetz *et al.* (124) screened all the 6000 predicted genes in the genome of *S. cerevisiae* and identified 692 interacting pairs. The Y2H method was also applied to *M. grisea* to study signalling pathways regulating appressorium development during infection of rice (74). Fang *et al.* (30) used a high-throughput Y2H-based system to identify protein-protein interaction maps of the signal transduction pathways in defence reaction of rice. Because of the availability of both the host and pathogen in *M. grisea*-rice system, large-scale Y2H screening studies could be undertaken to identify components of the 'interactome' between rice and the rice blast fungus. One key consideration has to be kept in mind while interpreting Y2H results, the number of false positives can be remarkably high. Thus, protein-protein interactions mapped by Y2H screening methods need to be verified by other techniques, such as protein cross-linking, co-immuno-precipitation, or affinity-column purification. In addition, this method is not suitable for dealing with membrane proteins.

Two-dimensional polyacrylamide gel electrophoresis (2D-PAGE) coupled with tandem mass spectrometry is, by far, the most widely used technique in proteomics research. One of the drawbacks of 2D-PAGE is that integral membrane proteins, low abundance proteins and small proteins cannot be separated by this method. Also, proteins with high molecular weight and extreme isoelectric points are not resolved well by 2D-PAGE. However, common post-translational modifications, such as phosphorylation and glycosylation, can be reliably detected.

Considerable advances in proteomic approaches were made possible by integration of two-dimensional gel electrophoresis with matrix-assisted laser desorption/-ionisation-time of flight-mass spectrometry (MALDI-TOF-MS). This technique enables high-throughput and accurate characterization of proteins (69, 95). Briefly, a protein in a spot from 2D-PAGE is subjected to trypsin digestion and, then digested peptide fragments are analysed by MALDI-TOF-MS. The observed peptide masses are used to search database of predicted masses derived from translation of genome sequence data of a given organism. Proteomic analysis enables the identification of proteins that are differentially produced during plant-fungus interactions and disease generation. Such proteins could serve as potential markers for disease and at the same time could be used as candidates for developing therapeutics in plant diseases.

There are several published studies on 2D-PAGE-based proteomic approaches in pathogenic fungi *viz.* to study morphogenesis of yeast to filamentous form and to determine multidrug resistance in *C. albicans* (75, 93), to identify GPI-anchored proteins in *A. fumigatus* (13), to study conidial germination, germ-tube growth and appressoria formation in *M. grisea* (70), and to determine mutualistic (symbiotic) association in arbuscular mycorrhizal fungi (53). These studies have shown the potential of applying proteomics approaches in understanding pathogenesis in pathogenic and non-pathogenic fungi.

## PERSPECTIVE

Studies on host-pathogen interactions started with landmark papers of Flor in middle of 20$^{th}$ Century. Application of forward genetic approaches, which once led to the gene-for-gene hypothesis, has now globally evolved through the use of genomic tools. As large amounts of DNA sequences data accumulate in public databases and increasingly powerful computer software is developed for sequence analysis, the biological functions of many of the pathogen and host encoded genes are beginning to emerge. In addition, the availability of the genome sequences of many hosts and their pathogens allow examination of the gene regulatory networks that exist in the hosts and the pathogens and how interplay between them determine the outcome of an interaction. Models are already in place for *Arabidopsis* - fungal systems and in the foreseeable future these models could be extended to test other plant-pathogen systems such as potato and *Brassica* and their associated pathogens as the genome sequencing of these hosts and the pathogens are underway (www.genomeso-nline.org). Rice and *M. grisea* genomes are now available for studying host-pathogen interaction at the genome level using the systems biology approaches. Large-scale insertional mutagenesis, transcriptional profiling, and proteomic analyses are in progress using the rice - *M. grisea* system to study fungal infection vis-à-vis host reaction. Comparative analysis of the genome sequence of fungal pathogens from diverse taxa will help identify genes unique to each host-pathogen interactions as well as genes common to pathogens across taxa. For example, comparative analysis of *F. oxysporum* (Deuteromycotina; perfect stage *Gibberella*: Ascomycota; Sordariomycetes) with *M. grisea* (Ascomycota; Sordariomycetes) or with the corn pathogen *C. carbonum* (Ascomycota; Dothideomycetes) by reverse genetics and insertional mutagenesis is offering fresh insights into attributes required for this soil fungus to be a successful pathogen. Availability of genome sequences from plant pathogenic and other fungi will therefore help to identify new antifungal drug targets. This will facilitate the design of new ways to control fungal pathogens of plants.

The wealth of genomic information that has accumulated since the sequencing of *H. influenzae* in middle 1990s (35) is enormous and growing at an unprecedented rate. Discussing the findings of each of these papers is far beyond the scope of this review. We, therefore, present a snap shot of the potential of genomics for elucidating the molecular basis of plant-fungus interactions with schematics of the o'mics approaches to the study of phytopathogenic fungi (Fig. 2). Using the growing arsenal of available sequence data, combined with bioinformatics and functional genomics, should lead to a greater understanding of the genetic networks activated during a pathogenic fungus's interaction with its plant host, in pathogen virulence, and host resistance. This revolution in genomics is allowing dissection of the classical gene-for-gene hypothesis at the molecular level. Merging of forward genetics approaches with the genomics information for the study of host-pathogen interactions will pave the way towards a global understanding of plant and pathogen biology. This will make major strides in further increasing agricultural production by improving plant health in the years ahead.

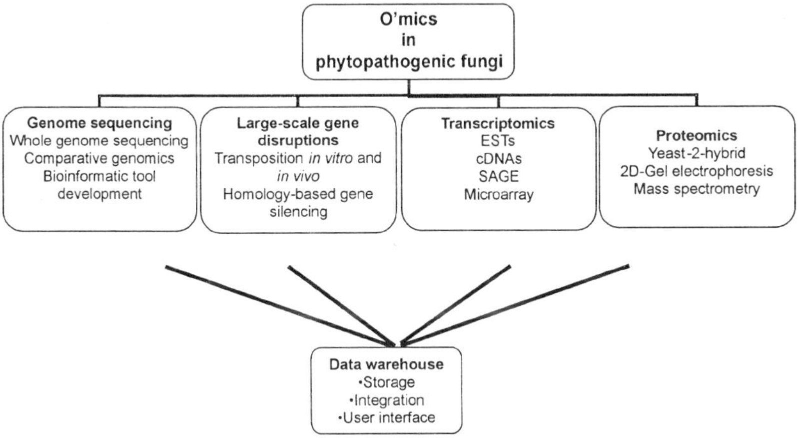


**Figure 2.** Schematics of the O'mics approaches to the study of phytopathogenic fungi

## *Literature Cited*


1. Adachi K *et al.* 2002. Efficient gene identification and targeted gene disruption in the wheat blotch fungus *Mycosphaerella graminicola* using TAGKO. *Current Genetics* 42: 123-127
2. Adams MD, *et al.* 1991. Complementary DNA sequencing: "expressed sequence tags" and the human genome project. *Science* 252: 1651-1656
3. Allard RW. 1991. The Genetics of host-pathogen coevolution: implications for genetic resource conservation. *Journal Heredity* 81: 1-6
4. Armstrong MR, *et al.* 2005 An ancestral oomycete locus contains late blight aviru-lence gene *Avr3a*, encoding a protein that is recognised in the host cytoplasm. *Proc. Natl. Acad. Sci. USA* 102: 7766-7771

5. Ashrafi K, *et al.* 2003. Genome-wide RNAi analysis of *Caenorhabditis elegans* fat regulatory genes. *Nature* 421: 268-272
6. Aubertot JN, *et al.* 2006. Improved resistance management for durable disease control: A case study of Phoma stem canker of oilseed rape (*Brassica napus*). *European Journal of Plant Pathology* 114: 91-106
7. Baldwin D, Crane V, Rice D. 1999. A comparison of gel-based, nylon filter and micro-array techniques to detect differential RNA expression in plants. *Current Opinion Plant Biology* 2: 96-103
8. Birren B, Fink G, Lander E. 2002. *Fungal Genome Initiative.* White paper, Whitehead Institute Centre for Genome Research, Cambridge, MA 02141, USA
9. Blattner FR, *et al.* 1997. The complete genome sequence of *Escherichia coli* K-12. *Science* 277: 1453-1474
10. Both M, *et al.* 2005. Transcript profiles of *Blumeria graminis* development during infection reveal a cluster of genes that are potential virulence determinants. *Molecular Plant Microbe Interaction* 18: 125-133
11. Braun BR, *et al.* 2005. A human-curated annotation of the *Candida albicans* genome. *PLoS Genet.* 1: 36-57
12. Brown JKM, Jessop AC, Rezanoor HN. 1991. Genetic uniformity in barley and its powdery mildew pathogen. *Proc. Royal Soc.* B246: 83-90
13. Bruneau JM, *et al.* 2001. Proteome analysis of *Aspergillus fumigatus* identifies glycosylphosphatidylinositolanchored proteins associated to the cell wall biosynthesis. *Electrophoresis* 22: 2812-2823
14. Burdon JJ. 1987. *Diseases and Plant Population Biology.* Cambridge University Press, Cambridge, UK
15. Burdon JJ. 1987. Phenotypic and genetic patterns of resistance to the pathogen *Phakopsora pachyrhizi* in populations of *Glycine canescens. Oecologia* 73: 257-267
16. Burdon JJ, Brown AHD, Jarosz AM. 1990. In *Pests, Pathogens and Plant Communities.* eds. JJ Burdon, SR Leather, Blackwell Publ., Oxford, UK, pp. 233-247
17. Burdon JJ, Jarosz AM. 1991. Host-pathogen interactions in natural populations of *Linum marginale* and *Melampsora lini.* I. Patterns of resistance and racial variation in a large host population. *Evolution* 45: 205-217
18. Burdon JJ, Jarosz AM. 1992. Temporal variation in the racial structure of flax rust (*Melampsora lini)* populations growing on natural stands of wild flax (*Linum marginale)*: Local versus metapopulation dynamics. *Plant Pathology* 41: 165-179

19. Christ BJ, Person CO, Pope DD. 1987. In *Populations of Plant Pathogens: Their Dynamics and Genetics.* eds. MS Wolfe, CE Carten, Blackwell Publ., Oxford, UK, pp. 7-19
20. Cogoni C. 2001. Homology-dependent gene silencing mechanisms in fungi. *Annual Review of Microbiology* 55: 381-406
21. Colebatch G, Trevaskis B, Udvardi M. 2002. Functional genomics: tools of the trade. *New Phytologist* 153: 27-36
22. Cottrell TR, Doering TL. 2003. Silence of the strands: RNA interference in eukaryotic pathogens. *Trends in Microbiology* 11: 37-43
23. Crute IR, Phelps K, Barnes A, Buczacki ST, Crisp P. 1983. The relationship between genotypes of three *Brassica* species and collections of *Plasmodiophora brassicae. Plant Pathology* 32: 405-420
24. De Backer MD, *et al.* 2001. Genomic profiling of the response of *Candida albicans* to itraconazole treatment using a DNA microarray. *Antimicrob. Agents Chemotherapy* 45:1660-1670
25. De Backer MD, *et al.* 2001. An antisense-based functional genomics approach for identification of genes critical for growth of *Candida albicans. Nature Biotechnology* 19: 235-241
26. Dean RA, *et al.* 2005. The genome sequence of the rice blast pathogen *Magnaporthe grisea. Nature* 434: 980-986
27. Dietrich FS, *et al.* 2004. The *Ashbya gossypii* genome as a tool for mapping the ancient *Saccharomyces cerevisiae* genome. *Science* 304: 304-307
28. Dykxhoorn DM, Novina CD, Sharp PA. 2003. Killing the messenger: short RNAs that silence gene expression. *Nature Review Molecular Cell Biology* 4: 457-467
29. English JJ, Mueller E, Baulcombe DC. 1996. Suppression of virus accumulation in transgenic plants exhibiting gene silencing of nuclear genes. *The Plant Cell* 8: 179-188
30. Fang Y, *et al.* 2002. Development of a high-throughout yeast two-hybrid screening system to study protein-protein interactions in plants. *Molecular Genetics Genomics* 267: 142-153
31. Fields S, Song O. 1989. A novel genetic system to detect protein -protein interactions. *Nature* 340: 245-246
32. Fire A, *et al.* 1998. Potent and specific genetic interference by double-stranded RNA in *Caenorhabditis elegans. Nature* 391: 806-811
33. Firnhaber C, Puhler A, Kuster H. 2005. EST sequencing and time course microarray hybridizations identify more than 700 *Medicago truncatula* genes

with developmental expression regulation in flowers and pods. *Planta* 222: 269-283

34. Firon A, Villalba F, Beffa R, d'Enfert C. 2003. Identification of essential genes in the human fungal pathogen *Aspergillus fumigatus* by transposon mutagenesis. *Eukaryotic Cell* 2: 247-255
35. Fleischmann RD, *et al.* 1995. Whole-genome random sequencing and assembly of *Haemophilus influenzae* Rd. Science 269: 496-512
36. Flor HH. 1942. Inheritance of pathogenicity in *Melampsora lini. Phytopathology* 32:653-669
37. Flor HH. 1953. Epidemiology of flax rust in the North Central States. *Phytopathology* 43: 624-628
38. Flor HH. 1956. The complementary genetic systems in flax & flax rust. *Adv. Genet.* 8: 29-54
39. Flor HH. 1971. Current status of the gene-for-gene concept. *Ann. Rev. Phytopathol.* 9: 275-296
40. Fraser AG, *et al.* 2000. Functional genomic analysis of *Caenorhabditis elegans* chromosome I by systematic RNA interference. *Nature* 408: 325-330
41. Galagan JE, *et al.* 2003. The genome sequence of the filamentous fungus *Neurospora crassa*. *Nature* 422: 859-868.
42. Galagan JE, *et al.* 2005. Sequencing of *Aspergillus nidulans* and comparative analysis with *A. fumigatus* and *A. oryzae*. *Nature* 438: 1105-15
43. Geiger HH, Schumacher AE, Billenkamp N. 1988. Frequencies of vertical resistances and virulences in the rye-powdery mildew pathosystem. *Plant Breeding* 100: 97-103
44. Goffeau A, *et al.* 1996. Life with 6000 genes. *Science* 274:546-567
45. Goldoni S, *et al.* 2004. Biologically Active Decorin Is a Monomer in Solution. *J. Biol. Chem.* 279: 6606-6612
46. Gorlach JM, McDade HC, Perfect JR, Cox GM. 2002. Antisense repression in *Cryptococcus neoformans* as a laboratory tool and potential antifungal strategy. *Microbiology* 148: 213-219
47. Graves PR, Haystead TAJ. 2002. Molecular Biologist's Guide to Proteomics. *Microbiol. Mol. Biol. Rev.* 66: 39-63
48. Guimil S, *et al.* 2005. Comparative transcriptomics of rice reveals an ancient pattern of response to microbial colonization. *Proc. Natl. Acad. Sci.* 102: 8066 – 8070
49. Gygi SP, Rochon Y, Franza BR, Aebersold R. 1999. Correlation between protein and mRNA abundance in yeast. *Molecular Cell Biology* 19: 1720-1730
50. Hamilton AJ, Baulcombe DC. 1999. A novel species of small antisense RNA in posttranscriptional gene silencing. *Science* 286: 950-952

51. Hammond SM, Caudy AA, Hannon GJ. 2001. Post-transcriptional gene silencing by double-stranded RNA. *Nature Review Genetics* 2: 110-119
52. Hamer L, *et al.* 2001. Gene discovery and gene function assignment in filamentous fungi. *Proceedings of National Academy of Sciences* 98: 5110 – 5115
53. Harrier L.A. 2001. The arbuscular mycorrhizal symbiosis: a molecular review of the fungal dimension. *J. Exp. Bot.* 52: 469-478
54. Harry I, Clarke D. 1987. The genetics of race-specific resistance in groundsel (*Senecio vulgaris* L.) to the powdery mildew fungus, *Erysiphe fischeri* Blumer. *The New Phytologist* 107: 715-723
55. Hensel M, *et al.* 1995. Simultaneous identification of bacterial virulence genes by negative selection. *Science* 269: 400-403
56. Irie T, Matsumura H, Terauchi R, Saitoh H. 2003. Serial analysis of gene expression (SAGE) of *Magnaporthe grisea*: genes involved in appressorium formation. *Mol. Gen. Genomics* 270: 181-189
57. Jansen R, Gerstein M. 2000. Analysis of the yeast transcriptome with structural and functional categories: characterizing highly expressed proteins. *Nucleic Acid Research* 28: 1481-1488
58. Jantasuriyarat C, *et al.* 2005. Large-scale identification of expressed sequence tags involved in rice and rice blast fungus interaction. *Plant Physiology* 138: 105-115
59. Jarosz AM, Burdon JJ. 1990. Predominance of a single major gene for resistance to *Phakospora pachyrhizi* in a population of *Glycine argyrea. Heredity* 64: 347-353
60. Jarosz AM, Burdon JJ. 1991. Host-pathogen interactions in natural populations of *Linum marginale* and *Melampsora lini*: II. Local and regional variation in patterns of resistance and racial structure. *Evolution* 45: 1618-1627
61. Jorgensen R. 1990. Altered gene expression in plants due to trans interactions between homologous genes. *Trend Biotechnology* 8: 340-344
62. Kadotani N, Nakayashiki H, Tosa Y, Mayama S. 2003. RNA silencing in the phytopathogenic fungus *Magnaporthe oryzae. Mol Plant Microb Interact* 16: 769-776
63. Kal AJ, *et al.* 1999. Dynamics of gene expression revealed by comparison of serial analysis of gene expression transcript profiles from yeast grown on two different carbon sources. *Molecular Cell Biology* 10: 1859-1872
64. Kamath RS, *et al.* 2003. Systematic functional analysis of the *Caenorhabditis elegans*

genome using RNAi. *Nature* 421: 231-237

65. Katinka MD, *et al.* 2001. Genome sequence and gene compaction of the eukaryote parasite *Encephalitozoon cuniculi. Nature* 414: 401-402
66. Kato-Maeda M, Gao Q, Small PM. 2001. Microarray analysis of pathogens and their interaction with hosts. *Cell. Microbiol.* 3: 713-719
67. Kazan K, Schenk PM, Wilson I, Manners JM. 2001. DNA microarrays: new tools in the analysis of plant defence response. *Molecular Plant Pathology* 2: 177-185
68. Kellis M, Patterson N, Endrizzi M, Birren B, Lander ES. 2003. Sequencing and comparison of yeast species to identify genes and regulatory elements. *Nature* 423: 233-234
69. Kersten B, *et al.* 2002. Large-scale plant proteomics. *Plant Molecular Biology* 48: 133-141
70. Kim ST, *et al.* 2004. Proteome analysis of rice blast fungus (*Magnaporthe grisea*) proteome during appressorium formation. *Proteomics* 4: 3579-3587
71. Kitamoto N, Yoshino S, Ohmiya K, Tsukagoshi N. 1999. Sequence analysis, overexpression, and antisense inhibition of a beta-xylosidase gene, xylA, from *Aspergillus oryzae* KBN616. *Appl. Environ. Microbiol.* 65: 20-24
72. Kiyosawa SA. 1982. Genetics and epidemiological modeling of breakdown of plant disease resistance. *Ann. Rev. Phytopathol.* 20: 93-117
73. Kuhn ML, *et al.* 2006. Genetic linkage maps and genomic organization in *Leptosphaeria maculans. European Journal of Plant Pathology* 114: 17-31
74. Kulkarni RD, Dean RA. 2004. Identification of proteins that interact with two regulators of appressorium development, adenylate cyclase and cAMP-dependent protein kinase A, in the rice blast fungus *Magnaporthe grisea. Mol. Genet. Genomics* 270: 497-508
75. Kusch H, *et al.* 2004. A proteomic approach to understanding the development of multidrug-resistant *Candida albicans* strains. *Mol. Genet. Genomics* 271: 554-565
76. Lagopodi AL, *et al.* 2002. Novel aspects of tomato root colonization and infection by *Fusarium oxysporum* f. sp. *radicis-lycopersici* revealed by confocal laser scanning microscopic analysis using the green fluorescent protein as a marker. *Mol. Plant-Microbe Interact.* 15:172-179
77. Langin T, Capy P, Daboussi MJ. 1995. The transposable element *impala,* a fungal member of the *Tc1-mariner* superfamily. *Mol. Gen. Genet* 246: 19-28
78. Larrainzar E, O'Gara F, Morrissey JP. 2005. Applications of autofluorescent

proteins for *in situ* studies in microbial ecology. *Ann. Rev. Microbiol.* 59: 257-277

79. Layton AC, Kuhn DN. 1988. The virulence of interracial heterokaryons of *Phytophthora megasperma* f. sp. *glycinea*. *Phytopathology* 78: 961-966
80. Leung YF, Cavalieri D. 2003. Fundamentals of cDNA micro-array analysis. *Trends Genet.* 19: 649-659
81. Lo C, Adachi K, Shuster JR, Hamer JE, Hamer L. 2003. The bacterial transposon Tn7 causes premature polyadenylation of mRNA in eukaryotic organisms: TAGKO mutagene-sis in filamentous fungi. *Nucleic Acids Res* 31: 4822-4827
82. Loftus B, *et al.* 2005. The genome of the protist parasite *Entamoeba histolytica*. *Nature* 433: 865-868
83. Lu Z, *et al.* 2004. *In vivo* study of *Trichoderma*-pathogen-plant interactions, using constitutive and inducible green fluorescent protein reporter systems. *Appl. Environ. Microbiol.* 70:3073-81
84. Machida M, *et al.* 2005. Genome sequencing and analysis of *Aspergillus oryzae*. *Nature.* 438: 1157-1161
85. Martinez D, *et al.* 2004. Genome sequence of the lignocellulose degrading fungus *Phanerochaete chrysosporium* strain RP78. *Nature Biotechnology* 22: 695-699
86. Matsumura H, *et al.* 2003. Gene expression analysis of plant host-pathogen interactions by SuperSAGE. *Proc. Natl Acad. Sci. USA* 100: 15718-15723
87. Mullins ED, Kang S. 2001. Transformation: a tool for studying fungal pathogens of plants. *Cell. Mol. Life Sci.* 58: 2043-2052
88. Mullins ED, *et al.* 2001. *Agrobacterium*-mediated transformation of *Fusarium oxysporum*: an efficient tool for insertional mutagenesis and gene transfer. *Phytopathology* 91: 173-180
89. Murad AMA, *et al.* 2001. Transcript profiling in *Candida albicans* reveals new cellular functions for the transcriptional repressors CaTup1, CaMig1 and CaNrg1. *Mol. Microbiol.* 42: 981-993
90. Nadimpalli R, Yalpani N, Johal GS, Simmons CR. 2000. Prohibitins, stomatins and plant disease response genes compose a protein superfamily that controls cell proliferation, ion channel regulation and death. *Journal Biological Chemistry* 275: 29579-29586
91. Nehls U. 2003. Ectomycorrhizal development and function – transcriptome analysis. *New Phytologist* 159: 5-7
92. Nierman WC, *et al.* 2005. Genomic sequence of the pathogenic and allergenic filamentous fungus *Aspergillus fumigatus*. *Nature* 438: 1151-1156

93. Niimi M, Cannon R, Monk B. 1999. *Candida albicans* pathogenicity: a proteomic perspective. *Electrophoresis*. 20: 2299-2308

94. Nof E, Dinoor A. 1981. The manifestation of gene-for-gene relationships in oats and crown rust. *Phytoparasitica* 9: 240

95. Nordoff E, *et al.* 2001. Large-gel 2-DE-MALDI-TOF-MS – an analytical challenge for studying complex protein mixtures. *Electrophoresis* 22: 2844-2855

96. Pain A. 2005. Fungi behaving badly. *Nature Reviews Microbiology* 3: 832-833

97. Pain A, Bohme U, Berriman M. 2006. Hot and sexy moulds! *Nature Reviews Microbiology* 4: 244-246

98. Parker MA. 1985. Local population differentiation for compatibility in an annual legume and its host-specific fungal pathogen. *Evolution* 39: 713-723

99. Park SM, Kim DH. 2004. Transformation of a filamentous fungus *Cryphonectria parasitica* using *Agrobacterium tumefaciens*. Biotechnology and Bioprocess Engineering 9: 217-222

100.Pawar MN, *et al.* 1985. Survey of the virulence of *Peronosclerospora sorghi* isolates from Ethiopia, Nigeria, Texas (USA), Honduras, Brazil and Argentina. *Phytopathology* 75: 1374

101.Persiel F, Lein H. 1989. Investigations on the resistance of China asters, *Callistephus chinensis*, to *Fusarium oxyspo-rum* f. sp. *callistephi*. *Journal Plant Disease Protection* 96: 47-59

102.Peter M, *et al.* 2003. Analysis of expressed sequence tags from the ectomycorrhizal basidiomycetes *Laccaria bicolor* and *Pisolithus microcarpus*. *New Phytologist* 159: 117-129

103.Pietravalle S, Lemarie S, van den Bosch F. 2006. Durability of resistance and cost of virulence. *European Journal of Plant Pathology* 114: 107-116.

104.Ramos HJO, *et al.* 2002. Monitoring *Azospirillum*-wheat interactions using the gfp and gusA genes constitutively expressed from a new broad-host range vector. *J. Biotechnol*. 97: 243-252

105.Ratcliff F, Harrison BD, Baulcombe DC. 1997. A similarity between viral defense and gene silencing in plants. *Science* 276: 1558-1560

106.Rauyaree P, *et al.* 2001. Genes expressed during early stages of rice infection with the rice blast fungus *Magnaporthe grisea*. *Molecular Plant Pathology* 2: 347-354

107.Rho HS, Kang S, Lee YH. 2001. *Agrobacterium tumefaciens*-mediated transformation of the plant pathogenic fungus, *Magnaporthe grisea*. *Mol. Cells* 12: 407-411

108.Schena M, Shalon D, Davis RW, Brown PO. 1995. Quantitative monitoring of gene expression patterns with a complementary DNA microarray. *Science* 270: 467-470

109. Schena M, Heller RA, Theriault TP, Konrad K, Davis RW. 1998. Microarrays: biotechnology's discovery platform for functional genomics. *Trends Biotechnology* 16: 301-306

110. Skinner DZ, Stuteville DL. 1985. Genetics of host-parasite interactions between alfalfa and *Peronospora trifoliorum*. *Phytopathology* 75: 119-121

111. Soanes DM, Skinner, Keon J, Hargreaves J, Talbot NJ. 2002. Genomics of phytopathogenic fungi and development of bioinformatic resources. *Molecular Plant Microbe Interactions* 15: 421-427

112. Soanes DM, Talbot NJ. 2006. Comparative genomic analysis of phytopathogenic fungi using expressed sequence tag (EST) collections. *Mol. Plant Pathology* 7: 61-70

113. Stachowiak A, *et al.* 2006. Frequency of avirulence alleles in field populations of *Leptosphaeria maculans* in Europe. *European Journal of Plant Pathology* 114: 67-75

114. Sweigard JA, Ebbole DJ. 2001. Functional analysis of pathogenicity genes in a genomics world. *Curr Opin Microbial* 4: 387-392

115. Takano Y, *et al.* 2003. Large scale parallel analysis of gene expression during infect-ion-related morphogenesis of *Magnaporthe grisea*. *Mol. Plant Pathol.* 4: 337-347.

116. Tani S, Kim KS, Judelson HS. 2005. A cluster of NIF transcriptional regulators with divergent patterns of spore-specific expression in *Phytophthora infestans*. *Fungal Genetics Biology* 42: 42-50

117. Thomas SW, *et al.* 2002. Transcript profiling in the barley mildew pathogen *Blumeria graminis* by serial analysis of gene expression (SAGE). *Molecular Plant Microbe Interactions* 15: 847-856

118. Thompson JN. 1989. Concepts of coevolution. *Trends Ecology Evolution* 4: 179-183

119. Thompson JN, Burdon JJ. 1992. Gene-for-gene coevolution between plants and parasites. *Nature* 360: 121-125

120. Titarenko E, Lopez-Solanilla E, Garcia-Olmedo F, Rodriguez-Palenzuela P. 1997. Mutants of *Ralstonia (Pseudomonas) solanacearum* sensitive to antimicrobial peptides are altered in their lipopolysaccharide structure and are avirulent in tobacco. *J. Bacteriology* 179: 6699-6704

121. Tunlid A. 2003. Exploring plant-microbe interactions using DNA microarrays. *New Phytologist* 158: 235-238

122. Tunlid A, Talbot N. 2002. Genomics of parasitic and symbiotic fungi. *Curr. Opin. Microbiol.* 5: 513-519

123. Tzung KW, *et al.* 2001. Genomic evidence for a complete sexual cycle in *Candida albicans. Proc. Natl. Acad. Sci. USA* 98: 3249-3253

124. Uetz P, *et al.* 2000. A comprehensive analysis of protein-protein interactions in *Saccharomyces cerevisiae. Nature* 403: 623-627

125. Villalba F, *et al.* 2001. Transposon *impala*, a novel tool for gene tagging in the rice blast fungus *Magnaporthe grisea. Molecular Plant Microbe Interaction* 14: 308-315

126. Voiblet C, Duplessis S, Nathalie Encelot N, Martin F. 2001 Identification of symbiosis-regulated genes in *Eucalyptus globulus–Pisolithus tinctorius* ectomycorrhiza by differential hybridization of arrayed cDNAs. *Plant J.* 25: 181-191

127. Walhout AJM, Vidal M. 2001. High-throughput yeast two-hybrid assays for large - scale protein interaction mapping. *Methods* 24: 297-306

128. Wan J, Dunning FM, Bent AF. 2002. Probing plant-pathogen interactions and downstream defense signalling using DNA microarrays. *Functional Integration Genomics* 2: 259-273

129. Wood V *et al.* 2002. The genome sequence of *Schizosaccharomyces pombe. Nature* 415: 871-80

130. Wren BW. 2000. Microbial genome analysis: insights into virulence, host adaptation and evolution. *Nature Genetics* 1: 30-39

131. Yamamoto M, Wakatsuki T, Hada A, Ryo A. 2001. Use of serial analysis of gene expression (SAGE) technology. *Jour. Immunological Methods* 250: 45-66

132. Yu JH, *et al.* 2004. Double-joint PCR: a PCR-based molecular tool for gene manipulation in filamentous fungi. *Fungal Gen. Biol.* 41: 973-981

133. Zheng XF, Kobayashi Y, Takeuchi M. 1998. Construction of a low-serine-type-carboxypeptidase-producing mutant of *Aspergillus oryzae* by the expression of antisense RNA & its use as a host for heterologous protein secretion. *Appl Microbiol Biotechnol.* 49: 39-4.

*Rev. Plant Pathol. Vol. 4, 2008*
*Indian Society of Mycology and Plant Pathology*
*Scientific Publishers (India), Jodhpur, pp. 29-82*

# INTEGRATED DISEASE MANAGEMENT AS A HOLISTIC APPROACH TO RICE PRODUCTION

***K. Muralidharan and C.S. Reddy***

Department of Crop Protection, Directorate of Rice Research,
Rajendranagar, Hyderabad 500 030; email: muralidharan_km@yahoo.com

KEY WORDS: *Oryza sativa*, rice, integrated disease management, resistance

## ABSTRACT

The approach of humans who settled to domesticate and raise crop for food and other uses was towards evolution of crop husbandry. Organic sources including plant and animal wastes were used to increase production and protect the crop for centuries. Yet, in the history of agriculture in India, only one disease, the brown spot disease devastated crop to result in famine and loss of human lives. Farmers practiced essentially subsistence agriculture till India's independence. Slowly, with the upsurge in population and need to find additional food, feed and other material, farmers started using chemicals either to protect or to boost production. Many old and newer diseases caused by biotic agents like fungi, bacteria, viruses, nematodes, insects, weeds, and birds appeared in rice farms. The abiotic agents like drought or flood induced by the vagaries of monsoon, and nutritional deficiency or toxicity also resulted in the emergence of additional disease problems. Insect pests as moving objects attract a greater attention of farming community. Cultivators unfortunately misunderstand that they pose a threat to production. Indian farmers depend on moneylenders and private dealers for credit and chemicals. Although the farmers in rainfed areas are handicapped to find finance, those in irrigated area resort to unwarranted use of pesticides. All farm science practitioners have also fallen prey to visible constraints that affect rice. It has been more of an integrated pest management (IPM) than an integrated disease management (IDM). The complexity in these disease problems and the options available in tackling each of the causative biotic and abiotic factors are discussed. The non-adoptability of some options like economic threshold and biological control in rice farms with today's knowledge and development is demonstrated. It is pointless to synthesize an IPM strategy for an ecosystem or a location or a season. In future, opting for the services of farm consultants might be the more appropriate solution to the challenges posed by ailments in rice production.

## EVOLUTION OF CROP HUSBANDRY AND PLANT PROTECTION

The history of humans is filled with attempts to gain increasing control over environment. About 12,000 years ago, women and men settled near water sources and learned how to cultivate grain crops (52). It was a remarkable advancement that changed the life of family groups or tribes from a nomadic existence to a settled and sedentary life. This subsistence agriculture was hampered by unfavorable changes in weather, low soil fertility and low-yielding cultivars and yet, it continued for hundreds of years. Through centuries, however, many cultural and physical practices evolved for protecting crop from plant diseases through trial and error. Some of these measures are burning or ploughing-under crop refuse, rotation of crops, planting in the most favorable time of the season, using healthy seed, isolating crops, pruning and management of water and fertilizer. Surapala in *Vrikshayurveda* recognized good nutrition supply as a measure to prevent incidence of ailments. They have been used for hundreds of years to improve the chances of making a profitable crop. Yet, India suffered the worst famine in 1943 due to a devastative epidemic of brown spot disease (78) that made the crop in Bengal dry without producing grains. Due to food shortage more than two million people lost their lives. Decades later, the *green revolution* ushered in an era of increased production and self-sufficiency in rice by

1978 (52, 65). Soon, disease problems became severe as a result of more complex agriculture involving changes in production practices and introduction of pathogens into new areas. The search was intensified for developing more effective management methods. Many highly effective agricultural chemicals developed and widely used in world agriculture were introduced in India. These chemicals gave farmers a new ability to reduce pest populations at a relatively low cost. The widespread adoption of these chemicals for managing pests, along with the development of disease-resistant cultivars, led to a general decline in the use and study of many basic cultural and biological tactics for reducing pathogen populations to a level where profitable yields are possible. Soon fertilizers and pesticides were used indiscriminately (72). It became apparent that unacceptable levels of crop loss from disease causing pathogens were still occurring despite tremendous improvements in crop production and protection methods (50). In addition, a variety of problems have caused concern about various agricultural technologies and future food production. In particular, the development of pathogen populations resistant to chemicals, crop losses and stress due to the phytotoxicity of pollutants, shifts in the prevalence of pathogen species, and the harmful effects of some pesticides in the environment and on non-target organisms including humans have all raised questions about the efficacy and advisability of certain disease and pest management practices.

Modern agriculture can overcome many of these problems, but its increased complexity and intensity demand unprecedented precision in the management of crop diseases (72). Furthermore, the realization that plant diseases are natural components of agriculture that must be dealt with continuously fortifies the concept that diseases can never be *controlled* or *eliminated.* Effectively limiting losses from plant diseases requires that they be managed in a process that incorporates the latest knowledge into the management program as rapidly as possible.

## DISEASES AFFECTING RICE

*Disease* means some factor that makes an organism ill-at-ease. In agriculture, the term is applied to any malady that affects the health of a crop. The un-healthy condition is described as a disease. Many external and internal factors influence rice crop health and growth. These factors include both biotic and abiotic factors that are widely prevalent in rice ecosystems. Pathogens - fungal, bacterial and viral, insect pests, and nematodes are the main biotic factors that upset rice crop growth and productivity. Any one of these factors can lead to a disease in rice. Although rodents, particularly rats normally take away the panicles, they can also cause disease in rice plants by cutting growing culms, leaves and panicles. There are a few other agents like weeds, birds, termites, elephants, pigs, etc., which also affect rice (19). All the insect pests that affect rice health have already been extensively discussed. The main abiotic factors that affect rice include nutrient deficiency from major and minor elements, ambient temperature regimes (high or low), and rainfall - insufficiency leading to drought, or excess leading to flood. In India, nutrition-induced physiological disorders have been reported from time to time (27). They were called by various names such as *pansukh*, *dakhina*, *khaira* or physiological root rotting. Until the introduction of high-yielding rice cultivars, fertilizer use was low, and this led to the appearance of several deficiency diseases. *Khaira* was the dominant and widely prevalent zinc-deficiency disease (76).

## APPROACHES TO INTEGRATED DISEASE MANAGEMENT

### Important Principles

- Plant diseases can never be eliminated, only managed at an economically acceptable level.
- Disease management requires knowledge and judgment: for this it is essential to know biotic constraints like pathogens and pests, know the abiotic constraints and environment, and know the crop.
- Provide the crop with every possible advantage. Use all available practices in integrated disease management.
- Continuous disease management is basic to efficient production.
- Epidemiology is basic to the development of a sound disease management strategy.

- Contain the disease at the least possible cost and minimize losses. Every rupee or dollar spent in disease management should return 3-4 rupees or dollars.
- There are three main objectives: avoid unfavorable abiotic conditions, reduce the initial inoculum or stress, and reduce the rate of increase and spread of biotic agent or abiotic causal factor.
- Use proper practices. Develop and employ suitable disease management procedures. These include chemicals; resistant cultivars; seed treatment; crop rotation; proper planting date, seeding rate, and depth; proper harvesting and storage; and ploughing-under of crop debris.
- Use effective logistics. Deploy resources at the proper time, in the correct amount, and in the correct way. Trained personnel are essential.
- Beware of pollution. Use chemicals wisely. Preserve the environment.

Many farmers are increasingly aware of the need for efficient farm management. Changes in farm structure, cultural practices, the price of food, farm inputs, interest rates, the limited choice of crops, special processing and marketing requirements, governmental restraints, and increased transportation costs - all affect farmers' decisions to use a certain practice. In addition, it is obvious that no strategy to increase food production on a sustainable basis can afford to disregard environmental and ecological principles. Efficient agricultural production in the short-term can not justify eventually destroying the ecological basis on which our entire food production system rests. The aim should be to increase agricultural production without environmental destruction.

Continuous disease management means efficient and profitable crop production. Every season of the year, and during certain seasons almost daily, the farmer needs to be actively fighting pathogens. All farmers must be taught to manage disease causing organisms or abiotic stress, for every farmer's practices affect the community of farmers. Some diseases can be held at bay only when all farmers in the community participate in a continuous causative factor(s) management program.

## Economic Thresholds

Economic thresholds depend, in part, on the market value of crops. In deciding whether to plant a given crop in a given year, farmers must be aware of economic developments not only in their own community but also in other states and countries. Market demands, government policies, and current production levels in other parts of the country and of the world - all affect the crop prices and, in turn, the level of disease intensity that warrants biotic or abiotic constraint management efforts. With rice diseases, determination of economic threshold level (70) has been made only on theoretical presumptions. For the polycyclic pathogens and insect pests, generally the onset of primary infection, the time taken to complete one generation, duration to crop maturity and reproductive ability are considered for determining the threshold levels. However, overlapping generations occur. With monocyclic pathogens, information on the degree of primary infection and the rapidity of spread within and between plants are considered in addition to the crop maturity duration. The climatic conditions, cultural practices and the host resistance or susceptibility can totally alter the estimated threshold values. The abiotic constraints can to some extent be nullified by soil test and climate forecasts. In certain cases, pre-emptive strike against a pathogen, an insect or other pest, or an abiotic constraint may be warranted as with neck blast, stink bug, or nutrient deficiency or toxicity. It has been clearly demonstrated that rice farmers tend to ignore forecast warning based on economic threshold values (70). The silent, unmoving and unnoticed disease causing nutrient deficiency or toxicity, and fungi, bacteria and virus pathogens do not attract the attention of farmers, as do insect pests. Insect pests show a definite and visible movement in the fields. Therefore, even if insects are incident at levels well below the economic thresholds, farmers jump to a conclusion that they pose a threat to production and immediately apply chemicals to control them.

## Diagnosis of Diseases

Diagnosis is the art or act of identifying a disease from the signs and symptoms. The signs refer to the actual presence of the pathogen. The symptoms refer to the visible effects of biotic or abiotic agent on the host. The first step in diagnosis of a plant disease is to carefully look at plants in the field and gather all additional information from the grower. Attempts should also be made to determine disease distribution in the field. This may help to clarify the question of whether the pathogen is air-borne, seed-borne, water-borne or soil-borne. Air-borne pathogens are blast, brown spot, banded blight (*R. solani* in cooler climate in north India) and sheath rot. Most of the insect pests are also air-borne. Bacterial leaf blight and bacterial leaf steak of paddy are seed-borne and water-borne and spread to plants along with irrigation. Soil-borne pathogens are sheath blight, stem rot and damping-off of paddy. If diseased plants occur regularly over the entire field at random this may point to the involvement of an air-borne pathogen or a deficiency or toxicity in nutrient supply. If diseased plants are found in small circular patches or aggregates the causal agent may be a soil-borne pathogen. The soil-borne pathogens may even be nematodes like *Hirschmanniella oryzae* (Van Breda de Hann) Luc and Goodey, *Heterodera oryzicola* Rao and Prakash, and *Meloidogyne* spp. that have been introduced inadvertently into the field through the use of externally contaminated seeds. If a disease affects the crop in a large geographical area, air-borne pathogens or vector-borne pathogens may be suspected. The best examples for vector-borne pathogens include green leafhopper vector-borne tungro virus and brown planthopper vector-borne grassy stunt virus or ragged stunt virus. Source of seed for raising a crop should also be determined particularly if it was not certified to be disease-free. Frequently, certain pathogens like *Pyricularia grisea*; *Drechslera oryzae* and bacterial leaf blight create havoc due to seed contamination. The symptoms of disease and the location of the symptoms on the plant give further clues on the nature of the disease. Quite often the lower leaves are attacked first and these leaves slowly dry-up. If only the top parts of plants are affected, air-borne pathogens can be suspected.

Examination of the plants for signs of biotic pathogen or abiotic causal agent is most helpful. One needs to look for insects like green leafhopper for tungro and brown planthopper for grassy stunt or ragged stunt. Sheath blight, stem rot and seedling blight pathogens can be diagnosed by looking at the sclerotial bodies on rice plants, weeds, irrigation water and soil. Sheath blight, foot-rot or stem-rot diseases may also be identified by the detection of mycelia on the surface of water, leaf or plant. Blast, brown spot, foot rot and sheath rot pathogens produce conidia and so can be diagnosed by observing the spores. Bacterial leaf blight or leaf streak can be detected by the yellowish exudates on the leaves or cut-ends of leaves. The signs and symptoms that are observed must be carefully compared with documented information on the respective disease for confirming the diagnosis of a rice disease.

## Available Options for Integrated Disease Management

In India, rice is cultivated in 42 million hectares under four major ecosystems *viz.*, irrigated (19 million hectares), rainfed lowland (14 million hectares), flood prone (3 million hectares) and rainfed upland (6 million hectares) ecosystems (84). As described earlier, rice is affected by a number of biotic and abiotic factors that cause disease and reduce yields. A few of them cause concern as epidemic outbreaks, either frequent as in blast, bacterial leaf blight, stem borer and rodents, or discontinuous and occasional as in sheath blight, tungro virus or brown planthopper, or rare as in false smut, foot rot or gall midge. Some pathogens pose problem in only one or a few farm fields where the entire crop may be affected. Even these sporadic epidemics that occur in an extremely restricted area pose a threat to farm production and farmers' livelihood. Once the causal agent of a disease has been correctly identified, it is possible to develop plans to manage the disease. During the past 100 years, much research has been conducted on pathogens, diseases, and management methods. Today we can draw on this vast store of knowledge to help us in our efforts to manage rice diseases. Intelligent rice disease management is an economic necessity. Together with need-based fertilizer and disease management, it is a form of insurance that helps to guarantee the farmer a profitable yield. It helps to prevent wide fluctuations in yield. It also helps to prevent disastrous epidemics and famines.

There are at least five categories for the management of plant diseases:

1) Avoidance of pathogens by *exclusion* of the pathogen from a geographical area, either voluntarily or by legislation, and by *evasion* by the host plant, to prevent the pathogen from coming into contact with the host;

2) Eradication or removal of the pathogen from the host, soil, from other reservoirs of inoculum, or from a geographical area – the actual result may be a reduction of inoculum rather than the total eradication of the pathogen;
3) Protection of plants by environmental changes, biological methods, or physical methods that produce conditions less favourable for disease development or by the use of pesticides;
4) Use of cultivars with genetic resistance that resist the biotic or abiotic constraint and disease development; and
5) Chemotherapy with systemic chemicals to kill pathogens that are already in the plant.

Our efforts have been aimed at understanding and combating rice diseases, primarily through host plant resistance breeding with selected donors. The options available for use of host plant resistance, development of disease epidemics and control measures for the management of some important disease are discussed in the following pages.

## BIOTIC CONSTRAINTS - FUNGAL DISEASES

### Blast

Blast disease is incited by *Pyricularia grisea* Sacc. {teleomorph *Magnaporthe grisea* (Hebert) Brarr}. It occurs at all stages of crop growth in rainfed, irrigated and hill rice, and severe incidence results in heavy or total loss in yield. Farmers in Andhra Pradesh, Himachal Pradesh, Tamil Nadu and in many other parts of India, had to abandon popular but susceptible cultivars because of the loss caused by blast. The sexual phase of blast pathogen has not been detected in nature. Infection by *Pyricularia grisea* leads to formation of spindle-shaped lesions with brownish margin and grayish center. The spots usually begin as small water-soaked, whitish, greyish or bluish dots. Fully developed lesions in 2-3 days reach 1.5 cm in length, and 0.5 cm in breadth. Adjacent lesions often coalesce under favorable conditions turning a major part or the entire leaf to produce a burnt appearance. Brown or black patches may appear around nodes and the nodes so infected often break apart. At a later stage, the fungus attacks the base of panicle at neck region and this infection is known as panicle blast or neck blast or spikelet blast. Neck infection causes the panicles to break and fall over, resulting in the loss of grains (57). Small brown to black spots formed by *P. grisea* may also be seen on the glumes.

The factors that influence blast epidemics are the susceptible variety, availability of inoculum to initiate the disease, excessive application of nitrogen fertilizer, low night temperature (24°C), high humidity, cloudy and drizzle weather or dew resulting in leaf wetness. When conditions are conducive, the pathogen multiplies rapidly to produce abundant conidia from lesions. The disease moves quickly from field to field by producing myriad number of spores that are disseminated by wind in all directions. These spores upon falling on rice plant, initiate further disease to progress rapidly through the entire field. The repeated cycles of spore production and infection continues throughout the crop growth. Under favourable conditions, the green lush crop growth is turned in to burnt up appearance (56). During mid-1980s, some farmers in Andhra Pradesh and Tamil Nadu committed suicide as they lost everything in rice crop failure following blast epidemics on IR 50.

Cultural practices exert a profound influence on blast development. Generally use of over aged seedlings and delay in planting of rice seedlings increase severity of disease incidence (121, 122). Dense planting is commonly advocated for obtaining maximum yields. Leaf and neck blast infections were found to increase significantly with the increase in the density of populations (122). Nitrogen supply induces a heavy incidence of blast (5) regardless of application of other fertilizers.

Rice blast outbreak rules have been developed (47, 54, 56) with which early and accurate warnings can be issued to farmers:

- For leaf blast outbreak - (1) moderate to high incidence of leaf blast in trap nurseries sown at weekly intervals with any locally susceptible cultivar under high plant population density and fertilizer, and (2) continued prevalence of low night temperatures (< 20°C) coupled with high relative humidity (> 90%) for 7 to 10 days, or 3) cloudiness, dewfall or light drizzle.
- For neck blast outbreak – (1) at least a low to moderate leaf blast incidence during tillering stage, or in

susceptible trap plants, and (2) low night temperatures (< 20°C) during panicle initiation to flowering stage, or (3) cloudiness, dewfall or light drizzle during the reproductive phase.

- If any one rule is not fully satisfied, more conducive changes in any other rule may compensate and lead to outbreak of leaf or neck blast. Example, intense dewfall, extended dew period, or frequent drizzle may compensate for high temperature (>20°C) regimes up to 26°C or more.

Nitrogen management with need-based application of fungicide may be adopted to control blast on a susceptible variety. It is advisable not to top dress with nitrogen fertilizer until fungicide is applied, if conditions are conducive for disease development. Fungicides like ediphenphos 50 EC (1 ml/l) or tricyclazole 75 WP (0.6 g/l) or carbendazim 50 WP (1 g/l), may be used to check further development of the disease (18). Yet, farmers find it difficult to adopt forecast-based chemical protection to control blast (66).

Blast continues to be a threat to realizing yield potential in cultivars in all the rice ecosystems (19). The methodology for uniform blast nursery (UBN) evaluations of germplasm accessions was developed at Central Rice Research Institute, Cuttack. Studies indicated that resistance in rice cultivars to blast disease was governed by genes that were either monogenic dominant or monogenic recessive or digenic or polygenic (79). In Japan, Kiyosawa (36) made gene analysis using seven stable blast isolates. Thirteen dominant resistance genes were identified at eight loci. By analyzing 132 isolates collected during 1962-68 from all-over the country, Padmanabhan *et al* (80) identified 31 races using international blast differentials *viz.,* Raminad Strain 3, Zenith, NP 125, Usen, Dular, Kanto 51, CI 8970 (s) and Calaro. After intensive investigations for more than a decade on the leaf blast lesion attributes, screening method, and purity in international differentials, Veeraraghavan (120) established the prevalence of only one race, IC-17 in India. Many fungicides like tricylazole control blast effectively (48).

Diseases in plants are caused by an interaction between host, pathogen and environment. Evidence on co-evolution of host and pathogen suggest for the presence of large variability in both host and pathogen. This variability must be dynamic i.e., continuously undergoing change, if the two organisms are to exist without eliminating each other. Results from investigations on rice and *Pyricularia grisea* Saccardo (= *Pyricularia* (*Piricularia*) *oryzae* Cavara (= telomorph *Magnaporthe grisea* (Hebert) Barr comb nov.) demonstrate variability as a dynamic weapon for the survival of the combatants in the environment. This variability can be equally used as a tool for humans to tilt the balance in the contest to favour the crop species. Based on such an understanding, pathologists and breeders used METs in India to successfully develop many varieties with durable resistance to blast (49).

From limited tests, Sridhar *et al* (112) reported the usefulness of blast resistance genes in combinations. During 2001, extensive multi-environment tests were made with near-isogenic lines carrying different blast resistance genes across the country. Tadukan carrying resistance gene *Pi-ta* showed small lesions infecting < 2% leaf area indicating a very high level of durable resistance to blast disease. They clearly demonstrated the expression of a high degree of resistance in A57 carrying three resistance genes (*Pi-1, Pi-2* and *Pi-4*). The performance of BL 245 with two resistance genes (*Pi-2* and *Pi-4*) and C101LAC (*Pi-1*) was comparable to A57 (58, 60). The performance of these NILs was marginally superior to the resistant checks (Tadukan, Rasi, Tetep and IR 64) and commercial cultivars like Triguna, Tulasi, Vikas, Vivekdhan 62, VL Dhan 221, VL Dhan 61 and VL Dhan 81. Besides these, different state variety release committees have so far released over 250 cultivars resistant to *Pyricularia grisea* (25).

## Sheath Blight

Sheath blight disease is caused by the fungus *Thanetophorus cucumeris* Frank Donk., anamorph *Rhizoctonia solani* Kuhn. It is widely prevalent, particularly in irrigated rice ecosystem but seldom assumes epidemic proportions. The asexual phase of the pathogen actually incites the disease in plants just above the water line in fields to cause damage to yields. The disease appears in moderate to severe intensity in a few states like Andhra Pradesh, Assam, Kerala, Orissa and West Bengal. Although sheath blight occurs, its impact on rice yields is demonstrated to be relatively less in the country when judged against blast. Losses from sheath blight in irrigated ecosystem may be around one tonne per hectare (62). *R. solani* as a pathogen causes lesions and necrosis in plant tissues. The profuse saprophytic external growth visually appears to be more threatening due to thick mycelia and numerous sclerotia. The pathogen affects all plant parts *viz.*, sheaths, internodes, upper leaves and panicles. On sheaths, *R. solani* infection leads to appearance of spots that are grayish in colour and ellipsoid or ovoid in shape. The leaf of the affected sheath dries up. In humid weather, white threads of fungal body can be seen all over the surface of leaf sheaths. The infected leaves and internodes turn grey to straw colour with lateral brown bands resembling snakeskin.

On the infected portions, initially white mycelial knots (1-2 mm) appear which later turn to brown and dark brown flattened visible structures called sclerotia. Many such sclerotia aggregate and appear in a large compact mass with dark brown colour. They adhere to plants, but are easily detached even by gentle wind. Once detached, the sclerotia are carried by water stream, and upon coming in contact with plants get adhered to the sheath and initiate the disease at the point of attachment. The infection then ascends to the upper part of the foliage and to flag leaves. Plants become more susceptible with increasing age apparently due to loosening of leaf sheath from the culm. Early sowing, early planting and dense plant populations encourage disease development. Application of nitrogenous fertilizers at high doses increases the severity of sheath blight disease. The progress of the disease is governed by ambient humidity and temperature. Sheath blight epidemics occur in highly humid conditions with a daily mean temperatures hovering at 30°C. Under favorable conditions, the disease spreads to top portions of the plant and chokes the emergence of panicles. *R. solani* invades spikelets causing sterility or improper grain filling. In southern parts of India, this pathogen does not produce any propagative air-borne spores. However, in cooler conditions, air-borne sexual spores are produced.

The pathogen has a wide host range and occurs on all grasses and broad-leaved weeds that grow on rice bunds. Similar symptoms and sclerotia are produced on all these hosts. The sclerotia fall into water in fields and initiate infections on rice crop. Even if leaves of rice plants come in contact with infected weeds on bunds, they pick-up infection and spread the disease.

The nutritional requirement for the mycelial growth in *R.solani* is so minimal that virtually on the surface of water, the mycelia grow and descend from bunds on to fields. Hence, keeping bunds clean of weeds will help in checking the disease spread from primary sources.

Sheath blight spreads very slowly and a change in environment from humid to dry weather will stop the disease progress. If the humid weather is likely to persist for a prolonged time and disease is noticed all along the peripheral areas in fields, a foliar spray application of either validamycin 3 L (2.5 ml/l) or hexaconazole 5 EC (2 ml/l) will prevent its spread. In addition to keeping bunds clean, soil solarization helps to reduce sheath blight incidence. Only T 141, OS 4, BCP 3, Saibham, Buhjan, Suduwee, Remadja, Ta-Poo-Cho-Z, Nangmons 4, Athebu, Phourel and ARC 15368 have been identified as donors expressing moderate resistance to sheath blight. So far no rice culture has been found to be resistant to *R. solani* infection. Pankaj, Swarnadhan and Vikramarya exhibit a good degree of tolerance to sheath blight. Transgenic IR 64 rice plants over-expressing rice chitinase, a pathogen-related protein, have been generated (16). The performance of these transgenic plants is yet to be tested in India.

## Brown Spot

The fungus, *Drechslera oryzae* Subramanian and Jain (=*Helminthosporium oryzae* Bred De Haan) is the causal agent of brown spot of rice. The perfect stage is not normally seen. Dastur (15) reported *Cochliobolus miyabeanus* (Ito & Kuribayashi) Drechsler ex. Dastur to be the perfect stage of the brown spot pathogen. In mid-1940s, serious damage from brown spot was generally associated with other primary problems such as nutrition deficiency and physiological disorder. With the introduction of dwarf or semi-dwarf high-yielding rice cultivars, fertilizers and irrigation, brown spot became a minor disease. In late 1990s, however, increased incidence of brown spot disease was reported particularly in rainfed uplands in Chotanagpur area and in rainfed hill areas in Uttaranchal (19). The characteristic symptom occurs on leaf as circular to oval spots that are dark brown to purplish brown. These spots frequently run together so that the entire leaf surface withers and dries up. The spots may also be distinct, isolated and scattered through out the leaf surface. In wet weather, dark olivaceous velvety mats, made up of creeping hyphae, erect condiophores and conidia of the fungus develop on these spots. Symptoms may also appear on the coleoptile, leaf sheaths, panicle branches and glumes. Black or dark brown spots appear on the glumes and entire surface of glumes may be covered with spots. This pathogen is well known to infect weeds like *Leersia hexandra* (Sw.), *Brachiaria mutica* (Stapf.), *Panicum repens* (L.) and *Echinochloa spp.* and wild rices.

The disease is seed-borne internally as mycelia. If affected seeds are to be sown, fungicide seed treatment (with copper oxychloride) will be useful in reducing the disease on seedlings. If brown spot symptoms are observed in the later growth stages of the crop, iprodione 50 WP (2 g/l) or mancozeb 75 WP (2.5 g/l) can be used to control the disease. Other methods that are suggested and practiced for the control of this brown spot disease include field sanitation, crop rotation, proper fertilization, good water management and use of soil amendments. Stubbles help to carry the pathogen and enable its survival during the absence of rice crop. Donors such as T 141, BAM 10, Ch 13, Ch 45, BJ 1, Sigadis and TKM 6 are resistant to brown spot pathogen. Tetep, ARC 7117, Ramtulasi, Manoharsali, Raminad strain 3 and Tadukan are tolerant to this disease. Among released cultivars, Rasi, IR 36, CNM 539, CNM 540 and Jagannath possess resistance to brown spot disease.

## Sheath Rot

Sheath rot was first reported to occur in India in 1973 (1). The pathogen *Acrocylindrium oryzae* Sawada was revised later as *Sarocladium oryzae* Gams and Hawksworth. Sheath rot has become widespread in Andhra Pradesh, Bihar, Kerala, Orissa, Tamil Nadu, West Bengal and northeastern states. Wounds and entry points created by leaf sap sucking pests facilitate early infection. Lesions begin to grow as an oblong or irregular spots with typically brown margins. The spots enlarge, coalesce and cover almost the entire sheath. Severe outbreaks caused considerable yield losses as in Nellore (55) and in Punjab (86). The pathogen causes dark brown to chocolate brown spots on upper sheaths resulting in either partial or incomplete emergence of the panicle along with seed sterility. The disease usually appears during heading to maturity stages. A whitish to light-pink powdery growth of the fungus may be found on the panicle inside the sheath. The young panicles may rot completely.

Grains inside the choked panicles and on the partially emerged panicles may be chaffy, light to dark-brown and covered by a white to light-pink mat of mycelium and spore mass. If the pathogen attacks after the panicle emergence, the grains may be partially or completely filled and the glume discoloration is caused. Availability of moisture in the form of rain or dew at flowering encourages an outbreak of sheath rot disease (55). Application of carbendazim (0.1%), twice at fortnightly intervals commencing from boot leaf stage is effective in preventing damage from *S. oryzae*. Thiophanate-methyl, ediphenphos (0.1%), hexaconazole, propiconazole and mancozeb are also useful in controlling the sheath rot infection. Many cultivars possessing resistance or tolerance to sheath rot are available (25).

**False Smut**

False smut of rice caused by *Ustilaginoidea virens* (Cooke) Takahashi (telomorph *Claviceps oryzae-sativae* (Cke) Tak.) is a well-known minor disease for more than a century in India. In fact this fungus was first described and named by Cooke in 1878 based on the specimen collected from Tamil Nadu. In the past, a few spikelts were infected by the pathogen, and big (1-2 cm) and dark green olivaceous smutted balls used to appear. Farmers always considered the appearance of the false smut incidence in their fields as a sign for bumper harvest (91). The conditions, which generally favour crop growth, also favour smut development (108). During 1980s, there was a change in the nature of infection caused by the fungus. Infection was noticed on almost all the spikelets in a panicle; bulk of the spikelets in panicles turned into small olive-green to black smutted balls (69). Even the cattle refused to feed on the straw from infected fields as it had a smear of colored spores fallen on to it from smut balls. This disease has been reported to be prevalent in Andamans, Andhra Pradesh, Assam, Haryana, Jammu and Kashmir, Karnataka, Madhya Pradesh, Maharashtra, Manipur, Orissa, Punjab, Rajasthan, Uttaranchal and West Bengal. False smut disease epidemics were observed in 1984 in Guntur district, Andhra Pradesh and in 1990 in Punjab. Damage to grain yields was estimated to be up to 70% (69).

The effects of the pathogen on the host plants are visible only after flowering when the fungal infection of the ovary in individual kernels transforms them into large, velvety, green smut balls (pseudomorphs). At first the smut balls are small and remain confined between the glumes. They gradually enlarge and become 1-cm or more in diameter enclosing the floral parts. Young spore ball that is flattened, smooth and light yellow in colour, changes to orange, yellowish green, green, olive-green and finally to greenish-black. Pseudosclerotia are covered by a membrane which bursts releasing the spore mass. In Uttaranchal at high altitudes in the hills, true sclerotia are found loosely attached to pseudosclerotia and become prominent after the dispersal of spores (110). The true sclerotia contain asci in perithesia embedded in a stroma. Pseudosclerotia retain their viability up to 7-months at room temperature (25-35°C). Primary infection is from pseudosclerotia hibernating in the soil and chlamydospores. Conidia play an important role in the secondary infection of the host. Chlamylospores are air-borne. *Ustilaginoidea virens* has been recorded on maize, *Oryza officinalis* Wall ex Watt., *Digitaria marginata* Link., and *Panicum* spp. which may serve as important reservoirs of primary inoculum. High humidity (>92%) and rainfall accompanied by cloudy days during flowering favours the disease incidence. Use of sclerotia–free seed and crop rotation are important to check disease incidence. At the time of harvest, the diseased plants should be removed, and destroyed to prevent sclerotia from falling on to the ground. The disease is also favoured by application of nitrogen fertilizer. Split application of nitrogen (each at 10 kg/ha) should be recommended in disease prone areas. The fungus is primarily soil-borne and the sclerotia in soil remain viable for a long period of time. The secondary infections occur with the help of air-borne sporidia. Spray application of mancozeb 75 WP (2.5 g/l) or chlorothalonil 75 WP (2 ml/l) around flowering by targeting spray on only emerging panicles, can help control the disease incidence (18). Bala, Cauvery, Sabarmati, Prakash and Pankaj have been reported to be free from false smut infection in rice fields.

## Grain Discolouration

Various organisms may infect rice grains before or after harvest, causing discolouration. The discolouration may appear externally on glumes or internally in kernels, or on both glumes and kernels. A high seedling mortality occurred when seed discoloration was in both embryo and endosperm regions (81). On glumes, symptoms vary depending upon the organism involved and the degree of infection. Sometimes there are distinct black dots, usually caused by fungal reporductive structures on normal or bleached areas of the glumes. Often the lesions are pale or greyish in centre with a dark brown margin. The kernels internally, and sometimes also externally are stained with distinct colours. *Monascus purpureus* (Went) causes a red colour. *Wolkia decolorans* Van der Wolk is the cause of yellow grain. *Penicillium puberulum* Bain stains the grain orange. *Erwinia herbicola* (Lohnis) Dye causes black rot. Pink spots are caused by *Fusarium*. On glumes green and blue colored spore appear due to infection with *Aspergillus* and *Penicillium* spp. The organisms associated with grain discolouration are *Drechslera oryzae* and *Alternaria padwickii* (Ganguly) Ellis, *Gibberella fujikuroi* (Sawada) Ito, *Nigrospora oryzae* (Berk. & Br.) Petch. *Epicoccum* spp., *Curvularia* spp., *Phoma sorghina* (Sacc.) Boerema *et al* and *Alternaria*. Some of the fungi could also develop as saprophytes on harvested grains during storage when conditions are abnormal (124). These storage fungi are *Aspergillus* spp., *A. flavus, Penicillium* and others. High atmospheric humidity and rainfall accompanied by cloudy days during flowering favors grain discoloration. When infection is by field fungi, the main effects are reduced viability and grain quality. Seedling blights and other diseases may occur when the seeds are planted. When the infection is by storage fungi, besides a reduction in viability and grain quality, there may also be production of toxins. Injuries from insects in the field, or from wind and rain storms during storage, may also increase grain discolouration. Spikelet sterility and grain discoloration in Andhra Pradesh was suspected to be associated with mite, saprophytic fungal species and white-tip nematode (92).

Grain discolouration is a serious problem in hybrid rice seed production plots. This might be due to the clipping-off of flag leaf and foliar application of gibberellic acid to encourage pollination and obtain a higher hybrid seed-set in cytoplasmic male sterile female parental lines. The hybrid seed production plots can be protected from grain dicolouration by application of propiconazole 25 EC (1ml/l) in the evening hours around panicle initiation to flowering period (26). This application can be limited to cover only emerging panicles.

**Fusarium Wilt or Bakanae**

Foot rot or *bakanae* disease is caused by *Fusarium moniliforme* Sheld (teleomorph *Gibberalla fujikuroi* (Sawada) Wr. It occurs in almost all the rice growing areas in the country. *Fusarium* wilt of seedlings in nursery has been reported to occur in Andhra Pradesh, Assam, Haryana, Karnataka, Manipur, Tamil Nadu, Uttar Pradesh and West Bengal (19). In Haryana and West Bengal, yield losses may vary from 3 to 95% due to foot rot disease. Diseased seedlings in nursery are lanky, pale yellow and taller than the healthy and uninfected plants. Such seedlings die either before or after transplanting. Diseased plants in fields show brown discoloration of tissues at lower nodes that facilitate an easy separation of the root system at collar region resulting foot rot symptoms. Infected plants have fewer tillers and their leaves dry one after the other, which wilt and die within a few weeks. Presence of white or pink mycelial growth, adventitious roots from the lower internodes, and wider leaf angles with stem are diagnostic features of the disease. The fungus also causes necrotic lesions on the leaves, stem, panicle or kernel leading to withering of growing shoots. The disease is both seed-borne and soil-borne. The panicle infection is caused by secondary air-borne conidia and ascospores discharged from diseased plants from heading to harvest. The fungus grows intercellularly in stigma and anthers and finally reach to cover the ovary. The fungus remains viable for 16-28 months in seeds and infected plants (119). Diseased debris also serves as a primary source of inoculum as the fungus can survive in plant debris for 10-28 months. Lower temperature (5-10°C) and humidity (RH 30-35%) conditions favor the survival of the pathogen in infected grains and stubbles. Cultural practices that aid in disease management include selection of healthy seed, late sowing of early maturing cultivars and crop rotation. Foliar application of benzimidazoles like carbendazim 50 WP or benomyl 50 WP (1 g/l) can control foot rot disease in fields. Seed treatment with carbendazim or benomyl (2 g/kg seed), checks the seed-borne infection. Aryan, HR 98, HR 101, HR 105, Co 18, Co 22, ADT 8, Ptb 7 and PR 106 are considered to be resistant to *Fusarium* (106). In glasshouse, resistance to foot rot pathogen was found in ADT

10, ADT 41, ASD 16, Amulya, Sabita, Akitphou, Eriemaphou, Prasanna, GR 4 and MTU 1010 (38).

**Udbatta Disease**

Udbatta disease is the infection by a *Ephelis oryzae* Syd. (teleomorph *Balansia oryzae-sativae* (Narasimhan & Thirumalachar) Hasioka) that makes the panicle to appear like incense stick. Sydow in 1914 first described this disease from India. Shivanandappa and Govindu (105) considered udbatta disease as an important disease in some areas in Karnataka causing losses. The symptoms are evident at the time of panicle emergence during October to January. The boot leaf of the affected plant is slightly distorted and a silvery lustre is observed on upper surface of the leaf blade. The infected panicles emerge as small cylindrical rods covered by white mycelium. The infected plants are stunted. Sometimes the mycelium and conidia form narrow stripes on the flag leaves along the veins before the panicle emerge. Usually all tillers of a plant are infected, indicating systemic infection. The fungus is internally seed-borne and infection is initiated at the time of emergence of the panicle (88). Seedling infection is the principle mode of infection. High soil temperature (28°C) and abundant soil moisture in the nursery bed are conducive to the development of disease (45). The fungus survives on several grass hosts, including *Aeschynomene indica* L., *Eragrostis tenuifolia* (A. Rich.) Hochst. ex Steud., *Arthraxon ciliaris* P. Beauv. fm. *glabrescens* Hack., *Sacciolepis indica* (L.) Chase, *Cynodon dactylon* (L.) Pers., *Pennisetum* sp. and rye grass. Host plant resistance is not yet discerned although CB11 was reported to be free from this disease. A hot water seed treatment at 54°C for 10 min controls the seed-borne infection (44).

**Leaf Scald**

The causal agent of leaf scald disease is *Microdochium oryzae* Hashioka and Yokogi Samuels (syns. *Gerlachia oryzae* (Hashioka & Yokogi) Gams.; *Rhinchosporium oryzae* Hashioka & Yokogi). The disease was first reported in 1973 from Kerala (103). The ascigerous stage of *Monographella albescens* (Thumen) Parkinson *et al*, has been reported from northern India (41). Zonate lesions typically appear on lower leaves starting usually from the tips or margins. Oblong or diamond-shaped water-soaked brown blotches appear initially. As the lesions become older, zonation fades in lesions that turn to light grayish-olive and finally to straw colour. Lesions coalesce covering the entire lamina, which in later stages dries up. Lesions on neck of panicles and on spikelets are also known. Leaf scald does not pose as a problem in rice cultivation. It appears only occasionally and that too near the crop maturity. Thiophanate-methyl 70 WP is very effective in controlling leaf scald (14). Tetep, TKM 6, Rasi, VL Dhan 221, IET 1192 and Jaya are considered as resistant, based on field observations.

## Stem Rot

The disease is caused by *Sclerotium oryzae* Catt. (anamorph *Helminthosporium sigmoideum* Cav. and telomorph *Magnaporthe salvinii* (Catt.) Krause and Webster). Although stem rot is detected often in Haryana (2) and a few other states, it is considered to be a weak pathogen. This pathogen at best may cause a premature lodging in crop. Stem rot infection starts with sclerotia in the soil or plant debris. The sclerotia are disseminated by irrigation. The first symptom appears as small, black, irregular lesions on the outer leaf sheath near the water line. The fungus penetrates mostly through wounds into the inner leaf sheath and finally rots the basal portion of the stem. Numerous black, round and shining bodies (sclerotia) are formed externally and internally on the affected sheaths and stems. The plants lodge and put-forth a number of green secondary tillers but the earheads produced are chaffy. TN 1, IR 8, Jaya, Latisail, Pankaj and Rasi are resistant to stem rot (107).

## Kernel Smut

Kernel smut also known as kernel bunt is a floral disease of rice that is caused by *Tilletia barclayana* (Bref.) Sacc. and Syd. (syn. *Neovassia barclayana* Bref., *T. horrida* Tak., *N. horrida* (Tak.) Padw & Khan). The disease occurs in most of the rice growing areas (10). During 1997, a widespread occurrence of kernel bunt was reported in Punjab where 80% of panicles were infected. National Seeds Corporation encountered a heavy incidence of kernel smut in seed multiplication plots in Orissa during 1980, because of which 35% of the seed lots were rejected. The disease appears after grain formation in rice. Generally, some of the grains in a panicle are attacked which may be partly or wholly infected. Minute black pustules or streaks burst through the glumes of severely infected grains. Black spores released from infected grains settle on apparently healthy grains or leaves which are more evident early in the morning. The fungus is seed-, soil- and also air-borne. Teliospores of the fungus fall on ground from infected grains and also reach soil along with the infected seed where they germinate and produce sporidia. Primary sporidia further produce secondary sporidia that are banana-shaped. They become air-borne and multiply generally on leaves of rice and other plants. After infection the hyphae remain in between aleurone layer and seed coat, digesting the endosperm and producing the sorus. Humid weather, frequent rainfall and temperatures ranging between 23 and 33°C during flowering are favourable for disease development. Teliospores constitute primary inoculum. The disease is more incident in sandy loam soils (109). Cultural practices that reduce disease severity include the use of disease-free seed, avoidance of excessive use of nitrogenous fertilizers and hot water treatment of seed. Spray application of benzimidozole (Benlate 50 WP or Thiophante methyl 70 WP-0.1%) during anthesis prevents damage from kernel bunt (10).

## BACTERIAL DISEASES

### Bacterial Leaf Blight

Symptoms of bacterial leaf blight disease appear from the tip or edges of leaves as yellow water soaked and undulating lesions parallel to the veins that later turn to straw yellow colour. The disease initially starts from either one or both sides of leaf margin. As the disease progresses, the drying spreads downwards and inwards of leaf blade causing drying and death of the leaf. Often amber coloured bead-like bacterial exudates are present on lesions. In systemic infection, seedlings wilt and die. Grains are either partially filled or become chaffy. Bacterial blight is caused by *Xanthomonas oryzae* pv. *oryzae* (Ishiyama) Swigs *et al.* The rod-shaped bacterium is gram negative, 1.75 x 0.60 μ in size, motile with a polar flagellum and is without spore forming capacity. Bacterial cells are surrounded by mucus capsules and are joined to form aggregated and stable mass. After the introduction of dwarf, high-yielding TN1, which was known for susceptibility to bacterial blight, the disease assumed greater importance in India.

Bacterial blight is associated with cyclonic storms during monsoon periods (77). Its incidence and severity is very much influenced by rainfall, winds, rainy days, susceptibility of the cultivar and nitrogen fertilizer application. The yield losses have been reported to vary from 2-74% in such epidemics (53, 93). Severe epidemics of bacterial leaf blight in two consecutive wet seasons (1979 and 1980) in northwestern India, reduced the grain yields drastically. Every year, bacterial leaf blight causes some damage to rice crop in several districts in India (18, 19). The disease occurs in two phases. *Kresek* or wilt phase usually occurs during active tillering stage and results in the death of tillers. Seedlings are killed by *kresek* if inoculum pressure is very high (89, 116). The leaf blight phase is common in all the rice growing areas and appears from maximum tillering to heading stages. Sometimes, farmers mistakenly identify the disease as crop drying due to unfavorable conditions; they expect the crop to recover. But under favorable conditions, the disease spreads quickly to devastate the entire crop. In diseased plants, panicle emergence is poor and spikelets are partially filled and discolored. Bacterial leaf blight causes more damage in rice fields where high doses of nitrogen fertilizers are used as in Punjab and Haryana. Epidemics of bacterial leaf blight occur frequently all along the Indian coast exposed to cyclonic storms and intense monsoon rains. Injuries to rice leaves caused by rain storms, strong winds, sap sucking insects and intercultural operations form entry points for the pathogen (53) besides other natural pores present in rice leaves. Once inside the leaf, the bacteria multiply quickly to produce millions of bacterial cells and cause blight. Milky dewdrops ooze out of the affected leaf lesions containing numerous bacterial cells.

The damage from natural epidemics of bacterial leaf blight (BLB) of rice caused by *Xanthomonas oryzae* pv. *oryzae* was examined in 400 farmers' fields in four rice-growing districts in India during wet seasons. One hundred contiguous farm fields were rated for BLB on a 0-9 scale. Disease prevalence at levels above the economic threshold (score >5) ranged from 7 to 39%. Studies were made with samples from representational farms taken at the rate of one farm for every 10% disease prevalence out of 100 farms per district. The spatial pattern of BLB development within and among rice hills was either random or uniform. On the representational farms, the mean BLB severity was 65-71%. The yield loss estimated on the attainable yield using regression models ranged from 17-31%. In calculating production loss, the actual production in the entire district is usually taken assuming damage from disease at all farms, and this leads to an over estimation. A modification was made to consider disease prevalence as a percentage of farms with disease above the economic threshold level in a district. This adjustment enabled the production loss estimate to be tailored to the actual area damaged by the pathogen. By multiplying district production estimates by the disease prevalence (%), production loss in a district was precisely derived. In BLB epidemics, the most conservative estimates for the production yield losses in districts were 3 to 16%. The production losses during the four epidemics were considerably different ranging from 92,000 to 105,000 tonnes in Nellore, 30,000 to 36,000 tonnes in West Godavari, and 46,000 tonnes in Karnal to 22,000 tonnes in Rangareddy district (87).

The studies on pathogenic variation from India deal with the leaf blight phase only, as this is the major problem encountered in rice fields. Generally, damage from *kresek* or seedling wilt is limited to Haryana and Punjab (19). Gupta *et al* (31) reported the presence of 11 virulence genotypes in *X. campestris* pv. *oryzae* populations from northwestern India based on differentials. Reddy and Reddy (101) collected 150 isolates from 25 locations in India and classified them into two pathotypes: pathotype I was avirulent on DV85 but virulent on Cemposelak and Java 14, and pathotype II was virulent on DV85 but avirulent on Cemposelak and Java 14. Pathotype I was further divided into two sub-groups *i.e.,* pathotype Ia and Ib based on the avirulence or virulence, respectively on another differential, IR20. They reported that pathotype Ia was prevalent in Punjab, East Uttar Pradesh, Tamil Nadu and Kerala, while pathotype II was limited to West Bengal. DNA finger printing of 67 isolates of *Xanthomonas oryzae* pv. *oryzae* collected during 1994 and 1995 from 18 locations in India belonged to a single lineage representing pathoype Ib (125). The resistance to bacterial leaf blight disease in some cultivars was considered to be due to a combination of two or more genes or to new genes that were often described as dominant, recessive, inhibitory, complementary or polygenic.

Near-isogenic lines (NILs) in the background of IR 24 possessing a single gene or multiple genes in various combinations were developed at International Rice Research Institute (IRRI), Philippines (34). Sridhar *et al* (122) found the effectiveness of individual resistance genes *xa8* or *Xa21* in NILs. In field inoculation tests, Shanti *et al* (104) found effectiveness of some individual genes with the broadest spectrum of resistance in *xa5* and *Xa7*. Sridhar (111) concluded that lines with two- or three-genes in various combinations in IR24 background (*Xa4* + *xa5*, *xa5* + *Xa21* and *Xa4* + *xa5* + *Xa21*) were resistant to all the pathotypes occurring in the Eastern India.

Rice genotypes carrying resistance genes to bacterial blight disease were evaluated (26) in multi-environment tests (METs). Differential culivar like Java 14 (*Xa1, Xa3* and *Xa12*) or DV 85 (*xa5* and *Xa7*), and cultivars IET 8320 and IET 8585 carrying several unknown resistance genes showed resistance at many of the test locations (18). Their stable performance over decades indicated durable resistance to bacterial blight. Near-isogenic lines (NILs) carrying different single genes for resistance to bacterial leaf blight *viz*., *Xa1, Xa3, Xa4, xa5, Xa7, xa8, Xa10, Xa11, xa13, Xa14, Xa18,* and *Xa21,* and two genes *viz*., *Xa4* + *xa5*, and *Xa4* + *xa13* were shown to be susceptible at 7-12 locations in the country (59). Yet, resistance genes *viz*., *Xa4, xa5, xa13,* and *Xa21* in other two-, three- or four-gene combinations were found to exhibit a high level of field resistance. This study also brought out the futility of identifying and classifying pathotypes based on the reaction of differentials to *Xanthomonas oryzae* pv. *oryzae* (61). The IR 64 lines with bacterial blight resistance genes (*xa5, xa13* and *Xa21*) developed at the Central Rice Research Institute, Cuttack through marker-assisted selection showed high disease resistance and yield potential (64).

Transgenic rice with cloned gene *Xa 21* in IR 72 developed at the International Rice Research Institute using particle bombardment was tested in a greenhouse possessing containment facility to meet with bio-safety regulations. Several isolates of the pathogen *viz. Xanthomonas oryzae* pv. *oryzae* derived from different rice growing areas in India were used in leaf clipping inoculation tests made at 18, 25 and 70 days after sowing. TN1 was highly susceptible while Ajaya was resistant. Transgenic rice TC-DRR 2 (IR 72 carrying cloned *Xa 21* gene and with *Xa 4* already present) was susceptible to several isolates (4-12), although its overall performance was rated as moderately susceptible. The performance of this transgenic line was poor compared to the near isogenic line IRBB 21 (IR 24 carrying *Xa 21* gene). The pathogen populations studied indicated nine isolates as more or less equally virulent, three moderately virulent, and six others as avirulent. Transgenic TC-DDR 2 (IR 72 with *Bt* gene) and cultivars IR 72, IR 24 and PR 106 were susceptible to all the virulent isolates. Only one isolate (DXO11) that was originally isolated from a small lesion on Ajaya in fields, when clip inoculated on to healthy plants, caused susceptible reaction on Ajaya and all other genotypes tested (51)

Rain splashes and wind aid in dissemination of the bacterium. Field to field irrigation also aids in the spread of the pathogen. There are no effective chemicals to protect the crop from bacterial leaf blight disease. Streptocycline and cow dung applications have been reported in production oriented surveys to contain the spread of this disease (19). Extensive studies, however, have shown their ineffectiveness in controlling the bacterial leaf blight disease (18). The options for control of bacterial blight include use of resistant cultivars and judicious nitrogen management. If favorable weather persists and disease is already incident on crop, it is advisable to withdraw application of nitrogen fertilizer. Some of the commercially cultivated resistant cultivars are Ajaya, PR 4141, PR 114, Swarna, MTU 4870, HKR 120, IR 36, IR 64 and Saket 4. Donors identified for resistance to bacterial leaf blight are BJ 1, TKM 6, Lacrose-Zenith-Nira, Java 14, Wase-aikoku and Sayaphal.

## Bacterial Leaf Streak

Symptoms of the disease appear on leaf as linear water soaked, translucent streaks of 1 to 4.5 cm in length. Bacterial leaf streak is caused by *Xanthomonas oryzae pv. oryzicola* (Fang *et al*) Dye. The streaks are dark green in colour at first and later turn yellowish-orange brown. The progress of the leaf streak is longitudinal. Several such steaks however, may coalesce causing total blight of the leaves. Yellowish droplets of bacterial exudates are often visible on the surface of leaf streaks. Generally, symptoms are more pronounced in *indica* cultivars. Infection of spikelets results in death of stamen and browning of glumes. The exudates with bacterial cells on infected leaves serve as inoculum for rapid secondary spread of the pathogen. The rod-shaped bacterium is Gram negative, 1.2 x 0.3-0.5 μ in size, motile by a polar flagellum and without any spore formation. Usually the bacterium is capsulated (17). The bacterium gains entry either through injured epidermal cells or directly through the stomata colonizing the sub-stomatal cavity and then invades intercellular spaces of the mesophill cells. Invasion of bacterium is confined to interveinal parenchyma, bundle sheath parenchyma and sclerenchyma extensions connecting the vascular bundles to lower and upper epidermis. Hence discrete streaks limited by veins are formed. The transmission of the disease through seed has been reported (3, 115). The pathogen is carried from field to field by irrigation. Pathogen is also known to survive in infected straw. Humid conditions, intermittent rainfall and leaf or plant injuries from storms or cyclones favor disease incidence. During 1988 and 1989 in the north coastal Andhra Pradesh and southern parts of Orissa, a severe outbreak of this disease was encountered (90). The incidence of bacterial leaf streak occurs sporadically and is often restricted to only a few fields. Occasionally, damage from bacterial leaf streak may be encountered on seedlings in nursery. Streptocycline spray application may check the disease from any further damage to seedlings. Three pairs of independent resistant genes were found to govern resistance in BJ 1 (75). BJ 1, IR 36, TN1, Ptb 10, CR 1001, Zenith and TKM 6 are known to be resistant to bacterial leaf steak pathogen.

## VIRAL DISEASES

### Tungro Virus Disease

Tungro virus disease occurs if a susceptible crop, virus inoculum and green leafhoppers (*Nephotettix virescens* Distant.) to carry virus as vectors, are available in a rice field. The disease attracted the public attention for the first time following an epidemic outbreak in the eastern parts of Uttar Pradesh, Bihar, and West Bengal (95). Later, the disease apparently moved towards peninsular India during 1977. Severe tungro virus disease outbreaks threatened rice production in Andhra Pradesh and Tamil Nadu during 1984. Regular production oriented survey records clearly established the discontinuous occurrence of tungro in the country (19).

Tungro virus disease epidemics occur frequently in some parts of the country as indicated by the production oriented surveys in rice fields. The aim was to quantify the levels of disease incidence, analyse the historical data on epidemic occurrence and estimate the grain yield losses. Tungro disease which appeared in north India for the first time during 1967, moved in 1977 to peninsular India. Tungro outbreaks were discontinuous within a district, state and country over the years. The outbreaks of this virus disease were restricted to irrigated and rainfed shallow lowlands. The loss from tungro epidemics steadily increased during 1979-1980. Three major epidemics in farmers' fields (1984, 1988 and 1990) caused severe quantitative and monetary losses. Each of the other two epidemics (1987 and 1998) led to a similar loss of about a million tonnes in rice production but showed a steady increase in loss in terms of real value. An epidemic outbreak of tungro during 2001 in only three districts of West Bengal caused a paddy or unmilled rice production loss of 0.5 million tonne that amounted to Rs. 2911 millions at current prices. This study clearly demonstrated that tungro epidemics could cause a maximum yield loss of 53% in a district, 23% in any one state and 2% in all-India rice production (63).

Stunting of infected plants and turning leaf colour from green to yellow to orange red characterize tungro disease incidence. Newly emerging leaves of infected plants are often pale with chlorotic intervenial areas. The leaf lamina is often twisted following virus attack. While orange-yellow colouration of the foliage is characteristic, variations often exist ranging from green to pale, or intensely orange, red and sometimes with brown spots. If the plants are infected in early growth stages, there is no flowering. If plants are infected late, there is a delayed and uneven flowering. Tungro virus reduces the number of spikelets in panicles and yield; it also decreases filling, weight and starch content in grains (11).

Two viral particles namely spherical (RTSV – an RNA virus) and bacilliform (RTBV- a DNA para retrovirus) particles were considered to be associated with the tungro disease (32, 102). Recent studies, however, raised the question of involvement of bacilliform virus as a pathogen in tungro disease (62). Viruliferous green leafhoppers *Nephotettix virescens N. nigropictus* Stahl. and *Recilia dorsalis* Motsch., introduce the virus into rice leaves when they probe to suck nutrients. Thus, a tungro-infected plant suffers from damage caused by both virus and the insect vector. The disease can occur at any stage from nursery onwards. The initial disease on rice crop is seen along the weedy border of rice fields which later spreads into the main field. The self-sown of seedlings in infected IR 64 in fields in coastal Andhra Pradesh led to an outbreak of tungro in the ensuing crop also during 1990. Tungro is found only in irrigated and rainfed lowland rice ecosystems. Applying any chemical cannot directly control tungro virus disease. But, the spread of tungro disease can be checked indirectly by controlling the vector with a pesticide application. A low dose application of imidachloprid 200 SL (100 ml/ha) in nurseries after a reported outbreak in the earlier crop can effectively control tungro from affecting the new crop. Practicing a fallow or introducing a pulse or oilseed crop can also break the continuous availability of virus inoculum in fields. Alternatively introduction of resistant cultivars like Nidhi, Vikramarya, IET 8565 and IET 8902 will prevent yield losses from tungro. Many other rice cultivars possessing resistance or tolerance to tungro virus in fields have been released by different states for commercial cultivation (62).

## Grassy Stunt Virus Disease

This virus disease causes severe stunting with profuse tillering. Leaves in infected plants are short, narrow and erect. They show numerous rusty spots or blotches, which spread giving a bronze colour to the plant. There is no flowering in affected plants, which present a bushy appearance. Suspected grassy stunt disease plants were first observed from India by Raychaudhari *et al* (95). Later, the first authentic report on symptoms and transmission properties was by Anjaneyulu (6). Brown planthopper, *Nilaparvata lugens* Stahl., transmits the causal virus. The insect acquires the virus in the nymphal stage and becomes infective at the adult stage. Most infective insects can retain their infectivity until death. The shortest acquisition access period may be 30 minutes; the shortest inoculation access period may be 5 minutes. The incubation period in the plant may be 10-19 days. Brown planthopper invasion in rice first appeared in Kerala during 1973-74, when incidence of grassy stunt virus was also recorded. This virus was responsible for heavy yield losses in about 15,000 ha in Trichur and Alleppy districts. This disease appears rarely and is restricted to Kerala and Tamil Nadu (6, 29). During 1988, a new strain GSV II was detected that showed susceptible reaction on all *Oryza nivara* Sharma et Shastry accessions except DWR 21004-5 (28).

## Ragged Stunt Virus Disease

Ragged stunt virus disease is found seldom in ratoons, particularly towards crop maturity in fields, where brown planthooper was incident earlier. Symptoms of the disease are stunting, twisting and curling of the leaves. The flag leaf is twisted around the panicle and presents a ragged appearance with serrated margins. Some times galls are present along leaf veins. Other symptoms include production of more tillers, incomplete panicle exertion and unfilled grains. Most of these symptoms are clear only at booting stage. The brown planthhopper (*Nilaparvata lugens*) is the vector for this virus.

## Rice Yellow Dwarf

Rarely in a few fields, one or two rice hills may be found to be infected with yellow dwarf virus disease. This disease can hardly cause any outbreak due to long incubation period of the pathogen in the vector. The disease is caused by a mycoplasma-like organism - MLO (46). The first symptom of yellow dwarf is a general chlorosis, especially on the newly emerged and young leaves. The infected leaf colour varies from yellowish-green to whitish-green. As the disease progress, the infected plants become severely stunted, tillering increases markedly, and leaves become soft and droop. The infected plants either do not produce any panicles or produce a few small panicles bearing mostly unfilled grains. *Nephotettix virescens* and *N. nigropictus* transmit yellow dwarf virus. The insect takes around 34 days to become infective and can transmit the pathogen within 2-10 minutes. The incubation period in plant varies from 23-30 days. Thus it takes around 57-64 days for the expression of the disease in the field. Seedling root-dip with oxytetracycline (500 ppm) prevented infection by MLO (96).

## NEMATODE DISEASES

### Stem Nematode

The stem nematode, *Ditylenchus angustatus* (Butler) Filipjev., caused complete failure of rice panicle emergence in rice crop spread over 2.5 m ha in Bengal (9). This nematode is locally called as *ufra* disease. The pathogen is soil-borne. Pre-adult nematodes are found in cottony-wool like mass at the base of stubbles and soil furrow during fallow season. When field is irrigated and planted with rice seedlings, nematodes break dormancy and climb to the foliage to feed on it ecto-parasitically causing leaf discoloration. Later, they accumulate in flag leaf to feed on growing panicle to cause distortion and poor or deformed emergence of panicle. At harvest, the nematodes once again congregate to form cottony-wool like masses or continue feeding on stubbles and ratoons. Semideep and deepwater rice cultivars, Rayada 16-06, Rayada 16-09 and Rayada B3 have been selected as resistant to stem nematode (12). Soaking seeds on triazophos or monocrotophos before sowing drastically reduced stem nematode incidence. Cultivar Padmapani completely escaped nematode infection (8). *Ufra* can cause damage to rice in *Indo-Gangetic* rainfed lowlands particularly in summer season. Therefore, burning stubbles and crop rotation to destroy cottony-wool masses controls *Ufra* effectively (13).

### White-Tip Nematode

In Chattisgarh, the white-tip nematode (*Aphelenchoides besseyi* Christie.) was found to be widely incident in 1935. Pre-adults of the nematodes are found in coiled quiescent stage on the lodicules and inside palea of rice seed. Sowing infected seeds in field activates nematodes which migrate to thin film of moisture present on the foliage. They feed on the growing point of the shoot and when leaves unfurl, whip like white-tips of 5 to 15 cm develop during vegetative stages of crop growth. During flowering the nematodes migrate to feed on reproductive parts of the plant (85). Due to this activity yield will be drastically reduced in affected plants. Adults coil and become quiescent in between seed coat and kernel during grain forming stage. Pre-soaking seeds at 28°C for 12-15 hours followed by hot water treatment at 50-55°C is effective in controlling white-tip nematodes (73).

## POST-HARVEST DISEASES

During post-harvest phase, an estimated 10% of food grains (nearly 20 million tones) are lost due to microbial spoilage and insect attack (123). Besides ambient temperature and humidity, certain fungal diseases, rodents and insects cause damage to stored grains. Proper storage of food grain is one of the most pressing problems in India. The stocks are kept in sheds, old buildings, unused space under cover of tarpaulin, black polythene or plastic sheets. As a consequence of seed-borne infection, grains are discolored and this discoloration degrades grains. The discoloration may be present externally on glume or internally on kernels or on both glumes and kernels. *Drechslera, Curvularia, Fusarium, Alternaria padwickii (Ganguly) Ellis.,* (Syn. *Trichochonis padwickii*) and *Aspergillus* have been found to infect harvested rice grains. Discoloration normally occurs when crop is left in the field itself. A pre-harvest spray application of copper oxychloride, particularly in humid area controlled the grain discoloration (30).

Insect pests that are encountered in stored rice include rice weevil *(Sitophilus oryzae* L.), red flour beetle (*Tribolium castaneum* Herb.), lesser grain borer (*Rhizopertha dominica* Fab.), khapra beetle (*Trogoderma granarium* Evert.), long headed flour beetle (*Latheticus oryzae* Waterhouse), cadelle beetle (*Tenebroides mauritanicus* L.), angoumois grain moth (*Sitotroga cerealella* Oliv.) and rice moth (*Corcyra cephalonica* Stainton). The prophylactic control measure is maintenance of sanitation in storage place, containers and other materials used. Spray application of malathion (0.5%), or pyrethrin (0.01%) on walls, ceilings and floor of storage area as well as gunny bags and other material prevents insects damage during storage. Mixing dried neem leaves in gunny bags with rice can effectively check pests from multiplying.

### Aflatoxin Contamination

Over 300 discoloured rice seed samples covering 220 varieties in 21 states in India were collected from areas exposed to rain. *Aspergillus* contamination was detected in most grain samples. *Aspergillus flavus, A. niger* and *A. ochracious* were identified after a critical examination of more than 600 cultures isolated from these samples. *Aspergillus flavus (*parrot green coloured) was the most dominant contaminant. The next major contaminants were *A. flavus* (olive green coloured), *A. niger and A. ochracious*. Norsoloronic acid (NOR), the first stable bright red-orange coloured intermediate in the aflatoxin biosynthetic pathway, with peaks at 280 and 360 nm was detected in several isolates. Such visual identification can help discern the contaminated grains quickly (99, 100). From temperate and cooler regions in the northern hemisphere, *Aspergillus oryzae* has been reported to produce aflatoxins $B_1$, $B_2$ and $G_2$. We have identified five isolates of *A. flavus* from discoloured rice samples collected in tropical fields as toxin producers. Only one fluorescent blue spot with $R_f$ 0.47 was identified on TLC that matched with that of authentic standard of aflatoxin $B_1$. It was purified in column chromatography and quantitatively estimated by developing TLC. The production of aflatoxin $B_1$ varied (307 to 333 $\mu$g/g grain substrate) with the isolates of *A.flavus* (98). The purified aflatoxin $B_1$ was precipitated with hexane and then re-crystallized, which showed mp 265-268°C. The identity of aflatoxin $B_1$ was further confirmed from change in colour of spot on TLC from blue to yellow on derivatization with $H_2SO_4$, HCl or trifluoroacetic acid in $HNO_3$, and from ultra violet spectrum (with peaks at 242, 265 and 361 nm). The proton NMR and $^{13}C$ NMR spectra corresponded closely with that of tetrahydrodeoxyaflatoxin $B_1$. The infrared spectrum showed carbonyl group at 1731 $cm^{-1}$ and other functional groups at 1655, 1635, 1597, 1557, 1125, 1072 and 1040 $cm^{-1}$ and the molecular weight was estimated at M/z 312 by mass spectrometry, which agreed with the aflatoxin $B_1$ composition, $C_{17}H_{12}O_6$. Bulk of the aflatoxins extracted from *A.flavus* in this study belonged to $B_1$ and neither $B_2$ nor $G_1$ or $G_2$ were detected (98). Aflatoxin contaminated seed produced attenuated or discoloured roots in some cultivars; in others, the growth of roots and shoots was affected. Among the biocontrol agents, culture filtrates of *Trichoderma virens* and *Pseudomonas fluorescens* (isolate, DRPf 005) at 200 ml/kg seed showed complete inhibition of Aspergilli on seed. Garlic bulb extract (5%) proved significantly effective with complete inhibition of all Aspergilli *in vitro*, while treatment of seed with clove extract was found highly effective @ 3g/kg in checking the *Aspergillus* contamination of seed. Of the fungicides, treatment of seed with carbendazim @ 3g/kg completely reduced the growth of *Aspergillus* spp. Among the chemicals evaluated, benzoic acid completely checked *A. flavus* and *A. niger* at 1g/kg and *A. ochraceus* at 4g/kg. Our study had shown that the seed treatment with biocontrol agents like *Trichoderma virens* and *Pseudomonas fluorescens* (DRPf 005 isolate), plant extracts like clove, fungicides like carbendazim and chemicals like benzoic acid prevent or drastically reduce the *Aspergillus* contamination in seed (100).

## ABIOTIC DISEASES

### Environmental Stress Induced Ill-health

The most important environmental factor that determines the health of rice crop is the water availability (27). It depends on monsoon rainfall and irrigation facility available. Variations in rainfall *viz.*, delayed on-set, prolonged breaks and abrupt retreat leading to drought, increased intensity or persistence for a longer period leading to excess, and cyclonic weather can devastate the crop stand, affect tillering and grain filling, and cause sterility. Floods and temporary water stagnation (10 – 30 days) and depth (30–140 cm) affect flowering and grain filling. Farmers cannot also adopt any cultural practices in flooded areas and simply leave the crop to nature in a state of helplessness due to the absence of a proper drainage system. Damage from water stress due to drought may be reversible or irreversible. Drought stress reduces leaf area, transpiration, photosynthesis and translocation, which affect rice yield. A timely application of nitrogen fertilizer at panicle initiation stage can overcome the damage from water stress during reproductive stage (24). Rice is sensitive to ambient low temperature regimes. In northwestern India during rabi, low temperatures fall to destructive or even lethal levels and prevent rice cultivation. If high temperatures prevail at flowering, spikelets become sterile. For inducing tolerance to temperature stress at anthesis, cysteine or ascorbic acid can be sprayed on the crop (22). Plants thus sprayed can withstand high temperatures (45°C) and survive for longer period (10-12 days) even when moisture is unavailable in the soil at anthesis. Diffused light conditions reduced both crop growth and biomass production but enhanced tiller mortality (23). Rice cultivars have been developed with host plant resistance to drought, flood, waterlogged or low light and temperature situations (25).

## Diseases Due to Nutrient Deficiency or Toxicity

When crops absorb little or excess of nutrients than their requirement, they develop abnormal appearances due to ill-health, otherwise called as nutritional disturbances. Symptoms of a nutritional disorder are not necessarily confined to leaves, but may occur on any part of the plant structure, including roots (35). Plant analysis for a particular nutrient or related nutrients, in conjunction with visual symptoms would greatly facilitate the diagnosis of nutritional problems correctly. The symptoms produced on plants by nutritional disturbances are very similar to symptoms produced by several other rice diseases. These symptoms include stunting of plants, poor tillering, uniform yellow or orange-yellow leaf-tips, narrow leaves showing senescence, advancement or delay in heading, and small panicles with poorly filled spikelets. Different nutrient deficiencies might show similar symptoms. Nutritional disorders in rice are caused by deficiencies in N, P, K, Fe, Mn, Zn, Si, or S. Toxicity of elements like Fe, Mn, Al and Bo can lead to ill-health of the crop. Toxicity of substances accumulated in the soil under reduced conditions is due to sulfides, organic acids, or $CO_2$. A high salt injury occurs in the presence of Na-salts. Ill-health of the crop is governed by the nutrient status of soils. Nitrogen deficiency may be encountered in all rice soils. Phosphorus deficiency occurs in lateritic soils, saline soils, acid sulfate soils, and heavy black soils. Potassium deficiency is seen in highly leached, light textured sandy or sandy loam soils. Iron deficiency is normally encountered in rainfed upland calcareous and alkaline soils or in soils with low organic matter and low available Fe. Manganese deficiency is not common in lowland rice but can be suspected when the iron content of soil solution is very high or when soil solution has very high $H_2S$. Zinc deficiency is found in soils with low available zinc status, calcareous saline and alkaline soils, soils with impeded drainage or high organic matter content. Silica would be deficient in lateritic soils and hill soils where due to laterization, all the silica from the surface is leached down. Sulfur deficiency is encountered in highly drained soils or due to use of sulfur free fertilizers such as urea.

Toxicity from iron occurs in acid sulfate soils, lateritic soils, soils with high organic matter or highly reducible iron content, acidic soils and saline soils. Manganese toxicity is found in highly acidic soils or in soils with poor aeration. Aluminum toxicity occurs in highly acidic soils where pH is below 3.5 in upland rice. Boron toxicity is found in highly saline soils and in fields irrigated with water containing a high concentration of boron. Sulfide injury occurs in soils containing sulfates, high organic matter, low Fe and low CEC. Toxicity due to organic acids occurs in peaty acid sulfate soils and in cold soils. Carbon dioxide ($CO_2$) accumulation occurs in calcareous saline soils or in soils with high organic matter; salt injury is encountered in saline-alkaline soils or alkaline soils. The critical limits of nutrients to classify soils as deficient or toxic are nitrogen and potassium < 250 kg/ha, phosphorus (Olsen P) < 5 ppm, calcium 90-100 kg/ha, magnesium < 45 kg/ha, sulfur 8-29 ppm, iron deficiency < 10 ppm, iron toxicity > 400 ppm, aluminum toxicity 300 ppm, salt injury (salinity) EC (electrical conductivity) > 4.00 deciSiemens, and alkalinity ESP (exchangeable sodium percentage) > 15%.

The available options to prevent ill-health due to deficiency or toxicity in nutrients include balanced fertilizer application based on soil test, providing a good drainage, addition of organic matter through green manure crop or pulse crop residue, liming acid sulphate soils (3-6 t/ha), or application of zinc sulphate fertilizer (15 kg/ha once in three crop seasons). Gypsum may be used (6 t/ha) in sodic soils with pH 9-10 to rapidly bring down ESP from 70 to 50% in the very first season (20, 21). In light-textured soils exposed to high rainfall and iron toxic soils, split application of potash (30 kg/ha) at planting, tillering and panicle initiation stages can increase grain yields (22).

## USEFULNESS OF BOTANICAL AND PLANT-DERIVED PESTICIDES

Many plant extracts, neem (*Azadirachta indica* Adr. Juss.) cake or water extracts have been screened *in vitro*. Neem coated urea (at 60 and 90 kg N/ha) was effective in reducing blast disease and increasing yield (33). Plant extracts tested maintained their fungitoxicity to *Rhizoctonia solani*, the sheath blight pathogen, following thermal treatment at 70°C for 10 minutes (39). Plant-derived products with or without azadirachtin as an active principle reduced blast or sheath blight severity in rice following foliar spray application on the rice crop (68). Application of Achook, Neemazal, Neemgold and Wanis reduced the severity of both leaf (38-51%) and neck blast (26-31%). However, with these biopesticide products, similar yield increases were recorded (29%). Achook, Neemazal, Neemgold, Spictaf and Wanis reduced both sheath blight disease score and tiller infection (13-23%). Marginal yield increases were recorded with these biopesticide products (1-13%). These plant-derived products give additional benefit as they also control other insect pests that harm the crop. Neem products inhibit reproduction in brown and white-backed planthoppers, deter oviposition in green leafhoppers and reduce white-ear incidence. Plant-derived products were found to be safe to *Cyrtorhinus lividipennis* Reuter. and *Trichogramma japonicum* Ashmead (18). Evidence from field experiments has clearly underscored the importance of recommending application of plant-derived products to control rice diseases and pests (63). These plant-derived products do not pose any threat to the farm worker applying them or to any biological control agent, which may be present in the field. They are more easily degradable in the environment. Compared to leaf blast, neck blast is known to cause more yield losses. Neck region of the rice panicle is a thin cylinder and has more wax coating. Therefore, improvements in plant-derived biopesticide products with wetting and sticking agents, and target spot application may be needed. Further work on the integrated use of plant-derived biopesticide products in rice disease control will be rewarding (68).

## BIOLOGICAL CONTROL

### Interactions between Rhizosphere microorganisms and Plant Pathogens

The rhizosphere is the first-line defense for roots against attack by pathogenic microorganisms. Complex and dynamic communities of microorganisms, including both beneficial and deleterious ones, heavily inhabit the rhizosphere region of plants. Of particular interest are the rhizosphere microorganisms with ability as antagonists to provide defense against soil borne plant pathogens. Fluorescent pseudomonads constitute a large population of the rhizosphere microflora because of their nutritional versatility, being able to utilize a large number of organic substrata. Specific strains of *Pseudomonas fluorescens* and *P.putida* readily colonize the growing roots and enhance the plant growth by reduction of root zone population of deleterious microflora and pathogens. These beneficial strains are also potential biocontrol agents of several soil-borne plant pathogens. The mechanisms by which these microorganisms provide defense include competition for nutrition and space sites, forms of hyperparasitism, induced resistance and by production of a variety of metabolites that include antibiotics, siderophores and other substances such as cyanide (97).

Use of antagonistic bacteria was suggested for the control on sheath blight disease (37). Treatments with *Pseudomonas* spp., *Trichoderma viride* Pers., *Gliocladium virens* Muller et al (117, 118) or *Glomus* spp. have been tried in an effort to control rice diseases, particularly blast and sheath blight. None of the *Pseudomonas* products tested showed any effect on rice blast disease in any of the experiments (67). Three selected bacteria *viz.*, fluoroscent *Pseudomonas* sp. (PF-9), *Bacillus* sp. (B-44) and chitinolytic bacterium (CH-1) were tested rigorously in glasshouse and fields (40). *Pseudomonas* sp. was effective in reducing severity either alone or in combination with one foliar application of carbendazim. Yet, there was no significant increase in grain yields due to use of *Pseudomonas* sp. The sclerotia produced by *Rhizoctonia solani* remain viable in the soil for several years. A repeated application of *Pseudomonas* is generally advocated to control sheath blight disease. *Pseudomonas* products marginally reduced the sheath blight disease incidence and severity. However, the grain yield differences in the treatments are not significant. Evidence from field experiments clearly demonstrated the futility of recommending *Pseudomonas* product application to control rice diseases (67).

## Using Organisms that Kill Insect Pests

Diseases in insect pests of rice initially incapacitate and then lead to death. Microbial control can be an alternative control measure that is safe to environment. There are scattered and limited reports on microbial pathogens in rice insects pests. It is well known that environment that is favourable of rice, partic- ularly moisture availability is also conducive for development of diseases in insects. Several diseases in rice insect pests are known to be caused by fungi, bacteria, viruses and nematodes.

*Beauveria bassiana* Balsamo, is well known to white moscaredine in many insect pests including beneficial insects like honey bees and silkworm. This pathogen is of common occurrence in most rice fields in the rainfed lowland, irrigated and rainfed hill ecosystems. It attacks yellow and pink stem borers, striped borer, rice skipper, leaf rollers or folders, green leafhoppers, brown planthoppers, green horned caterpillars, grass hoppers, rice bug and rice weevil. The fungal pathogen produces abundant conidia in diseased insects that are released into air. The disease is also transmitted by contacts between insects in the field. The pathogen is externally borne on eggs of several insects and causes disease soon after egg hatching. Hypodermis is destroyed at the site of invading pathogen hypha. The toxin produced by the pathogen causes a general paralysis and death of insect host. *Beaveria bassiana* was mass-produced in rice-husk supplemented with 2% dextrose. It caused more than 90% mortality in *Dicladispa armigera* Oliver (43).

*Entomophthora fumosa* Speare causes a high mortality in brown planthoppers. Five species of entomophthora namely, *E. virulenta* Hall and Dunn, *E. ignobilis* Hall and Dunn, *Entomophaga* sp., *Zoophthora radicans* (Brefeld) Batko and *Erynia delphacis* Hori., cause infection on brown planthopper. *Metarhizium anisopliae* Metch., is also known to infect brown planthopper. The infection initially starts through the transmission on contact with diseased insect. The pathogen produces spores that are air-borne and water-borne and cause secondary infection. *M. anisopliae* can be mass produced successfully in Czapeck's Dox agar (82). *Beauveria brongniartii* Sacc., is highly pathogenic on larvae of all stem borer species and rice skippers. *Fusarium proliferatum* (syn *Fusarium moniliforme* var. *intermedium* Neish & M. Legg.) was found to infect nyphs and adults of *Nilaparvatha lugens* (114). *F. pallidoroseum* (Cooke) Sacc. was reported (42) to cause infection in *Cnaphalocorcis medinalis* Guenee. Natural infection of green leafhopper by *Pandora delphasis* (Hori) Humber., has been reported (67). Using PCR-based method (4) detected in gall midge infection by *Wolbachia* sp.

Several bacteria also cause infections in rice insect pests. Rice green horned caterpillar (*Melanitis ledaismene* Cramer) is infected by the non-spore-forming facultative bacterial pathogen *Serratia marceseans* Bizio. Application of *Bacillus thuringensis* Berliner controlled leaf folders and stem borer. It is also useful in controlling stored grain pests. Most of the lepidopteran rice insect pests are victims of *Bacillus subtilis* (Ehrenberg) Cohen. *B. thuringensis* and *B. subtilis* h ave better applicability due to their ability to form spores. Although commercial formulations (Delfin 85%, Biolep, BTK II, and Dispel 3.5%) are available on a limited scale, their use is extremely restricted and the benefits from use is often doubtful (94). The gene coding the delta-toxin of *Bacillous thuringiensis* (Bt) has been transferred to rice. Preliminary reports indicated a reduced damage from stem borer in these transgenic IR 64 lines (74).

Some viruses are potential biological control agents because of their high virulence, remarkable durability, specificity and relative safety. Wide spread occurrence of rice skipper (*Parnars mathias* Faber.), nuclear polyhedrosis virus (NPV) on larvae of leaf folder or stem borer in rice fields is well known. Viron-H is a commercial formulation of *Heliothis* NPV that is effective in infecting eggs, larvae, pupae and adults of stem borer.

Mermithids constitute a group of insect parasitising nematode species. In rice, only a few species are reported to cause infection. They include *Agamermis* spp., *Heamermis* and *Mermis* spp. on stem borer. Merthids are very slender organisms no longer than 10-20 cm in length. Another group of entomogenus nematodes in the Rhabditida are those in the genus *Neoaplectana*. The DD 136 strain of *Neoaplectana carpocarsae* Dutsky, is generally associated with a bacterium, *Achromobacter nematophilus* Pioenary et Thomas. Spray application of DD 136 nematodes caused infection in stem borer moths and reduced white earhead damage due to stem borer insect pest. Another entomophilic nematode *Parasitorhabditids* sp. infects many rice insect pests. Both *N. carpocarsae* and *Parasitorhabditids* sp., can be easily multiplied in artificial media.

## FUTURE OF INTEGRATED DISEASE MANAGEMENT (IDM)

### Sources of Information

In the complex world of modern agriculture, farmers' dictum might well be - not to be the first by whom the new is tried, nor yet the last to lay the old aside. Successful farmers must undertake continuing education because every year sees a remarkable increase in the output of scientific data relevant to agriculture. The prompt acquisition and use of this information is one of the most vital components of farming, especially pest management, today. There is an unceasing supply of agricultural information available. Both public and private agencies make great efforts to assemble, digest, and dispense information to farmers and other interested citizens. Many states now have regular programs on television and radio devoted to agriculture both from public and private enterprises. Newspapers publish agricultural sections, and numerous magazines are devoted to agriculture. Experiment stations, universities, and agricultural firms regularly publish articles, bulletins, circulars, handbooks, manuals, and periodicals. Television and radio programs on farm topics are broadcast daily in many states. The extension services in some states have *teletips* that enable citizens to phone in and listen to a 3- to 5-minutes discussion on any of a large number of agricultural topics. Short courses, on-farm demonstrations, fairs, machinery exhibits, field days, and tours are designed to convey the latest information to anyone who cares to come, see, and listen. It is only common sense that successful farmers should seek out the latest information on efficient agricultural practices.

### Sustaining Agriculture With IDM

A growing concept has appeared that includes both a philosophy of farming and a set of agricultural practices. It is called alternative agriculture with low input sustainable agriculture. Three major economic and environmental issues motivate this approach. The first is the possible health risks from pollution of air, water, and foods by agriculture chemicals. The second is soil erosion and depletion of natural resources. The third is the effect of conventional agricultural production systems on future generations' access to an abundance of food, clean water, and a decent environment. Therefore, this new approach to integrated management includes two central ideas: reduction in agricultural chemical input for the sake of food safety; and environmental awareness, including resource conservation, and guarding farm land, and water quality. These basic concepts need to be incorporated into farm policy and research should be funded on ways to reduce manufactured inputs in agriculture. For example, crop rotations and cultivation might be substituted for herbicides, fungicides and pesticides. Similarly manure, crop residues, and legumes might substitute in the crop rotation of inorganic fertilizer. All farmers in the rainfed lowlands in Eastern India practice low input rice culture (71). But, in irrigated rice high input use is widely adopted. A high priority must be given to educate farmers to prevent excessive use of nitrogenous fertilizers and unwarranted dumping of chemicals in rice fields to control insect pests. Several factors tend to make sustainable agriculture less profitable, although some farmers have achieved considerable success with low input systems. First, farm programs do not give preference to low-input farming. Second, yields are likely to be lower because of both the problems of managing diseases and pests without chemicals and the use of land in the low-value rotation crops required for pathogen control and soil improvement. Third, low-input agricultural practices often require more management time and skill than conventional farming practices. Finally, reduced reliance on inorganic fertilizers, fungicides and pesticides requires that farmers have more knowledge about the complex relationships among crops, weeds, disease causing pathogens, insects, and soils.

Integrated disease management includes many of these low input aspects of sustainable agriculture and has demonstrated that wise biotic and abiotic problem management often translates into increased grower profits. In future, land and water resources will reduce. To ensure continued self-sufficiency in rice, production and productivity must be increased. Host plant resistance is the first choice to ensure crop protection from biotic and abiotic stresses. But, host resistance may not be available to check all diseases at all times. This is why judicious and increased chemical input use will become a basic necessity. The adoption of integrated disease management by farmers in the operational research project area (21) indicates that farmers are willing to reduce pesticide use if shown that it is economical to do so. We must protect our environment and increase the conservation of our resources for the future. However, agriculture must remain productive and profitable. Consequently it will take the combined efforts of scientists and farmers in various ecosystems to improve Indian rice crop husbandry and keep the environment fit for human habitation.

## A Successful IDM Program

- Select season, photoperiod sensitivity or insensitivity and maturity duration for the mega-environment (53) based on experience, location and monsoon data.
- Select high-quality seed that is pesticide-treated. Use resistant cultivars.
- Assay soil for nutrient status, fungal propagules like sclerotia, spores or other resting bodies or nematodes, and practice soil solarization.
- Determine planting date based on weather forecast.
- Use proper row spacing and seeding rates.
- Apply fertilizers based on soil test results.
- Always practice split application of fertilizer in small doses at proper time.
- Use techniques of biological control, if available.
- Apply pesticides properly at the correct rate and correct time. Wherever possible, use plant-derived pesticides.
- Harvest crop efficiently; store it properly.
- Destroy crop residues if found infected or infested, otherwise plough the field and recycle organic matter as much as possible.
- Rotate crops, grow legume crop.
- Keep abreast of latest developments.

Rice crop suffers yield losses as a result of damage from pests and diseases only when the causal organisms attack and feed in large numbers. Often the climatic conditions during crop growth stage encourage rapid multiplication and spread of a pest or pathogen population resulting in either restricted or wide spread damage. Imbalance or excessive application of fertilizers, particularly nitrogen leads to the development and expression of all diseases. Therefore, the primary goal of IDM should be to educate and motivate farmers to use the correct dose of appropriate fertilizers at the right time when plants need them. Timely intervention to control pests reduces the loss substantially. Regular and periodic information on the intensity of occurrence of diseases and insect pest at different location are gathered through survey and surveillance program. This information helps to identify and recommend effective control measures. Advance planning is necessary for production and deployment of such items as equipment, fertilizer, pesticides, seed, and water. Safe, efficient, and effective management practices must be carried out. The farmer needs to know what each particular practice will do, how and under what circumstances it is to be used, and its relative effectiveness.

## AGRICULTURAL CONSULTANTS TO AID IN IDM

With the *Green revolution* in the county, extension personnel from state departments of agriculture have provided invaluable service and given good advice to farmers. Subject Matter Specialists in extension service are trained in scientific agriculture and, as public employees, deliver and administer programs supported cooperatively by each state's agricultural extension service. Traditionally, they have brought the latest information at no cost to farmers and advised them in the most efficient agricultural practices. These extension agents seek out farmers to tell them of the newest discoveries. One reason why agriculture is strong in India is the service of thousands of extension workers who have devoted their professional lives to helping farmers. However, there are too many farmers and not enough extension personnel available to supply the continuing advice and information necessary to meet each individual farmer's needs.

Modern farmers or farm managers cannot hope to fight all their battles alone. They need individual help from personnel trained in the various tactics of rice disease management. There is a scope to introduce and make available another source of help -

private agricultural consultants who can lease their expertise to farmers. These consultants can perform more efficiently such tasks as selecting proper cultivars, assaying the soil, monitoring pest populations, deciding where and if pesticides are needed, and correcting harvesting, storage, and marketing practices.

The agricultural consultant is a trained agronomist, economist, entomologist, pathologist, or member of a firm composed of skilled agriculturists, who help the farmers plan a production program. These consultants give advice on such diverse decisions as cultivar selection, planting dates, soil assays, monitoring for pathogen, insect and weed incidence, disease management practices, irrigation, harvest dates, proper storage, and marketing. Usually the consulting firms can charge on a per-hectare basis for their services and/ or receive a bonus based on yield increases. A farmer can pay for consultant services by omitting one unnecessary pesticide application or an unnecessary or inappropriate use of fertilizer, based on information furnished by the consultant. As modern agriculture becomes more sophisticated, and farm enterprises increase in size and complexity, private consultants will assume a vital role in helping agribusiness firms to avoid unnecessary risks and to be aware of the cost-benefit considerations involved with every decision made in growing a rice crop.

## *Literature Cited*


1. Agnihothrudu V. 1973. *Acrocylindrium oryzae* Sawada – Sheath rot on paddy. *Kavaka* 1:69-71.
2. Ahuja SC and Shrivastava MP. 1990. Stem rot of paddy – A review. *Agric Rev* 11:61-78.
3. AICRIP. 1971. *Annual Reports*. All-India Coordinated Rice Improvement Project, Hyderabad.
4. Ambethgar V. 2002. Record of entomopathogenic fungi from Tamil Nadu and Pondicherry. *J Entomol Res* 26:161-167.
5. Amin KS, Venkatarao G. 1979. Rice blast control by nitrogen management. *Phytopath Z* 96:140-145.
6. Anjaneyulu A. 1974. Identification of grassy stunt, a new virus disease of rice in India. *Curr Sci* 43:416-417.
7. Behura SK, Sahu SC, Mohan M, Nair S. 2001. *Wolbachia* in the Asian rice gall midge, *Orseolia oryzae* (Wood-Mason): correlation between host

mitotypes and infection status. *Insect Mol Biol* 10:163-171.

8. Bhagawati B, Bora LC. 1993. Managing ufra disease in deepwater rice (DWR) in Assam, India. *Int Rice Res Notes* 18:30.
9. Butler EJ 1913. Ufra disease of rice. *Agric J India* 8:205-220.
10. Chahal SS. 1998. *Kernel smut of rice*. Tech Bull. National Agricultural Technology Information Centre, Ludhiana 141001.
11. Chowdhury AK, Mukhopadhyay S. 1975. Effect of tungro virus on yield components. *Int Rice Comm Newsl* 24: 75-76
12. Das D, Sarma NK, Borgohain R, Das P. 2000. Rayada B3 - a high-yielding, ufra-resistant deepwater rice for Assam, India. *Int Rice Res Notes* 25:17.
13. Das P. 1996. An integrated approach for management of rice stem nematode, *Ditylenchus angustus* in deep water rice in Assam. *Indian J Nematol* 26:222-225.
14. Das SR. 1995. Field-testing of fungicides for the control of leaf scald disease of rice. *Indian Phytopath* 48:182-184.
15. Dastur JF. 1942. Notes on some fungi isolated from black point affected wheat kernels in the central provinces. *Indian J Agric Sci* 12:731-32.
16. Datta K, Tu J, Oliva N, Ona I, Velazhahan R, Mew TW Muthukrishnana S, Datta SK. 2001. Enahanced resistance to sheath blight by constitutive expression of infection-related rice chitinase in transgenic elite *indica* rice cultivars. *Plant Sci,* 160:405-414.
17. Devadath S, Dath P. 1970. Histopathological studies of rice leaves infected by *Xanthomonas transluscens* f. sp. *oryzae. Indian Phytopath* 7(2):1-4.
18. DRR 1965-2005. *Annual Reports*. All-India Coordinated Rice Improvement Project. Directorate of Rice Research, Hyderabad, India.
19. DRR 1975-2005. *Production Oriented Survey. Annu Report*, Directorate of Rice Research, Hyderabad, India
20. DRR 1987. *Research Highlights* (1986). Muralidharan K, Pasalu IC (eds) Directorate of Rice Research, Hyderabad 500 030, India, pp 62.
21. DRR 1987-1993. Final Report of Operational Research Project on Integrated pest management in rice. Directorate of Rice Research, Hyderabad, India, pp28.
22. DRR 1990. *Research Highlights (1989-90)*. Muralidharan K, Prasad GSV and Pasalu IC (eds) Directorate of Rice Research, Hyderabad 500 030, India. pp38.

23. DRR 1991. *Research Highlights (1990-91).* Muralidharan K, Prasad GSV, Pasalu IC (eds) Directorate of Rice Resea-rch, Hyderabad 500 030, India. pp40.

24. DRR 1994. *Research Highlights (1993-94).* Muralidharan K, Prasad GSV, Pasalu IC (eds) Directorate of Rice Research, Hyderabad 500 030, India.

25. DRR 2001. *High-yielding rice varieties in India-2000.* Directorate of Rice Research, Hyderabad, India, p102.

26. DRR 2002. *Progress Report 2002.* Vol 2. Entomology & Plant Pathology, All-India Co-ordinated Rice Improvement Project. Directorate of Rice Research, Hyderabad, India

27. Ghose RLM, Ghatge MB, Subramanian V. 1966. *Rice in India.* Indian Council of Agricultural Sciences (ICAR), New Delhi, pp474.

28. Ghosh A, Vankataraman S. 1994. New strain of rice grassy stunt virus in India. *Int J Trop Pl Dis* 12:59-66.

29. Ghosh A, John VT, Rao JRK. 1979. Studies on grassy stunt disease of rice in India. *Plant Dis Reptr* 63:523-525.

30. Govindarajan K, Kannaian S. 1982. Fungicidal control of grain discoloration. *Int Rice Res Newsl* 7:1.

31. Gupta AK, Sharma SC, Saini RG. 1986. Variation in pathogenicity of some Indian isolates of *Xanthomonas campestris* pv. *oryzae. Phytopathology* 76:881-883.

32. Hibino H, Roechan M, Sudarsman S. 1978. Associa-tion of two types of virus particles with *Penyakit Habang* (Tungro disease) of rice in Indonesia. *Phytopathology* 68:1412-16.

33. Hooda KS, Srivastava MP. 1998. Impact of neem coated urea and potash on the incidence of rice blast. *Plant Dis Res* 13:28-34.

34. Huang N, *et al.* 1997. Pyramiding of bacterial leaf blight resistance genes in rice marker assisted selection using RFLP and PCR. *Theor Appl Genet* 95:313-320.

35. IRRI 1965. *The mineral nutrition of rice plant.* Johns Hopkins Press, Maryland, Baltimore, pp 494.

36. Kiyosawa S. 1981. Gene analysis for blast resistance. *Oryza* 18:196-203.

37. Krishanamurthy K, Gnanamanickam SS. 1997. Biological control of sheath blight of rice: induction of systemic resistance in rice plant associated *Pseudomonas* spp. *Curr Sci,* 72:331-334.

38. Krishanveni D, Laha GS, Muralidharan K. 2002. Evalu-

ation of rice cultivars against *Bakanae* disease of rice. November 14-15, 2002. *Plant Prot Assoc India, Natl Sem p*54-55. National Bureau of Plant Genetic Resources, Regional Station, Hyderabad.

39. Kurucheve V, Ezhilan JG, Jayaraj J. 1997. Screening of higher plants for fungitoxicity against *Rhizoctonia solani* in vitro. *Indian Phytopath* 50:235-241.

40. Laha GS, Venkataraman S. 2001. Sheath blight management in rice with biocontrol agents. *Indian Phytopath* 54: 461-464.

41. Maiti S, Singh AK, Saha HS. 1979. Occurrence of the perfect state of *Rhinchosporium* oryzae in India. *Pl Dis Reptr* 63:773-774.

42. Manisegaran S, Letchumanan S. 1995. Occurrence of *Fusarium pallidoroseum* (Cooke) Sacc. as a fungal pathogen of rice leaf-folder, *Cnaphalocrocis medinalis* (Guenee) in Karaikal region. *J Biol Cont* 9:133-135.

43. Mazumder D, Puzari KC, Hazarika LK. 1995. Mass production of *Beauveria bassiana* & its potentiality on rice hispa. *Indian Phytopath* 48:275-278.

44. Mohanty NN. 1975. Efficacy of different seed treatment for control of udbatta disease of rice. *Indian Phytopath* 28:521-523.

45. Mohanty NN. 1976. Influence of some meteorological factors on the development of udbatta disease of rice. *Indian J Agric Sci* 46:399-403.

46. Muniyappa V, Raychaudhuri SP, Maramorosch K (eds) 1988. Rice yellow dwarf disease. Mycoplasma diseases of crops - Basic and applied aspects. p 233-284.

47. Muralidharan K. 2000. Modelling for rice blast disease development. *In*: Dang JK. and Rishi N (eds) *Epidemiological principles in plant disease control.* Department of Plant Pathology, CCS Haryana Agricultural University, Hissar, India and Indian Council of Agricultural Research, New Delhi.

48. Muralidharan K. 2005a. Resistant varieties as a design factor in the field evaluation of fungicides for rice blast control. *J Mycol Pl Pathol* 35(2): 277-282.

49. Muralidharan K. 2005b. Variability, the weapon in the combat between plant and pathogen as illustrated with rice and blast. In: *Plant Protection in New Millennium* Pp 69-108. Satish Serial Poublishers, New Delhi 110 033. India.

50. Muralidharan K, Pasalu IC. 1996. Assessments of yield losses in rice. pp.35-38. In:

*National agricultural research project: basic research subproject on integrated pest management 1992-96*. National Centre for Integrated Pest Management (ICAR) New Delhi 12.

51. Muralidharan K, Reddy CS. 2004. Transgenic Rice with *Xa 21* is Susceptible to Bacterial Leaf Blight Pathogen Populations Prevalent in India. *J Mycol Pl Pathol* 34(3)773-777

52. Muralidharan K, Siddiq EA. 2000. Agricultural sciences in India – From 1200 B.C. to 2000 A.D. National Agricultural Sciences Museum, Indian Council of Agricultural Research, New Delhi 110 001.

53. Muralidharan K, Venkatarao G. 1979. Bacterial blight (*Xanthomonas campestris* pv. *oryzae*) on rice in Nellore district, Andhra Pradesh. *Indian Phytopath* 32:483-485.

54. Muralidharan K, Venkatarao G. 1980a. A simple method of forecasting outbreak of rice blast. *Indian Phytopathol* 33(4): 560-564.

55. Muralidharan K, Venkatarao G. 1980b. Outbreak of sheath rot *(Acrocylindrium oryzae)* on rice at Nellore. *Int Rice Res Newsl* 5:7-8.

56. Muralidharan K, Venkatarao G. 1987. Forecasting out break of rice blast disease. pp. 111-137. *In:* Kannaiyan S (ed), *Advances in Rice Pathology*, Tamil Nadu Agric Univ, Coimbatore, India.

57. Muralidharan K, Venkatarao G. 1993. Synchronized testing of rice leaves and panicles for blast resistance. pp 59-62. *In*: Muralidharan K and Reddy CS (Eds) *Plant disease problems in central India*. Proc Symp Central Zone, Indian Phytopathol Soc, Directorate of Rice Research, Hyderabad, India.

58. Muralidharan K, *et al.* 2003a Evaluation of blast resistance genes in India. *Rice Genet Newsl* 20:94-96.

59. Muralidharan K, *et al.* 2003b. Appraisal of bacterial leaf blight resistance genes in India. *Rice Genet Newsl* 20:96-98.

60. Muralidharan K, *et al.* 2004a. Performance of rice blast resistance genes in multi-environment test in India. *Indian Phytopath* 57:260-266.

61. Muralidharan K, *et al.* 2004b. Expression of bacterial blight resistance genes in rice genotypes in India. *J Mycol Plant Pathol* 34:365-371.

62. Muralidharan K, Krishnaveni D, Neeraja CN, Prasad GSV. 2003c. Reflections on the complexity of rice tungro virus disease. *Annu Rev Pl Path* 2: 379-436.

63. Muralidharan K, Krishnaveni D, Rajarajeswari NVL, Prasad

ASR. 2003d. Tungro epidemics and yield losses in paddy fields in India *Curr Sci* 85:1143-1147.

64. Muralidharan K, *et al.* 2005. Field performance of genotypes generated by molecular-assisted breeding to bacterial leaf blight in India. *Rice Genet Newsl* 21: 29-31.

65. Muralidharan K, Ravi V, Siddiq EA, Paroda RS. 1996. *Indian Agriculture - A saga of success in crop improvement and challenges ahead*. 2nd Int Crop Sci Cong: Crop Productivity and sustainability - shaping the future. November 17-23, 1996. pp. 84. National Academy of Agricultural Sciences and Indian Council of Agricultural Sciences, New Delhi.

66. Muralidharan K, *et al.* 1991. Farmers' constraints in the adoption of recommended rice blast control measures. *Oryza* 18:229-231.

67. Muralidharan K, *et al.* 2004c. Field application of fluorescent *Pseudomonas* products to control blast and sheath blight diseases in rice. *J Mycol Pl Pathol* 34: 411-414.

68. Muralidharan K, Reddy CS, Krishnaveni D, Laha GS. 2003e. Evaluation of plant-derived commercial products for blast and sheath blight control in rice. *Indian Phytopath* 56: 206-210.

69. Muralidharan K, *et al.* 1990a. *Ustilaginoidea virens* - A possible threat to increased rice production. *Indian Phytopath* 43(2): 257.

70. Muralidharan K, Shinde JE, Siddiq EA. 1990b. *Technology for Rice Production.* 2nd Ed. Directorate of Rice Research, Hyderabad. pp. 200.

71. Muralidharan K, *et al.* 1988. Rice in Eastern India - A reality. *Oryza* 25:213-245.

72. Nagarajan S, Muralidharan K. 1995. *Dynamics of Plant Diseases*. Allied Publishers, New Delhi, pp247.

73. Nandakumar C, Prasad JS, Rao YS, Rao J. 1975. Investigations on the white-tip nematode (*Aphelenchoides besseyi* Christie) 1942 rice (*Oryza sativa*). *Indian J Nematol* 5:62-69.

74. Nayak P, Basu D, Das S, Basu A, Ghosh D, Ramakrishnan NA, Ghosh M, Sen SK. 1997. Transgenic elite *indica* rice plants expressing *CryIAc* delta-endotoxin of *Bacillus thuringiensis* are resistant against yellow stem borer (*Scirpophaga incertulas*). Proc Natl Acad Sci USA 94:2111-2116.

75. Nayak P, Reddy PR, Mishra RN. 1975. Pattern of inheritance of bacterial leaf streak resistance in rice. *Curr Sci* 44:600-601.

76. Nene YL, Gairola C. 1965. Some preliminary observations on the *Khaira* disease of paddy. *Res Paper No1* Exp Stn, UP Agric Univ, Pantnagar, Nainital.

77. Ou SH. 1985. Rice diseases, 2nd Ed, pp109-200. Commonwealth Mycological Institute, United Kingdom.

78. Padmanabhan SY. 1973. The Great Bengal famine. *Annu Rev Phytopathol* 11:11-23.

79. Padmanabhan SY. 1975. *Inheritance of disease resistance in rice.* Final Tech Reprt US PL 480 Project. Central Rice Research Institute, Cuttack, India.

80. Padmanabhan SY, Chakrabarti NK, Mathur SC, Veeraraghavan J. 1970. Identification of pathogenic races of *Pyricularia oryzae* Cav. in India. *Phytopathology* 60:1574-1577.

81. Pandey V, Agarwal VK, Pandey MP. 2000. Location and seed transmission of fungi in discolored seeds of hybrid rice. *Indian Phytopath* 53:45-49.

82. Patel KC, Yadav DN, Dube HC, Patel RJ. 1990. Laboratory and mass production studies with *Metarhizium anisopliae*. *Ann Biol Ludhiana* 6:135-140.

83. Prasad GSV, Muralidharan K, Rao CS. 2001a. Stability and yield performance of genotypes: A proposal for re-grouping world rice area into mega-environments. *Curr Sci* 81:1337-1346.

84. Prasad GSV, Rao UP, Rani NS, Rao LVS, Pasalu IC, Muralidharan K. 2001b. Indian rice varieties released around the world. *Curr Sci* 80:1508-1511.

85. Prasad JS, Panwar MS, Rao YS 1992. Nematode pests of rice, pp 43-61. *In:* Bhatti DS and Walia RK (eds) *Nematode pests of crops*. CBS Publishers & Distributors, New Delhi.

86. Raina GL, Singh G. 1980. Sheath rot outbreak in Punjab. *Int Rice Res Newsl* 5(1):16.

87. Rajarajeswari NVL, Murlidharan K. 2006. Assessments of farm yield and district production loss from bacterial leaf blight epidemics in rice. *Crop Prot* 25:244-252 (in Press).

88. Ranganathan KG., Safeeulla KM. 1994. Seed-borne nature and mode of infection of *Ephelis oryzae* Syd. incitant of *udbatta* disease of rice. *Mysore Agril Sci.* 28:60-63.

89. Rangareddy P. 1987. *Kresek* syndrome of bacterial leaf blight. Pp. 320-329. *In*: Kannaiyan S (ed), *Advances in rice pathology*, Tamil Nadu Agric Univ, Coimbatore, India.

90. Rao CS, Muralidharan K, Rao AV. 1993. A sampling methodology for estimating yield losses due to bacterial leaf streak of rice. pp. 81-85. *In*: Muralid-

haran K and Reddy CS (Eds) *Plant disease problems in central India.* Proc Symp Central Zone, Indian Phytopathol Soc, Directorate of Rice Research, Hyderabad, India.

91. Rao KM. 1964. Environmental conditions and false smut incidence of rice. *Indian Phytopath* 17:110-114.

92. Rao PRM, Bhavani B, Rao TRM, Reddy PR. 2000. Spikelet sterility/grain discoloration in rice in Andhra Pradesh, India. *Int Rice Res Notes* 25:40.

93. Rao PS, Kauffman HE. 1977. Potential yield losses in dwarf rice varieties due to bacterial leaf blight in India. *Phytopathol Z* 90:281-284.

94. Rath PC. 1999. Evaluation of some *Bacillus thuringiensis* & neem formulations against yellow stem borer of rice under rainfed lowlands. *Oryza* 36:398-99.

95. Raychaudhuri SP, Mishra MD, Ghosh A. 1967. A preliminary note on the trasmission of a virus disease resembling 'tungro' of rice in India and other virus-like symptoms. *Plant Dis Reptr* 51:300-301.

96. Reddy AV, Jeyarajan R. 1990. Influence of number of hoppers on incubation period and symptom expression of rice yellow dwarf (RYD) pathogen. *J Mycol Plant Pathol* 20: 3, 277-278.

97. Reddy CS, Laha GS, Muralidharan K. 2005a. Interactions between Rhizos-phere microorganisms and Plant Pathogens. pp163 – 218. *In: Frontiers in plant sciences.* Mukergi KG, Tilak KVBR, Reddy SM, Gangwane LV, Prakash P, Kunwar IK (eds). I. K. International Pvt. Ltd., New Delhi – 110016, India.

98. Reddy CS, Reddy KRN, Kumar NR, Muralidharan K. 2005c. Characterization of aflatoxin $B_1$ produced by *Aspergillus flavus* isolated from discoloured rice grains. *J Mycol Pl Pathol* 34(3): 321-326.

99. Reddy CS, Reddy KRN, Kumar NR, Muralidharan K. 2005a. Norsolorinic acid production by *aspergillus* spp. isolated from discolored rice grains. *J Mycol Pl Pathol* 35(1): 53-58.

100. Reddy CS, Reddy KRN, Kumar NR, Laha GS and Muralidharan K. 2004. Exploration of aflatoxin contamination and its management in rice. *J Mycol Pl Pathol* 34( 3) 786-790.

101. Reddy MTS, APK Reddy. 1992. Occurrence of pathotypes of *Xanthomonas campestris* pv. *oryzae* in India. *J Mycol Pl Pathol* 22:205-206.

102. Saito Y, Iwaki M, Usugi T. 1976. Association of two types of particles with tungro group of disease in rice. *Ann*

*Phytopathol Soc Japan* 43: 375 (in Japanese).

103. Shanmugam SN, Gopalan N, George KM, Gopalakrishnan R. 1973. Leaf scald of rice (*Rhinchosporium oryzae* Hashioka (Yokogi). *Curr Sci* 42:582-583.

104. Shanti ML, George MLC, Cruz VCM, Bernando Ma, Nelson RJ, Leung H, Reddy JN, Sridhar R. 2001. Identification of resistance genes effective against rice bacterial leaf blight pathogen. *Plant Dis* 85:506-512.

105. Shivanandappa N, Govindu CH. 1976. Direct and indirect effects of udbatta disease on total number of panicles per hill in different paddy cultures. *Int Rice Res Newsl* 1:12.

106. Singh R, Sunder S. 1997. Foot rot and Bakanae of rice: retrospect and prospects. *Int J Trop Plant Dis* 15:153-176.

107. Singh R, Kumar A, Jalali BL. 2002. Variability, predisposing factors and management of stem rot of rice caused by *Sclerotium oryzae* Catt.: An overview. *Annu Rev Pl Pathol* 1:275-289.

108. Singh RA. 1974. Meteorological factors influencing the occurrence of false smut of rice. *Indian J Agric Sci* 44:718-721.

109. Singh RA. 1975. Sources of inoculum and the epidemiology of bunt of rice. *Il Riso* 25:77-80.

110. Singh RA. 1980. *False smut of rice*. Final Report ICAR Adhoc Scheme. GB Pant Univ. Agic. Tech. Pantnagar 263 145. Nainital, Uttaranchal.

111. Sridhar R. 2002. Use of molecular tools for understanding bacterial blight pathogen populations of rice and usefulness of disease resistance genes. *Indian Phytopath* 55: 417-429.

112. Sridhar R, Reddy JN, Singh UD, Agarwal PK. 1999b. Usefulness of combinations of bacterial leaf blight genes at Cuttack, Orissa, India. *Int Rice Res Notes* 24:24-25.

113. Sridhar R, *et al.* 1999a. Usefulness of blast resistance genes and their combinations in different blast-endemic locations in India. *Int Rice Res Notes* 24:22-24.

114. Srinivas PR, Pasalu IC. 1990. Occurrence of a fungal disease on brown planthopper of rice. *J Biol Cont* 4:67.

115. Srivastava DN. 1969. Bacterial blight and streak diseases of rice in India *Indian Phytopath* 22:163-164.

116. Srivastava DN. 1972. Bacterial blight of rice. *Indian Phytopath* 25:1-16.

117. Sudhakar R, Reddy CS, Nagarajrao HS. 1993. Effect of biological agents sheath blight of rice. *J Mycol Pl Pathol* 23: 108-109.

118. Sudhakar R, Reddy CS, Rao KC. 1997. Studies on antagonism of *Trichoderma* and *Gliocladium* species against *Rhizoctonia solani* of rice. *Indian J Plant Prot* 25:199.

119. Sunder S, Satyavir 1997. Survival of *Fusarium moniliforme* in soil, grains and stubbles of paddy. *Indian Phytopath* 50:474-481

120. Veeraraghavan J. 1975. Stability in *Pyricularia oryzae*. Cav. *Proc Indian Acad Sci* B 84:127-131.

121. Venkatarao G, Muralidharan K. 1982a. Effect of meteorological conditions on the incidence and progress of blast disease of rice. *Z Pflakrant Pflalshutz* 89:219- 223

122. Venkatarao G, Muralidharan K. 1982b. Plant population density and severity of rice blast disease. *Indian Phytopath* 35:318-320.

123. Vir D. 2002. Management of post harvest diseases of grains and seeds. *Annu Rev Pl Pathol* 1:123-135.

124. Waghray S, Reddy CS, Reddy APK. 1988. Seed mycoflora and aflatoxin production in rice. *Indian Phytopathol* 41:492-494.

125. Yashitola J, Krishanaveni D, Reddy APK, Sonti RV. 1997. The genetic diversity within the population of *Xanthomonas oryzae* pv. *oryzae* in India. *Phytopathology* 87:760-765.

*Rev. Plant Pathol. Vol. 4, 2008*
*Indian Society of Mycology and Plant Pathology*
*Scientific Publishers (India), Jodhpur, pp. 83-104*

# BIOLOGY, EPIDEMIOLOGY AND MANAGEMENT OF APPLE LEAF SPOT DISEASES-A REVIEW

***J.N. Sharma, Ravinder Kumar and L.N. Bhardwaj***

Department of Mycology and Plant Pathology, Dr. Y.S.Parmar University of Horticulture and Forestry Nauni, Solan - 173 230 Himachal Pradesh, India

KEY WORDS : Apple, leaf spot, biology, epidemiology, disease management

## ABSTRACT

Apart from apple scab, Alternaria leaf spot (*Alternaria alternata)*, frog eye leaf spot (*Sphaeropsis malorum)*, leaf blotch (*Phyllosticta mali)* and many other diseases caused by the species of *Ascochyta, Coniothyrium, Pestalotia, Mycosphaerella* and *Leptosphaeria* are major foliar diseases of apple. Onslaught of Marssonina blotch caused havoc in apple plantations by mid-season defoliation in the last decade in Himachal Pradesh. In Jammu & Kashmir, Alternaria leaf spot is reported to destroy the apple foliage and reduce fruit yield as a result of it. Most of these pathogens survive either in the lesions on leaves on grass or bare ground, in lesions on terminals, or as conidia in bud scales. Outbreaks of *Alternaria alternata* f. sp. *mali* were influenced by air, temperature, relative humidity, rainfall and number of rainy days. Empirical models have been developed for the prediction of Alternaria leaf spot based on various environmental factors. *Azadirachta indica* and *Dodonia viscosa* extract, and *Mentha arvensis* oil gave the maximum growth inhibition of *Alternaria alternata*. Among the bio-control agents, *Trichoderma hamatum* exhibited the maximum inhibition followed by *T. harzianum* and *T. viride*. EBI fungicides hexaconazole, difenocanozole and tebuconazole, and iprodione gave cent per cent inhibition of growth as well as spore germination of *A. alternata in vitro*. Propineb, dodine, Shield and fluquinconazole were also effective in inhibiting the growth and spore germination, whereas benzimidazole fungicides carbendazim, thiophanate methyl and benomyl did not show any significant inhibition. Under field conditions, difenoconazole, iprodione + carbendazim, hexaconaxole, propineb, dodine and Shield gave fairly good disease control in the field. A spray schedule comprising the sprays of dodine (0.075%), carbendazim (0.05%), hexaconazole + mancozeb (0.35%) and tebuconazole (0.05%) at different stages (from fruit set to pre-harvest stage) was recommended.

## INTRODUCTION

Apple *(Malus domestica* Borkh.) is the most important fruit crop grown extensively in temperate regions all over the world. In India, apple is commercially cultivated in the states of Jammu and Kashmir, Himachal Pradesh and Uttaranchal on an area of 241.6 thousand hectares with annual production of 11.58 lakh tonnes. Its cultivation has revolutionized the socio-economic condition of hilly farmers in Himachal Pradesh and has become the number one commercial fruit crop which is being grown over 82,000 ha with annual production of 5.55 lakh tonnes (2). The area under apple cultivation in this state has increased during the last decade but the production has not increased proportionally. Also, the productivity of apple per unit area is very less in comparison to advanced apple growing countries of the world. The reasons for low apple productivity could be many, but one of the reasons is surely the damage caused by insect pests and a variety of fungal, bacterial and viral diseases. Of these, fungal diseases are of utmost economic importance and are responsible for causing huge losses to the growers.

Leaf spot diseases are the most frequent and could be seen even in the best managed orchards. The disease incidence and damage inflicted however, differ from orchard to orchard and area-to-area depending upon the plant protection measures adopted and the prevalent weather conditions. A typical leaf spot is definitely a delimited necrotic lesion, often with brown, some with white centre and dark margins (91). The spots gradually increased and formed large dead areas known as 'blight' or 'blotch'. Many leaf spots probably cause far more damage than they appear to. Some leaf spots apparently cause damage beyond the mere reduction in the active photosynthetic leaf area. They may cause more or less severe defoliation, depending on the kind of spot, the host plant and the number, size and position of spots on the leaves. These spots diminish the photosynthetic efficiency and also disturb the physiological activity of the host. The early leaf fall and in extreme cases almost entire defoliation may occur much before the normal time. In Himachal Pradesh, there has been an increase in the incidence of many leaf-spotting fungi over the years. Onslaught of Marssonina blotch caused havoc in apple plantations by mid-season defoliation in the middle of last decade (61,63,65,67). Inspite of the recommended practices, many more leaf spot diseases are appearing in the commercial orchards. In this review, a critical account on the work done on various leaf spots of apple world over is presented.

## GEOGRAPHICAL DISTRIBUTION

Leaf spots in apple are incited by a number of fungi, the most important genera being *Alternaria, Ascochyta, Marssonina, Coniothyrium, Phyllosticta, Pestalotia, Mycosphaerella* and *Leptosphaeria*. *Alternaria* species have been reported from several countries attacking apple fruit and leaves. Roberts (53) identified the fungus as *A. mali*, while Taylor (81) thought it to be *A. tenuis*. Sawamura (57) and Tweedy and Powell (82) did not mention the species of *Alternaria*. It is well investigated fungus and only two species viz. *A. mali* Roberts and *A. tenuis* Auct. have been found widely distributed all over the world. Gupta and Agarwala (27) reported *A. mali* Roberts from Himachal Pradesh causing leaf blight of apple leaves in Kotgarh apple belt. Different species and the fungal isolates causing leaf spots of apple trees in Russia have been identified. The pathogen has been identified as *Alternaria alternata*. Inoculation of the apple leaves with conidial suspension of the Krasnodar strains resulted in the appearance of necrotic spots characteristic of 'alternariosis' (22). Very recently, Sharma *et al.* (66) reported the occurrence of Alternaria leaf spot (*Alternaria alternata*) on apple from some of the fairly sprayed orchards in

Kotkhai and surrounding areas. Clinton (12) reported that a leaf spot of apple in Illinois was due to *Sphaeropsis malorum*. Ivanoff (28) reported *Phyllosticta briardi* Sacc. on apple leaves from Bulgaria. Wenzl (90) observed *Phyllosticta angulata* causing a new leaf spot in nursery apple plants in more humid regions of Austria. Moore (47) also observed angular spot disease of apple leaves in 1937 at Lingfield, Surrey caused by *Phyllosticta angulata* Wenzl as in Austria. Adams (1) reported the attack on apple leaves with circular spots caused by *Phyllosticta pirina* in Delaware and later Peregrine and Watson (50) reported *Phyllosticta pirina* as a new record from Tanganyika on pear leaves. Martsikh (44) observed that out of the fourteen fungi isolated from apple leaves causing brown spots, *Sphaeropsis malorum* and *Phyllosticta mali* were the main fungi responsible for causing the leaf spots in Maldovia. McDonald (45) reported *Phyllosticta pirina* Sacc. on apple leaves from Kenya as a new host record. Voglino (88) reported *Phyllosticta prunicola* from the neglected apple trees in the vicinity of Susa in North-West Italy. Govi (20) reported the widespread and severe attack in the provinces of Kmilia and Rome by a leaf spot caused by *Phyllosticta prunicola* (Opiz.) Sacc. Sibilia (72) reported that apple leaves at Belzana were found infected with *Phyllosticta prunicola*.

Furstenberg (21) reported that untreated apple trees in the same district in Germany were severely attacked by leaf spot fungi in the month of July or August and also reported the white spot infection caused by *Mycosphaerella sentina* on pear trees. Vukovits (89) observed the leaf spot diseases of fruit trees to be severe in Austria during humid weather. The important leaf spots recorded were due to *Phyllosticta mali* Prill. and Del. on apple and *Mycosphaerella sentina* Schroet on pear. Khabiri (32) described *Mycosphaerella sentina* on apple and pear from Persia. Gandineau (23) studied biology, host range and methods of control of the principal leaf spot diseases of apple and pear in France. Viennot-Bourgin (87) recorded *Mycosphaerella tulasnei* (Janes) Both. and *Cladosporium herbarum* Lk. ex Fr. from France as major causes of leaf spot diseases of apple. Scott and Rorer (58) and Crabill (13) established the fact that *Sphaeropsis malorum* was the primary cause of the typical frog-eye leaf spot and other fungi such as *Alternaria, Coniothyrium, Coryneum* and *Phyllosticta* species were secondary invaders. The detailed taxonomic study of these fungi was made by Stevens (75) and concluded that it was limited only to apple. Brien and Dingley (6) reported *Leptosphaeria* and *Coniothyrium* species invading rose and also apple in Russia, however, the incidence was benign.

Viennot-Bourgin (86) reported *Leptothyrium pomi* on apple leaves from Iran. Kaul (30) reported first time in India that *Leptothyrium pomi* caused leaf spot disease in apple trees. Rothers (54) reported *Pestalotia breviseta* Sacc. causing spots on apple leaves in Russia. Petrak and Esfandiari (51) reported *Ascochyta piricola* Sacc. on apple leaves from Iran. Sokolov (74) reported from Russia that apple leaves were attacked by *Phyllosticta mali*, *Pestalotia* and *Hendersonia* species during the years 1956-58 in epiphytotic conditions. Zhukova (99) stated that *Phyllosticta mali* was widespread in the Southern districts of Krasnoyarsk region. Mundkur and Ahmad (48) reported *Linospora ochracea* and *Cercospora mali* on apple leaves as new records to India. Kaul (31) reported *Cercospora rubsotincta* on *Prunus avium* from Srinagar. Chaudhary *et al.* (9) reported *Pestalotia hartigii* for the first time from Kashmir valley. In India, the prevalence of Marssonina blotch on apple caused by *Marssonina coronaria* was first recorded in 1992 from Kotkhai area of Himachal Pradesh, and subsequently it was reported from all apple growing areas of the state. This disease appeared in epiphytotic proportions in all the apple growing districts of Himachal Pradesh in 1995 and jolted the apple industry in the state by mid-season defoliation of trees (60, 64, 67). First recorded in 1904 and 1905 seasons from Japanese orchards by Miyake (46), Marssonina leaf spot has been reported from South Korea, Canada, Rumania, Taiwan, China, Brazil and Italy (41, 80, 85). Sooty blotch (*Gloeodes pomigena*) and Phyllosticta leaf spot [*P.* (*Asteromella) mali*] were detected on apples in experimental plots in Ahrensburg, Germany (37). Bulajic *et al.* (7) reported *Alternaria mali* on apples in Yugoslavia. Filajdic and Sutton (16) reported *A. mali* on strains of Delicious apple (*Malus domestica*) in the summers of 1987 and 1988 in Western North Carolina, U.S.A.

In Himachal Pradesh, different leaf spot diseases were prevalent in moderate to severe form in all the locations surveyed in 2003 and 2004 seasons (39). Most serious leaf spot (Alternaria leaf spot) was caused by *Alternaria alternata* (Fr.) Keissler. The maximum disease severity of this leaf spot was observed at

Chaithla in Shimla district followed by Nauni in Solan district. The next serious in hierarchy was Marssonina blotch caused by *Marssonina coronaria* (Ell. & J.J. Davis) J.J. Davis with maximum disease severity at Nauni in Solan district followed by Pragatinagar in Shimla district. Other leaf spots were however, recorded in poorly managed orchards in Nauni with a disease severity of 19.50 per cent.

## SYMPTOMATOLOGY

Different fungi produce varied leaf spot symptoms and few diseases have been named on the basis of symptoms like shot hole, frog-eye, blotch and blight etc. In other cases, where the symptoms are not specific, the generic name of the fungus precedes the leaf spot for classification and better understanding. Lewis (42) reported a large number of fungi occurring on wood and leaves of apple. The leaf spots appeared as small purple spots, scattered over the affected leaves, the spots increased in size and were usually somewhat circular in outline and the centre became dead and brown in colour. He found little black fruiting bodies of fungi on dead spots. The fungi most frequently associated were *Phyllosticta limitata* Pk., *Coniothyrium pirina* (Sacc.) Sheldon and *Sphaeropsis malorum* Pk. Dennis and Wakefield (14) studied the leaf spots of apple and observed circular whitish spots of dead tissues ranging from 2 mm to 6 mm in diameter. The black pycnidia borne on dead surface irrespective of the leaf spot size were externally the same but a microscopic examination identified the two separate organisms viz., *Phyllosticta mali* Prill. & Del. and *Coniothyrium pirinum* (Sacc.) Sheldon producing same type of symptoms. Moore (47) observed angular spot disease of apple leaves in 1937 at Lingfield, Surrey caused by *Phyllosticta angulata* Wenzl. Small angular spots of about 1-2 mm diameter were defined as diagnostic symptoms of this disease. These spots were in groups, though not restricted to any other portion of the leaf. Adams (1) reported the attack on apple leaves with circular spots caused by *Phyllosticta pirina* in Delaware. Voglino (88) reported from the neglected apple trees in the vicinity of Susa in North-West Italy, a leaf spot circular and black in colour, which bore the pycnidia of *Phyllosticta prunicola* (Opiz.) Sacc. and also spherical pycnidia of *Phyllosticta briardi* Sacc. Unamuno (83) found that

*Phyllosticta mali* produced round to elongated, pale to bleached spots upto 10 mm in diameter on upper surface of apple leaves seriously damaging the trees.

Dowson (15) reported certain apple varieties (e.g. Cox's Orange Pippin, Jonathan, Scarlet and Sturmer Pippin) in Tasmania to be liable to a leaf spot which may appear early or late in the season. The spots were of various tinges of grey, brown and purplish and were circular, round or irregular. Some were found in the middle of the leaves, while others were confined to the edge and tips besides spotting of the surface, the leaf margins darkened to purple or bronze and curl upwards. The grey spots were due to *Phyllosticta pirina* and some of the rounded brown spots were caused by *Sphaeropsis malorum*, but both fungi were commonly found growing on leaves injured by sprays or on trees in a bad condition of health.

Yu *et al.* (97) reported leaf spot diseases of apples in Gansu, China. Wheel spot, grey spot, brown spot and circular spots were identified, with wheel spot being identified as the main disease. Bulajic *et al.* (7) isolated and identified *A. mali* from leaf spots on Red Delicious apples collected from 7 locations in Yugoslavia. The spots were circular, tan to brown, 2-5 mm diameter and often bordered by a purple ring. Leaf abscission led to 50 per cent defoliation in some orchards. Glaser and Kaiser (25) isolated *Alternaria tenuis* [*alternata*] from leaves of Golden Delicious with necrotic spot symptoms. Nema (49) observed that inoculation of fruits with *Alternaria alternata* reduced the concentration of sugars to a greater extent in cv. Delicious than in American. Recently, Sharma *et al.* (66) reported the occurrence of a brown spot disease caused by *A. altanata* from some of the orchards in Kotkhai which produced 1-3 mm diameter reddish brown necrotic spots on young foliage of apple in early summer months later coalescing to form bigger blotches and giving blighted appearance followed by defoliation. Dark brown to black elongated lesions also appeared on the petioles and mid-ribs leading to severe yellowing and defoliation of infected trees.

Kudela *et al.* (38) reported a severe outbreak of calyx end fruit rot of apples at harvest time in the gardens of Southern Bohemia, Czech Republic in 1992. *A. alternata* and *Botrytis cinerea* were isolated from the diseased calyx end of the fruit. On inculation, *A.*

*alternata* produced symptoms identical to those observed in the field. They suggested that the disease outbreak was caused by *A. alternata* and was also related to the decrease in apple scab (*Venturia inaequalis*) control.

A study conducted to identify the species of fungal isolates causing leaf spots of apple trees in Russia reaffirmed the pathogenicity of *Alternaria alternata*. Inoculation of apple tree leaves with conidial suspension of the Krasnodar strains resulted in the appearance of necrotic spots characteristic of 'alternariosis' (22). A hitherto unknown disease Marssonina blotch (*Marssonina coronaria*) appeared in all apple plantations in Himachal Pradesh and caused widespread mid-season defoliation in mid 1990's (60,65,67). This disease first appeared in the form of 5-10 mm size brown spots on the upper surface of the mature leaves which later coalesced and the surrounding areas turned yellow followed by severe leaf yellowing and defoliation, while the fruits were still hanging on the trees (63,68). Infected areas on the leaves were distinctly covered with numerous pinhead-like fungal fruiting bodies - the acervuli.

Glaser and Kaiser (25) isolated *Alternaria tenuis* [*alternata*] from apple leaves with necrotic spot symptoms. Four types of isolates were distinguished and all 4 produced spots on artificially infected leaves but were unable to cause fruit rot.

Filajdic and Sutton (17) isolated *Alternaria mali* from apple leaves of cv. Red Delicious. A host-specific toxin was isolated from *A. mali*. Red Delicious and lines obtained from it showed varied degree of susceptibility to artificial inoculation. Wu *et al.* (92) also proved the pathogenicity of two isolates of *Alternaria alternata* f.sp. *mali*. Of the two isolates $A_2$ was more pathogenic. Gagkaeva and Levitin (22) isolated *Alternaria alternata* from apple leaves. Inoculation of apple tree leaves with conidial suspension of the Krasnodar strains resulted in the appearance of necrotic spots. Culture suspensions with strains of high aggressiveness when applied on the leaves, also caused typical symptoms of 'alternariosis'.

## EPIDEMIOLOGY AND DISEASE DEVELOPMENT

Tak *et al.* (79) found that the spore germination of *Alternaria alternata* (Fr.) Keissler was maximum at 30°C, 100 per cent RH and 5.5 pH. While studying the influence of pH and incubation period on enzymes produced by *A. alternata* (Fr.) Keissler, they observed that polymethylgalacturonase activity was maximum in culture filterates from glucose potassium nitrate medium with 1% pectin after 9 days of incubation at pH 4 and 5. Polygalacturonase activity was maximum in a non-pectin medium after 12 d at pH 6. Cellulolytic enzyme activity was maximum in a medium with 1 per cent pectin after 9 d at pH 6. Selmaoui and Douira (59) found microsclerotia in *A. alternata* collected from lesions on different apple varieties in Morocco. Wu *et al.* (92) isolated two strains ($A_1$ and $A_2$) of the pathogen of the apple leaf spot disease. On PDA cultural medium, $A_1$ was black-green in colour with very few aerial hyphae but with a lot of conidia, and $A_2$ was grey in colour with many hyphae and few conidia. Growth of $A_1$ and $A_2$ was maximum at 30° and 28°C, respectively and the most suitable pH was 7.

Furstenberg (21) reported that untreated apple trees in Germany were severely attacked by leaf spot fungi in the month of July or August and also reported the white spot infection caused by *Mycosphaerella sentina* on pear trees. Vukovits (89) observed the leaf spot diseases of fruit trees as severe in Austria during humid weather. The important leaf spots recorded were *Phyllosticta mali* Prill. and Del. on apple and *Mycosphaerella sentina* (Fr.) Schroet on pear. Van Poeteren (84) reported that Star apple trees at Beugen were heavily attacked by leaf spot diseases resulting in complete defoliation before the middle of June. Sokolov (74) reported from Russia that apple leaves were attacked by *Phyllosticta mali, Pestalotia* and *Hendersonia* species during the year 1956-58 in epiphytotic conditions. Cold spring and heavy summer rainfall were the reasons for the widespread occurrence of various leaf spots. Zhukova (99) stated that *Phyllosticta mali* was widespread in the southern districts of Krasnoyarsk region. Mass infection occurred in the middle of June and pycnidia appeared usually at the end of July or beginning of August. Wet and warm weather was favourable for the development of the fungus.

Filajdic and Sutton (19) examined the survival of *A. mali* in lesions on leaves on grass or bare ground, in lesions on terminals, and as conidia in buds of Delicious apple trees in 3 distinct orchard environments in North Carolina, U.S.A. However, no lesions were observed on the terminals. Leaves on the ground were a more important overwintering site than buds at all 3 locations. Yoon *et al.* (95) reported that the meteorological factors influencing the outbreaks of *A. alternata* f.sp. *mali* were temperature, relative humidity, rainfall and number of rainy days. High temperature and relative humidity before July had negative effects on the disease, while in August; relative humidity, rainfall and number of rainy days had a positive effect. Filajdic and Sutton (95) studied the effect of combinations of 9 different temperatures from 4 to 36°C and 8 wetness periods from 2 to 48 h on the infection of Delicious apple seedlings by *A. mali*. Disease severity increased with increased wetness duration over all the temperatures tested and was highest at 12 to 28°C. The influence of temperature and wetness duration on infection of apple seedlings by *A. mali* was described by an equation:

Yij = -6.4580 + 0.1853Ti + 0.0912 Wj - 0.0033 T2i - 0.0030 W2j + 0.0194 TiWj - 0.0005T2iwj, in which Y = $\log_{10}$ (percentage of leaf area covered with lesions + 0.01), T = temperature (°C), W = wetness duration (h), i = individual temperature treatment and j = individual wetness duration treatment. The predicted optimum temperature for infection was 23.5°C. At this temperature, 5.1 h of wetness was required for light infection (0.2% leaf area covered with lesions). The model derived from lab data was tested in the field with healthy seedlings surrounded by inoculated ones.

A minimum of 6 h of leaf wetness duration was required at 15, 20 & 25°C for initiation of *Alternaria* infection on young leaves; whereas at 30°C, at least 18 h of leaf wetness was essential. On the mature leaves, disease initiated after 12 h of leaf wetness at 25°C; whereas at 15 and 20°C temperature, a minimum of 18 h leaf wetness was required. At 30°C temperature, the disease initiated after 24 h of leaf wetness duration. Among the different weather parameters, relative humidity was directly correlated with disease development. Regression analysis further showed that about 50 per cent contribution in disease development was due to temperature, rainfall and relative humidity.

## EVALUATION OF APPLE GERMPLASM

Zhao *et al.* (98) carried out an investigation on 15-20 years old trees of 136 apple varieties in the Institute of Pomology's apple germplasm collection. The 21 varieties found to be resistant to apple leaf spot (*A. mali*) included White Rozmalin, Dongguo, Iranpeach, Tsugaru, Xiangjiang Pingguo, Shennong, Jerseymac and Chicheng. The most susceptible varieties were Magu, Moscow Transparent, Sweet Antonovka and Mutsu. All the strains of cv. Fuji, except susceptible Nagaro 6, were moderately resistant. Shin *et al.* (71) rated foliage of 60 cultivars for resistance to *A. mali* on the basis of natural infection and inoculation. Gala, Honey Gold, Jonagold, Magnolia Gold and Mollie's Delicious were highly resistant. Maeno *et al.* (43) reported that AM toxins I and II caused necrosis and resulted in a marked increase in potassium release from green tissues (leaves, petioles, sepals and pericarp of young fruit) in the susceptible apple variety Red Gold. In the moderately resistant varieties Jonathan and American Summer Pearmain, green tissues and petals, styles, filaments and fruit flesh were affected. Resistant varieties did not react to either toxin. Twenty one apple varieties were assigned to three groups according to necrotic response to the toxins.

Saito *et al.* (55) observed that out of six crosses involving the varieties Golden Delicious, Starking Delicious, Fuji and Ralls; those three which had Starking Delicious as a parent had a low percentage (32-46%) of individuals in the bottom resistance category 0 and a broad range of segregation up to the top category 4. The three other crosses had 72-82 per cent of individuals in category 0 and 18-26 per cent of those remaining in category 1. Genetic control of resistance to *Alternaria mali* in Starking Delicious was, therefore, thought to differ from that in the other varieties. Sasnauskas *et al.* (56) reported apple cultivars Sylvia, Lavia, Rubin, Lodel, Staris, Slavianin, Gloster and Pamiatj Isajeva to be highly resistant to blotch. Bandaravicius *et al.* (3) reported Co-op 10 to be resistant to apple blotch and scab. Gelvonauskis and Gelvonauskiene (24) in their study used three scab resistant apple cultivars (Arbat, Florina and Prima) as male and eight cultivars (Tellissaare, Sylvia, Staris, Auksis, Aktiubinskoye, Discovery, Lodi and Quinte) as female and found significant differences in resistance to scab, powdery mildew and apple blotch [*Phyllosticta mali* (*Asteromella mali*)] among the crosses.

In Himachal Pradesh, evaluation of apple cultivars for resistance against alternaria leaf spot revealed that Fuji, Gala,

Red Fuji and Golden Delicious were resistant to this disease and all the commercial cultivars viz., Royal Delicious, Red June, Top Red and Starkrimson were highly susceptible (39).

## DISEASE MANAGEMENT

### Biological Control

Chen *et al.* (10) observed in lab tests that fluorescent *Pseudomonas* strain P420-4, *Bacillus* strain B526-7 and their mixture were highly antagonistic to *Physalospora piricola, Glomerella cingulata* and *Alternaria mali.* Fu *et al.* (20) reported *Bacillus* spp. isolated from apple trees to be effective in controlling *Alternaria solani* f. sp. *mali* and *Botryosphaeria berengeriana* on apple trees. Xin *et al.* (93) reported under laboratory tests that the inhibitory action of *Bacillus subtilis* towards *A. alternata, Trichothecium roseum* and *Fusarium* sp. was 100 per cent. Among different plant extracts and plant oils tested *in vitro,* neem (*Azadirachta indica*), mehandu (*Dodonia viscosa*) extract and mentha (*Mentha arvensis*) oil gave the maximum growth inhibition of *Alternaria alternata*. Among the bio-control agents, *Trichoderma hamatum* exhibited the maximum inhibition (68.76%) followed by *T. harzianum* (63.18%) and *T. viride* (62.62%).

### Chemical Control

A great deal of work on the control of leaf spot diseases caused by *Alternaria* species have been done at various places. Mycelial growth of *A. alternata* on Richard's agar medium was completely inhibited by propiconazole (Tilt) and tridemorph (Calixin-M) at 30 ppm (33). In descending order of effectiveness, mancozeb + metalaxyl (ridomil), chlorothalonil (daconil), cobox, carbendazim (bavistin), benomyl (benlate), propineb (antracol), and thiophanate significantly reduced mycelial growth at 10, 20 and 30 ppm concentration. Propiconazole was the most effective and thiophanate the least effective. Among the six fungicides tested against *Alternaria alternata*, mancozeb was found highly effective in inhibiting the mycelial growth followed by copper oxychloride and iprodione at 1000, 2000 and 3000 ppm. Mancozeb was also found effective in reducing the disease intensity under pot culture condition at 0.2 per cent concentration.

Ciferri (11) compared diathane Z-78 (0.3%) with other materials against *Alternaria mali* causing leaf spot of apple in different parts of Italy from 1949-1953 and reported it to be effective. Sharma and Agarwala (70) reported aureofungin at 20-40 μg/ml to be superior to other antibiotics tested in inhibiting the growth of sooty blotch of apple caused by *Gloeodes pomigena*. Sud and Agarwala (77) reported six fungicides belonging to the groups of copper and dithiocarbamates for their toxicity against *A. mali* Roberts. Ziram (0.1%) was more effective and closely followed by ferbam, zineb and captan (@ 0.15% each) in controlling the disease amongst the carbamates and none was effective out of the copper fungicides tested. Chakstynya (8) reported that the copper and sulphur containing fungicides used against *Phyllosticta mali* proved ineffective and Bordeaux mixture stimulated the development of the disease.

Lee and Kim (40) studied the effects of fungicides on sporulation of apple leaf spot (*Alternaria mali* Roberts). The maximum inhibition of sporulation on inoculated leaf discs was obtained with rovral (iprodione) at all dosages tested, while antracol (propineb), diathane M-45 (mancozeb) and dikar (mancozeb + dinocap) were effective only at the higher recommended concentrations. Polyoxin and difolatan (captafol) were less effective on leaf discs from sprayed orchards than those from unsprayed. Bayleton (triadimeton) was the least effective fungicide tested and benlate (benomyl) stimulated sporulation. Kim *et al.* (35) reported that captafol, rovral (iprodione), thalonil (chlorothalonil), dithane M-45 (mancozeb) and polyoxin B sprays controlled *Alternaria* leaf spot of apple. The optimum spraying time was from 10 days after petal fall to the end of the rainy season.

Tak *et al.* (78) reported that out of a number of fungicides, growth regulators, fats and oils tested for the control of *A. alternata*, the most effective were Rovral (iprodione), maleic hydrazide and hydrogenated ground nut oil as both pre- and post-inoculation treatments. Glaser and Kaiser (25) recommended that repeated orchard spraying with 0.3 per cent dithane M-45 (mancozeb) or with mancozeb + a solution of mineral fertilizer mixture (8 treatments), both reduced incidence of *Alternaria tenuis* (*alternata*) on apple cultivar Golden Delicious; but the combined spray programme (mancozeb + fertilizer) gave the best results with 93% control.

Yoon *et. al.* (94) studied the inhibitory effects of calcium compounds on the outbreak of alternaria leaf spot caused by *A. alternata* f.sp. *mali*. The Ca content of apple leaves resistant to *A. alternata* f.sp. *mali* [*A. mali*] was higher than that of susceptible leaves. Foliar sprays of Ca compounds [CaO, $Ca(OH)_2$ and $CaCO_3$] inhibited infection, application prior to inoculation being more effective than post-inoculation. Jiang *et al.* (29) tested the standard fungicides Bordeaux mixture and carbendazim, and 1.5 per cent duokangmeisu, 50 per cent puhaiin and 70 per cent mancozeb for the control of apple leaf spot diseases caused by *Marssonina mali* and *Alternaria mali*. Spraying twice in the spring and once in the autumn at the shoot growing stage with 1.5 per cent duokangmeisu gave the best disease control, followed by Bordeaux mixture. Yousuf and Ahmad (96) evaluated six fungitoxicants and one nitrogenous fertilizer, applied as single postharvest pre-leaf fall autumn spray for their potential to reduce conidial production of *A. mali* on overwintered apple leaves and intact buds. The ergosterol biosynthesis inhibiting fungitoxicants penconazole and hexaconazole followed by fenarimol proved to be the most effective in reducing conidial production on leaves by 61.40 to 66.46 per cent and on intact buds by 69.94 to 71.06 per cent. Urea recorded about 50.0 and 47.5 per cent conidial production on leaves and buds, respectively. Mancozeb was the least effective. The results of the study conducted by Reuveni *et al.* (52) indicated that a control programme based on spray applications of difenoconazole or polyoxin B, during bloom period can effectively reduce *Alternaria* on Red Delicious apples. Bassler and Hausladen (4) while evaluating fungicidal efficacy and first control application for *Alternaria alternata* and *A. solani*, found that out of 4 starting dates for the application of fungicides in June-July, the best results were obtained when treatment started on 8$^{th}$ June (first appearance of symptoms).

EBI fungicides hexaconazole, difenocanozole and tebuconazole, and iprodione + carbendazim gave cent per cent inhibition of growth as well as spore germination of *A. alternata in vitro* (39). Propineb, dodine, shield and fluquinconazole were also effective in inhibiting the growth and spore germination, whereas benzimidazole fungicides carbendazim, thiophanate methyl and benomyl did not show any significant inhibition. Under field conditions, difenoconazole (0.025%), iprodione + carbendazim (0.15%), hexaconaxole (0.05%), propineb (0.3%), dodine (0.075%) and Shield (0.25%) gave fairly good disease control ranging from 92.11 to 98.51 per cent. a spray schedule comprising the sprays of dodine (0.075%), carbendazim (0.05%), hexaconazole + mancozeb (0.35%) and tebuconazole (0.05%) at different stages (from fruit set to pre-harvest stage) was highly effective in controlling the disease and provided 99.21 per cent disease control in the field.

## CONCLUSION

Once regarded as minor/weak pathogens, leaf spot fungi are taking a heavy toll in apple orchards owing mainly to the changing climatic scenario and intensive diverse cultivation practices. Site-specific fungicidal sprays directed against one pathogen is inadvertently leading to the emergence of another problem, thereby necessitating a different control strategy against the latter. There is an urgent need to evolve an integrated spray programme against all the foliar diseases of apple including leaf spots. More forecasting procedures need to be developed along with the use of bio-pesticides in the form of plant extracts, essential oils and biocontrol agents. Regular monitoring of leaf spot in apple is compulsory for taking well timed action to avoid indirect losses in terms of premature defoliation and reduced fruit quality and yield.

### *Literature Cited*


1. Adams JF. 1929. Report of the plant pathologist for 1928. *Quart. Bull. State Board of Agric.*, Delaware 19: 30.
2. Anon. 2004. http://www.fao.org.
3. Bandaravicius A, Gelvonauskiene D, Sasnauskas A. 2001. Investigation of biologically and economically important traits of introduced apple cultivars. *Sodininkyste-ir-Darzininkyste* 20(1): 3-15.
4. Bassler E, Hausladen H. 2003. *Alternaria* : evaluation of fungicidal efficacy and first control application. *Kartoffelbau* 6: 232-234.
5. Bhardwaj LN. 1972. *Studies on leaf spot diseases of apple (Malus pumila* Mill) *and their control under field conditions.* M.Sc. Thesis, Himachal Pradesh University, Agricultural

Complex Colelge of Agriculture, Solan. 76p.

6. Brien RM, Dingley J M. 1957. Third supplement to a revised list of plant diseases recorded in New Zealand, 1955-1957. *N.Z.J. Sci. Tech.* 38: 777-781.

7. Bulajic A, Filajdic N, Babovic M, Sutton TB. 1996. First report of *Alternaria mali* on apples in Yugoslavia. *Plant Dis.* 80: 709.

8. Chakstynya T. 1964. The effect of the fungicides on apple leaf spot caused by *Phyllosticta mali. Abs. in Referate. Zh. Rasten.* 10: 483.

9. Chaudhary KCB, Puttoo BL, Ashraf SS. 1987. Seed borne nature of *Pestalotia hartigii* on apple seeds. *Int. J.Tropical Plant Dis.* 5: 111-112.

10. Chen GY, Zheng TM, Mao QY. 1993. Experiments on the control of apple leaf and fruit diseases with fluorescent *Pseudomonas* sp. *Chinese J. Biol. l Cont.* 9: 163-166.

11. Ciferri R. 1953. Five years of experiments with fungicides with a zinc ehtylene bisdithiocarbamate base. *Notiz. Malatt. Piante* 1953. 23: 3-19.

12. Clinton GP. 1902. Black rot, *Sphaeropsis malorum* Berk. **In**: *Apple rots in Illinois. III.* Agri. Expt. Sta. Bull. 69: 192-193.

13. Crabill CH. 1915. The frog-eye leaf spot of apples. *Va. Agr. Exp. Sta. Bull.* 209: 3-16.

14. Dennis RWG, Wakefield EM. 1946. New or interesting British fungi. *Trans. Brit. Mycol. Soc.* 29: 141-166.

15. Dowson WJ. 1931. Apple leaf spot. *Tasman. J. Agric.* 2: 79-80.

16. Filajdic N, Sutton TB. 1991. Identification and distribution of *Alter-naria mali* on apples in North Carolina and susceptibility of different varieties of apples to Alternaria blotch. *Plant Dis.* 75: 1045-1048.

17. Filajdic N, Sutton TB. 1992a. First occurrence of Alternaria blotch on apple in the USA and susceptibility of different varieties to a new disease. *Zaschita Bilja* 43: 27-33.

18. Filajdic N, Sutton TB. 1992b. Influence of temperature and wetness duration on infection of apple leaves and virulence of different isolates of *Alternaria mali. Phytopathology* 82: 1279-1283.

19. Filajdic N, Sutton TB. 1995. Overwintering of *Alternaria mali,* the causal agent of Alternaria blotch of apple. *Plant Dis.* 79: 695-698.

20. Fu XC, Yan ZN, Xu WM, Mei RH, Chen C. 1997. Study on the biological control of *Alternaria solani* f.sp. *mali* and *Botry-*

*osphaeria berengeriana* of apple trees. *China Fruits* 3: 7-10.

21. Furstenberg, C. 1925. Advice on the control of some fruit pests & diseases. *Deutsche Obst -u. Gemusebauzeet* 21: 67-68.
22. Gagkaeva TY, Levitin MM. 2000. Identification of the pathogen causing leaf spots on apple trees in Krasnodar province orchards. *Mikologiya-i-Fitopatologiya* 34: 58-62.
23. Gandineau MM. 1947. Diseases of apple and pear trees. Some present day problems. *C.R. Synd. Arboric. Adour.* 6.
24. Gelvonauskis B, Gelvonauskiene D. 2003. Inheritance of resistance to powdery mildew and apple blotch progenies of scab resistant apple culivars. *Biologija* 1: 73-76.
25. Glaser T, Kaiser HU. 1986. Studies on the etiology and control of necrotic leaf spot of apple cultivar Golden Delicious. *Rocz. Akad. Rolniczej-w-Poznaniu, Ogrodinictwo* 165: 49-62.
26. Govi G. 1957. Parasitic spotting of apple leaves. *Ital. Agric.* 94: 630-632.
27. Gupta GK, Agarwala RK. 1968. Alternaria blight of apple. *FAO Plant Prot. Bull.* 16: 32-33.
28. Ivanoff B. 1924. Diseases of pome fruits and measures for their control. *Rev. Inst. Rech. Agron. en Bulgarie.* 3: 57-65.
29. Jiang RD, Lu SJ, Wang S, Wang JQ, Li Q. 1997. Control of apple leaf spot diseases by spraying. *China Fruits* 2: 51.
30. Kaul TN. 1957. Outbreaks and new records. *F.A.O. Plant Prot. Bull.* 5: 93-96.
31. Kaul TN. 1962. Occurrence of *Cercospora rubrotincta* Ell. and Ev. on Cherry in India. *Curr. Sci.* 27: 201-202.
32. Khabiri K. 1952. A contribution to the mycoflora of Iran. *First List. Rev. Mycol.* 17: 154-157.
33. Khosla K, Sharma IM, Kumar J. 2004. Management of leaf blotch (*Marssonina mali*) disease of apple plants in nursery. *Indian Phytopathol.* 57: 366-367.
34. Kim CH, Cho D, Kim SC. 1986. An empirical model for forecasting Alternaria leaf spot in apple. *Korean J. Plant Prot.* 25: 221-228.
35. Kim YB, Yiem MS, Jang HI. 1982. Studies on the chemical control of Alternaria leaf spot in apple trees. Res. Rep. Office of Rural Development, South Korea Horticulture 24(12):77-82.
36. Kovacs J, Fischl G, Nagy P. 1999. Effect of sugar forms and sugar concentration on *Alternaria alternata* (Fr.) Keissler under *in vitro* conditions. *Novenyvedelem* 35(2): 53-56.

37. Kruger J. 1991. Appearance of sooty blotch and Phyllosticta leaf spot in unsprayed fields at Ahrensburg. *Erwerbsobstbau* 33: 204-205.

38. Kudela V, Kovacikova E, Muska J. 1993. A calyx end rot of apple fruit caused by *Alternaria alternata. Ochrana Rostlin* 29: 319-225.

39. Kumar R. 2004. Studies on leaf spot diseases of apple and their management. M. Sc. Thesis. Dr. Y.S. Parmar University of Horticulture and Forestry, Nauni, Solan, 71p.

40. Lee CU, Kim MH. 1980. Effects of fungicides on sporulation of apple leaf spot *Alternaria mali* Roberts. *Korean J. Plant Prot.* 19: 169-174.

41. Leite Junior RP, Tsuneta M, Kishino AY. 1986. Apple leaf spot caused by *Marssonina coronaria. Fitopathol. Brasil.* 11: 725-729.

42. Lewis CE. 1912. Inoculation experiments with fungi associated with apple leaf spot and canker. *Phytopathology* 2:49-62.

43. Maeno S, Kohimoto K, Otani H, Nishimura S. 1984. Different sensitivities among apple and pear cultivars to AM-toxin produced by *Alternaria alternata* apple pathotype. *Journal of the Faculty of Agriculture, Tottori University* 19: 8-19.

44. Martsikh Zh G. 1965. Aetiology of brown spot of apple in Maldovia. *Infekts Zabolevaniya kul turn rast. Kishiner* 4: 72-81.

45. McDonald J. 1931. Annual report of the Senior Mycologist for 1930. *Ann. Rept. Dept. of Agric. Kenya for the year ending 31st December.* 1930: 206-216.

46. Miyake I. 1907. Ueber einige pilz-krantheiten unserer Nutzpflanzen. *Bot. Mag.* 21: 39-44.

47. Moore WC. 1939. New and interesting plant diseases I. *Trans. Brit. Mycol. Soc.* 12: 3-4.

48. Mundkur BB, Ahmad. 1946. Revision of and addition to Indian fungi. II. *Mycol. Pap. Imp. Mycol. Inst.* 18: 11.

49. Nema DK. 1986. Changes in sugars of apple due to *Alternaria alternata* infection. *Indian Phytopathol.* 36: 626-629.

50. Peregrine WTH, Watson DRW. 1964. Annual Report of the Plant Pathology Section, Department of Agriculture, Tanganyika, 21st Sept. 1963-30th Sept. 1964: 10.

51. Petrak F, Esfandiari E. 1941. Beitrage zur Kenntnisder iransschen pilzflora. *Ann. Mycol. Berl.* 39: 204-228.

52. Reuveni M, Sheglov D, Sheglov N, BenArie R, Prusky D. 2002. Sensitivity of Red Delicious apple fruit at various phenotypic stages to infection by

*Alternaria alternata* and moldy-core control. *European Journal of Plant Pathology* 108: 421-427.

53. Roberts JW. 1914. Experiments on apple leaf spot fungi. *J. Agric. Res.* 2: 57-66

54. Rothers BV. 1928. Outline of plant diseases in the district of Sotchi. *La Defense des Plantes. Leningrade* 6: 962-967.

55. Saito K, Nakayama R, Takeda K. 1974. Seedling screening for resistance to Alternaria leaf spot in apple. *Agriculture and Horticulture Nogyo Oyobi Engei* 49: 923-924.

56. Sasnauskas A, Gelvonauskiene D, Bandaravicius A. 2001. First results of biologically and economically important traits of introduced apple cultivars. *Sodininkyste-ir-Darzininkyste* 20: 317-324.

57. Sawamura K. 1962. Studies on spotted diseases of apple I. Causal agent of Alternaria blotch. *Bull. Tohoku Agric. Exp. Sta.* 23: 163-175.

58. Scott WM, Rorer JB. 1908. Apple leaf spot caused by *Sphaeropsis malorum. U.S. Dept. Agr. Bur. Plant Ind. Bull.* 121: 45-54.

59. Selmaoui K and Douira A. 1999. Microsclerotia in *Alternaria alternata. Phytopathol. Mediterr.* 38: 43-46.

60. Sharma JN. 1999. Marssonina blotch - a new disease causing premature leaf fall in apple. *Indian Phytopathology* 52: 101-102.

61. Sharma JN. 2001. Diagnosis and control of premature leaf fall problem in apple. *J. Mycol. Plant Pathol.* 31: 305-310.

62. Sharma JN. 2004. *Biology, epidemiology and integrated management of Marssonina blotch causing premature leaf fall in apple.* Final Report of ICAR Adhoc Scheme. Department of Mycology and Plant Pathology, Dr. Y.S. Parmar University of Horticulture and Forestry, Nauni, Solan, 41p.

63. Sharma JN, Sharma A, Sharma P. 2004. Out-break of Marssonina blotch in warmer climates causing premature leaf fall of apple and its management. *Acta Hortic.* 662: 405-409.

64. Sharma JN, Gautam DR. 1997. Studies on premature leaf fall in apple - A new problem. *Indian Journal of Plant Protection* 25: 8-12.

65. Sharma JN, Verma LR. 1999. Premature leaf fall problem in apple : Diagnosis and control. pp. 81-88. In: *Diseases of Horticultural Crops-Fruits.* (Eds. Verma, L.R. and Sharma,

R.C.). Indus Publishing Co., New Delhi.

66. Sharma JN, Gupta D, Bhardwaj LN, Kumar R. 2005. Occurrence of Alternaria leaf spot (*Alternaria alternata*) on apple and its management. In: *Integrated Plant Disease Management* (Eds. Sharma, R.C. and J.N. Sharma). Scientific Publishers, Jodhpur (Raj.). Pp 25-31.

67. Sharma JN, Sharma PK, Sharma, RC, Bhardwaj LN. 2004. The genus *Marssonina*: Its biology, epidemiology and management. *Ann. Rev. Plant Pathol.* 3: 271-292.

68. Sharma JN, Sharma RD, Verma KD. 2002. Diagnosis and control of premature leaf fall problem of apple in Himachal Pradesh - A success story. *Information Bulletin.* Dr. Y.S. Parmar University of Horticulture and Forestry, Nauni, Solan. 8 pp.

69. Sharma PK. 2003. Studies on Marssonina blotch (Marssonina coronaria Ell & J.J. Davis) J.J. Davis, causing premature leaf fall problem in apple and its management. M.Sc. Thesis. Dr. Y.S. Parmar University of Horticulture and Forestry, Nauni, Solan, 62 p.

70. Sharma VC, Agarwala RK. 1968. *In vitro* studies of the effect of antibiotics on *Dematophora necatrix. Indian Phytopathol.* 20: 291-293.

71. Shin YU, Kang SJ, Kim MS. 1986. Studies on resistance to Alternaria leaf spot in apple cultivars. *Res. Rep. Rural Development Administration, Horticulture, Korea-Republic* 28(2): 39-45.

72. Sibilia C. 1949. Review of the most noteworthy phytopathological records observed in 1949. *Bull. Staz. Pat. Vege. Roma.* 8: 261-294.

73. Sidlauskiene A, Rasinskiene A, Surviliene F. 2003. Influence of environmental conditions upon the development of *Alternaria* genus fungi *in vitro. Sodininkyste-ir-Darzininkyste* 22(2): 160-166.

74. Sokolov AM. 1961. Resistance of apple to spot. *Agrobiology, Moscow* 4: 625-626.

75. Stevens NE. 1933. Two apple black rot fungi in the United States. *Mycologia* 25: 536-548.

76. Sud VK. 1968. *Study of the fungi causing leaf spot diseases in pome and stone fruit trees.* M.Sc. Thesis. Himachal Agricultural College and Research Institute, Solan, 74 p.

77. Sud VK, Agarwala RK. 1970. Laboratory and field evaluation of fungicides for the control of Alternaria blight of apple. *Indian Phytopathol.* 24: 201-204.

78. Tak SK, Verma OP, Pathak VN. 1985. Control of Alternaria rot of apple fruits by post-harvest application of chemicals. *Indian Phytopathol.* 38: 471-474.

79. Tak SK, Verma OP, Pathak VN. 1986. Facotrs affecting conidial germination of *Alternaria alternata* (Fries) Keissler causing fruit rot of apple. *Korean J. Plant Prot.* 24: 129-133.

80. Takahashi S, Sawamura K. 1990. Marssonina blotch. **In**: *Compendium of apple and pear diseases* (Eds. Jones, A.L. and Aldwinckle, H.S.). APS Press. St. Paul, Minnesota, pp. 23.

81. Taylor J. 1966. Ghost spot of apple leaves caused by *Alternaria tenuis. Phytopathology* 56: 553-555.

82. Tweedy BG, Powell D. 1960. Alternaria rot of apple.. *Phytopathology* 50: 657.

83. Unamuno LM. 1932. Some new or little known micromycetes of the spanish flora. *Bot. Soc. Espanola Hist. Nat.* 32: 439-449.

84. Van Poeteren N. 1928. Report on the activities of the phytopathological service in the year 1926. Verslag en Meded. *Plantenziektenkundigen Dienst te Wageningein* 51: 100.

85. Verneau R. 1987. Report of Marssonina panattoniana on Cichorium intybus. Ric. Fitop. Compan. 13-14: 45-48.

86. Viennot-Bourgin G. 1958. A contribution to the knowledge of the parasitic fungi of Iran. *Ann. Epipht.* 9: 97-210.

87. Viennot-Bourgin G. 1961. Notes on plant pathology. *Ann. Epiphyt.* 12: 115-119.

88. Voglino P. 1930. The ochraceous spots of apple (*Phyllosticta briardi* Sacc.). *La Difesa delle Piante.* 7: 1-3.

89. Vukovits G. 1959. 1959, a year for leaf spot diseases. *Pflanzenarzt* 12: 124-126.

90. Wenzl H. 1936. A new leaf spot disease of the apple (*Phyllosticta angulata*). *Phytopathology* 9: 349-356.

91. Westcott C. 1959. *Plant Disease Handbook. Second Edition.* D. Yan Nostrand Company, 1 NC, New York.

92. Wu GB, Wang YZ, Wang PS, Wang JQ, Gong BY. 1999. Study on the differentiation and biological characteristics of *Alternaria alternata* f.sp. *mali. China Fruits* 4: 11-15.

93. Xin YC, Qin SL, Li BD, Yin SC, Ding XH, Lei CX. 2000. Control efficacy and inhibitory action of *Bacillus subtilis* XM16 strain on disease and pathogens of apple mould core. *Acta Phytopathol. Sinica* 30: 66-70.

94. Yoon JT, Lee JT, Choi DU, Shon SG. 1989a. Inhibitory effects of calcium compounds on the outbreak of Alternaria leaf spot caused by *Alternaria alternata* f.sp. *mali*. *Korean J. Plant Pathol.* 5: 306-311.

95. Yoon JT, Lee JT, Park SD, Park DO. 1989b. Effects of meteorological factors on the occurrence of Alternaria leaf spot caused by *Alternata alternata* f.sp. *mali*. *Korean Journal of Plant Pathol.* 5: 312-316.

96. Yousuf V, Ahmad QN. 2001. Impact of single autumn application of chemicals on primary inoculum production of *Alternaria mali,* the causal agent of Alternaria blotch of apple in Kashmir. *Appl. Biol. Res.* 3: 23-27.

97. Yu ZG, Liu YS, Xie M Zhang YH. 1994. Occurrence and control of apple leaf spot diseases. *Bull. Agricl. Sci. Technol.* 7: 29.

98. Zhao JC, Gong X, Liu WJ, Wang K, Hong YM. 2001. Investigation of leaf spot disease of apple varieties. *South China Fruits* 36(6): 58-59.

99. Zhukova VD. 1964. Brown spot of apple and its control. *Trudy Krasnoyar Nauchnoissled. Inst. Sel. Khoz.* 2: 163-164.

*Rev. Plant Pathol. Vol. 4, 2008*
*Indian Society of Mycology and Plant Pathology*
*Scientific Publishers (India), Jodhpur, pp. 105-138*

# PATHOGENESIS-RELATED PROTEINS IN PLANT DEFENCE

***B.N. Chakraborty and M. Sharma***

Immuno-Phytopathology Laboratory, Department of Botany, University of North Bengal, Siliguri 734 013, Darjeeling, India

KEY WORDS: Pathogenesis-related proteins, disease resistance, induced systemic resistance, systemic acquired resistance, immunological procedures, electron microscopy

## ABSTRACT

One particularly well-studied aspect of plant resistance is the pathogenesis-related (PR) protein response. These proteins were initially identified in tobacco (*Nicotiana tabacum* L.) leaves responding hypersensitively to tobacco mosaic virus (TMV) and have since been identified in a range of other species where their synthesis may be induced by microbial infection (viruses, viroids, fungi, or bacteria) or by chemical elicitors. The introduction of polyacrylamide electrophoresis by which proteins could be separated on the basis of their combination and charge has been an important innovation in unearthing PR proteins. At least seventeen PR protein families are now known and few have various biological activities, including enzymic activity as ß-glucanases and chitinases, chitin binding, and membrane permeabilization. Resistance can derive from a combination of characteristics, which act additively or synergistically. Both constitutive and inducible resistance mechanisms may play roles in the resistance of plants to pathogens and may also form the basis for future attempts to confer resistance by genetic engineering. The antifungal proteins (AFP) are directly involved in attacking or killing fungi and reducing their growth while disease resistance proteins (DRP) are signal mechanisms that activate different resistance mechanisms including the levels of AFP genes. Besides proteins newly defined mRNAs (cDNAs) are often considered as additional members of the existing families where shown to be induced by pathogens or specific elicitors.

## INTRODUCTION

Plants are compelled to withstand stresses of all kinds, be it biotic, abiotic or anthropogenic as a consequence of their immobility. Higher plants protect themselves from various stresses, such as pathogen attack, wounding, application of chemicals including phyto-hormones and heavy metals, air pollutants like ozone, ultraviolet rays, and harsh growing conditions. Such constraints lead to production of a wide array of defense compounds, which are either induced or preformed. One of the ways in which plants respond to biotic and/or abiotic stress factors are the accumulation of various novel proteins collectively referred to as pathogenesis-related proteins or PR-proteins. Such protective plant proteins specifically induced in pathological or related situations have been intensively studied from the agricultural interest. On the other hand, many of the reserve proteins accumulated in seeds and fruits take the constitutive defense function against microbial pathogens and invertebrate pests in addition to their storage function. These inducible or constitutive defense mechanisms of higher plants are relatively conserved in the course of evolution. Accordingly, most plants produce or accumulate similar proteins under certain situations irrespective of their morphological differences. These proteins were first observed in tobacco, *Nicotiana tabacum*, cultivars reacting hypersensitively to tobacco mosaic virus (105, 43). The occurrence of these proteins was not pathogen specific but determined by the type of reaction of the host plant, indicating that these proteins were of host origin. PR proteins were found in small amounts in senescing leaves of flowering plants and in relatively larger quantities when necrosis was more severe. This led to the assumption that these polypeptides were stable proteolytic breakdown products of larger leaf proteins. By definition, these proteins abbreviated as 'PR' proteins, were described as "proteins coded for by the host plant but induced specifically only in pathological or related situations" (105). PR proteins are most often of low molecular weight, selectively extractable in low pH, highly resistant to proteolytic degradation/ or endogenous proteases and localized predominantly in the intercellular space (57). PR proteins have been found in several plant species belonging to various families (Table 1).

**Table 1.** Plant species in which pathogenesis-related proteins have been identified

| Family | Plant species |
|---|---|
| Amaranthaceae | *Gomphrena globosa* |
| Chenopodiaceae | *Chenopodium amaranticolor, C. quinoa, Beta vulgaris* |
| Compositae | *Gynura aurantiaca, Helianthus annuus* |
| Cruciferae | *Arabidopsis thaliana, Brassica nigra, B. juncea, B. napus, B. rapa, Raphanus sativus* |
| Cucurbitaceae | *Cucumis sativus, C. melo, Cucurbita maxima, C. pepo* |
| Gramineae | *Hordeum vulgare, Zea mays, Avena sativa, Oryza sativa, Triticum aestivum* |
| Malvaceae | *Gossypium hirsutum* |
| Papilionaceae | *Medicago sativa, Phaseolus vulgaris, P. lunatus, Cicer arietinum, Vigna unguiculata, V. radiata, Arachis hypogea, Lablab purpureus, Pisum sativum, Glycine max* |
| Rutaceae | *Citrus sinensis* |
| Solanaceae | *Capsicum annuum, Petunia, Solanum demissum, S. nigrum, S. tuberosum, S. dulcamara, Nicotiana debneyi, N. glutinosa, N. rustica, N. sylvestris, N. langsdorfii, N. plumbaginifolia, N. tomentosiformis, N. tabacum, Lycopersicon esculentum* |
| Theaceae | *Camellia sinensis* |
| Umbelliferae | *Petroselinum crispum, Apium graveolens* |
| Vitaceae | *Vitis vinifera* |
| Pinaceae | *Picea abies* |

## OCCURRENCE AND TYPES

The introduction of polyacrylamide electrophoresis by Davis (31) and Ornstein (80) by which proteins could be separated on the basis of their combination and charge was a chief innovation in unearthing PR proteins – the new protein components induced by TMV in hypersensitively reacting tobacco. The protein extracts from *N. tabacum* revealed thirty bands and additional separation was achieved with varying concentrations of the acrylamide in the gels and the electrophoretic pattern was represented by densitometric tracings (105). Moreover, it was observed that the same proteins were induced in tobacco by different viruses, whereas different proteins appeared upon infection in *N. glutinosa*. Aided by selective extraction methods (101, 44, 82) it was established that not only viruses but also fungi and bacteria were able to induce similar protein components in various plant species particularly in incompatible combinations resulting in hypersensitive necrosis. Proteins that are induced exclusively during disease development in compatible host-pathogen combinations have hardly been considered as PR proteins. PR proteins have been grouped into families on the basis of similarities in molecular weights, amino acid composition and serological properties confirmed by nucleotide sequencing of corresponding cDNAs (100). The classification has set a convenient standard for other plant species, in which these family numbers now similarly designate PR proteins with properties homologous to tobacco PR proteins. Initially the PR proteins were grouped in to five main classes consisting of the 10 major acidic PR proteins of tobacco characterized both by biochemical and molecular biological techniques and designated as PR-1 to -5 (44, 102). A unifying nomenclature was proposed based on their grouping into eleven families, classified for tobacco and tomato, sharing amino acid sequences, serological relationships and/ or enzymatic or biological activity. The criteria for inclusion of new families of PR proteins were (i) protein(s) must be induced by a pathogen in tissues that do not normally express the protein(s), and (ii) induced expression must have shown to occur in at least two different plant-pathogen combinations, or expression in a single plant-pathogen combination must have been confirmed independently in different laboratories (106). Each PR protein family is numbered and the individual family members are assigned lower case letters in the order in which they are described. In accordance with the recommendations of the Commission for Plant Gene Nomenclature, PR-genes are designated as *ypr*, followed by the same suffix as of the family. Later on three more peptides, which were capable of inducing defense responses of plants, were identified. These three families (PR-12, 13 and 14) comprise the pathogen induced plant defensins (PR-12) (40) thionins (PR-13) (3) and lipid transfer proteins or LTPs (PR-14) (42). So far, 17 families of PR proteins have been recognized (Table 2). However, the properties of all these proteins have not yet been elucidated.

**Table 2.** Recognized families of pathogenesis-related proteins

| Families | Type member | Properties | Gene symbol |
|---|---|---|---|
| PR-1 | Tobacco PR-1a | antifungal | *Ypr1* |
| PR-2 | Tobacco PR-2 | β-1,3-glucanase | *Ypr2*, [*Gns2* ('*Glb*')] |
| PR-3 | Tobacco P, Q | chitinase type I,II, IV, V, VI,VII | *Ypr3*, *Chia* |
| PR-4 | Tobacco 'R' | chitinase type I, II | *Ypr4*, *Chid* |
| PR-5 | Tobacco S | thaumatin-like | *Ypr5* |
| PR-6 | Tomato Inhibitor I | proteinase-inhibitor | *Ypr6*, *Pis* (*Pin*') |
| PR-7 | Tomato $P_{69}$ | endoproteinase | *Ypr7* |
| PR-8 | Cucumber chitinase | chitinase type III | *Ypr8*, *Chib* |
| PR-9 | Tobacco 'lignin-forming peroxidase' | peroxidase | *Ypr9*, *Prx* |
| PR-10 | Parsley 'PR1' | 'ribonuclease-like' | *Ypr10* |
| PR-11 | Tobacco 'class V' chitinase | chitinase, type I | *Ypr11*, *Chic* |
| PR-12 | Radish Rs-AFP3 | defensin | *Ypr12* |
| PR-13 | Arabidopsis THI2.1 | thionin | *Ypr13*, *Thi* |
| PR-14 | Barley LTP4 | lipid-transfer protein | *Ypr14*, *Ltp* |
| PR-15 | Barley OxOa (germin) | oxalate oxidase | *Ypr15* |
| PR-16 | Barley OxOLP | 'oxalate oxidase-like' | *Yrp16* |
| PR-17 | Tobacco PR27 | unknown | *Yrp17* |

Besides proteins newly defined mRNAs (cDNAs) are often considered as additional members of the existing families where shown to be induced by pathogens or specific elicitors. However, because PR proteins are generally defined by their occurrence as protein bands on gels, and classified within each family once the protein has been characterized, cDNA or genomic sequences without information on the corresponding protein cannot be fitted in to the adopted nomenclature. Thus for naming it is necessary to gather information at both the nucleic acid and the protein level when dealing with a stress-related sequence falling within the definition of PR proteins. Conversely, homologies at the cDNA or genomic level may be encountered without information on the expression or characteristics of the encoded protein. Such sequences obviously belong to the PR-type families, but yet cannot, be considered to correspond to pathogen-induced PR proteins and named accordingly. In more than a few situations, it is difficult to distinguish PR proteins from related proteins/ mRNAs that are present in some organs or appear during specific developmental stages. Homologous proteins/ mRNAs in healthy tissues in which no induction by pathogen infection has yet been demonstrated, are to be termed PR-like proteins (PRLs) and their genes *yprl* (104).

## STRUCTURE AND FUNCTIONAL PROPERTIES

PR proteins are, as such, a collective set of novel proteins which a plant produces in reaction to a pathogen mainly in incompatible interactions and thus impedes further pathogen progress. The "related situations" in which PR proteins were found to be induced, seem to prove the point: application of chemicals that mimic the effect of pathogen infection or induced some aspects of the host response, as well as wound responses that give rise to proteins that are also induced during infections, can induce both PR proteins and acquired resistance. The occurrence of homologous PR proteins as small multigene families in various plant species belonging to different families, their tissue-specificity during development and consistent localization in the apoplast as well as in the vacuolar compartment and their differential induction by endogenous and exogenous signaling compounds suggest that PR proteins may have important functions extending beyond their apparently limited role in plant defense. During the hypersensitive reaction cellular damage and death is a major stress to the plant, as exemplified by high increases in abscisic acid and ethylene (103). It is possible, therefore, that PR proteins are stress proteins directed to alleviate harmful effects of cellular degradation products on thus far untouched neighboring cells. Both acidic and basic PR proteins may be induced by physiological necrosis or wounding or by high concentrations of ethylene (12). Such induction in the absence of pathogenic attack might be taken to indicate protection of cellular structures, either physically to stabilize sensitive membranes or macromolecules, or chemically to keep potentially harmful saprophytic microorganisms on tissue surfaces or in intercellular spaces in check. In virtually any natural stress condition *e.g.,* heat, cold, drought, osmotic stress, water logging, anaerobiosis, metal toxicity, *etc.,* plants are known to react by the synthesis of novel, and sometimes partly overlapping, sets of proteins (111). The various conditions under which PR proteins occur are reminiscent of those under which heat-shock proteins (HSP) are induced. These proteins are ubiquitous in living organisms and associated with the acquisition of thermotolerance to otherwise lethal temperatures, but a causal connection is not evident. Interestingly, the promoters of all three tobacco PR-1 genes that are expressed, as well as of a different type of PR in parsley, contain a heat shock regulatory element, but the proteins are not induced to detectable levels by heat shock. Nevertheless, PR proteins might have an analogous function, though quite different, chaperonin-like function: unlike PR proteins, HSP are intercellular proteins that do not accumulate during heat shock. However, the specific occurrence of individual PR proteins in some floral organs, but not in others, points to other, more specific roles (30).

Tobacco PR-1-type proteins are encoded by a small gene family comprising at least 8 members. The acidic PR-1 proteins remain soluble at pH 3, whereas most other plant proteins are denatured under these conditions. The relative resistance to the action of endogenous and exogenous proteases, suggest that PR-proteins have a low turnover rate. The estimated half-life of the PR-1 proteins is 40 to 70 hours. In other plants, extracellular proteins corresponding to the acidic PR-1 proteins of tobacco do not necessarily have a low pI. For example, the tomato P4 and P6 proteins purified from the apoplastic fluid are basic (55).

Major interest has been devoted to plant hydrolases, β-1,3-glucanases (E.C.3.2.1.39) (PR-2) and chitinases (EC 3.2.1.14) (PR-3), as they are capable of cleaving fungal cell walls resulting in pathogen growth inhibition (77), and moreover, the products of the hydrolysis can act as elicitors of further defence responses (9). Both β-1,3-glucanases and chitinases are highly abundant proteins in plants involved in diverse physiological and developmental processes. They can act either alone or in combination strengthening their antifungal activity. Their accumulation is not restricted only to resistant plants but is often observed in compatible plant-pathogen interactions or even non-pathogenic combinations. The relative ineffectiveness of PR proteins in determining resistance to pathogens does not preclude an involvement in defense. As first proposed by Mauch and Staehelin (74) acidic, extracellular PR proteins might be involved in recognition processes, releasing defense-activating signal molecules from the walls of invading pathogens. This would hold particularly for chitinases and glucanases that could liberate elicitor-type carbohydrate molecules from fungal and bacterial cell walls. Thus, a β-1,3-glucanase induced in soybean seedlings by infection or chemical stress releases elicitor-active fragments from cell wall preparations of the fungus *Phytophthora megasperma* f.sp. *glycinea*. Such elicitors stimulate defense responses in adjacent cells and thus accelerate and enhance these reactions, as well as induce acquired resistance to further infection (45). PR-3 proteins are monomeric enzymes ranging from about 25 to 35 kDa in size. They act as endochitinases and release chitooligosaccharides of varying degrees of polymerization, of which the smallest are usually chitobiose and chitotriose. A PR-3 type chitinase was first isolated from bean leaves but similar enzymes have now been purified from about 20 different monocot and dicot plants. The presence of PR proteins in tea plants with PR-2 and PR-3 activity was reported by Chakraborty *et al.* (22). Association of PR-proteins with resistance in tea plants triggered by the fungus *Exobasidium vexans* (23) and a pest *Helopeltis theivora* has been established (21). The occurrence of almost all types of PR proteins in various floral tissues also suggests specific physiological functions during flower development rather than a role in general defense against pathogen invasion. This notion is supported by the presence in floral organs of additional PR-like proteins, glucanases (28) and thaumatin-like proteins (88). In Petunia flowers, chitinase activity is localized in the petals (about 15%) and stigma (about 85%). In the stigma it increases about five-fold following anther dehiscence, strongly suggesting that the chitinase has a specific function in reproduction (77). Demonstration that the PR-2 family are β-1,3-endoglucanases and the PR-3, 4, 8, and 11 families consist of chitinases with or without lysozyme activity, immediately suggested that these PR proteins are directed against cell walls of fungi and bacteria. Homology of the thaumatin-like proteins of the PR-5 family with a bifunctional α-amylase/trypsin inhibitor from maize seeds seemed consistent with a role in protection against phytophagous insects. However, no proteinase inhibitor activity has been demonstrated for PR-5 proteins. Resistance to insect attack is taken to be conferred primarily by wound-inducible proteinase inhibitors, which have now been grouped into the family PR-6 (48). A role of PR proteins as specific internal signal generating enzymes would be consistent both with their occurrence in specific organs and with their induction during the development and in response to stressful pathogen infections. The major chitinase of bean leaves first described by Boller *et al.* (10) to be induced by ethylene and located in the vacuole, appears to be also induced in abscission zones at the stem petiole-junction together with a PR-1-like protein, two isoforms of β-1,3-glucanase, other chitinases, and a thaumatin-like protein (94). However, the natural substrates for chitinases in higher plant cell walls remain to be determined.

Elucidation of the biochemical properties of the major, pathogen-inducible PR proteins of tobacco and subsequent cloning of their cDNAs and/ or genes revealed that proteins with substantial similarity to the classical PR proteins, which are mostly acidic and extracellular proteins, the homologous counterparts are mostly basic and localized intracellularly in the vacuole. As far as it has been possible to deduce, they possess the same type of enzymatic activities, but their substrate specificity and specific activity may be rather different (71). Newly defined mRNAs (cDNAs) may be added to existing families when shown to be induced by pathogens or pathogen specific elicitors, like are the, thionins (5) and defensins (14) both families are of small basic, cysteine-rich polypeptides, and qualify for inclusion as new families of PR proteins. The only non-plant proteins reported so far to display similarity to plant PR-2-type proteins are fungal β-1,3-glucanase (51). Plants contain several types of enzymes that have the ability to hydrolyze polysaccharides from microbial cell walls. Endo-β-1,3-glucanases, for instance, are enzymes that specifically hydrolyze β-1,3-ether linkages in β-1,3-glucans and β-1,6-glucans, resulting in a range of oligosaccharides down to disaccharides (113). Most endo-β-1,3-glucanases are structurally related and form a family known as PR-2-type proteins which catalyze endo-type hydrolytic cleavage of the 1,3-β-D-glucosidic linkages in β-1,3-glucans. The β-1,3-glucanses are abundant highly regulated enzymes widely distributed in seed plants (52). Although the major interest in β-1,3-glucanases stems from their possible role in the response of plants to microbial pathogens, there is strong evidence that these enzymes are also implicated in diverse physiological and developmental processes in the uninfected plant, including cell division, microsporogenesis (17), pollen germination and tube growth, fertilization (79) embryogenesis, fruit ripening, seed germination (69), mobilization of storage reserves in the endosperm of cereal grains (38), bud dormancy, and responses to wounding, cold, ozone and UV-B (50). Basic glucanases belong to class I, whereas class II accommodates acidic β-1,3-glucanases including the two acidic 41 kDa stylar β-1,3-glucanase isoforms (79) and class III β-1,3-glucanases harbor an acidic tobacco enzyme slightly different from the class II enzyme. The products of three tandemly arranged *Arabidopsis* β-1,3-glucanase genes also fall into this class (34). There is considerable indirect evidence that, in analogy to tobacco class I chitinases and barley lectin, the C-terminal extension contains a signal for targeting to the vacuole. Results obtained with cultured tobacco cells provide strong evidence that vacuolar chitinases can be secreted into the medium via a novel pathway (75). In contrast to β-1,3-glucanase I, the class II and III members of the PR-2 family are secreted into the extracellular space (93). The tobacco class II β-1,3-glucanase PR-2a, PR-2b, PR-2c and the class III β-1,3-glucanase PR-2d, also known as PR-2, PR-N, PR-O, PR-Q′, respectively, are acidic proteins without the C-terminal extension present in the class I enzymes ranging in apparent size from approximately 34 to 36 kDa in denaturing gels (107). Induction of β-1,3-glucanases I has been observed during germination. β-1,3-glucanases is exclusively induced in the micropylar region of the endosperm where the radicle will penetrate but prior to endosperm rupture. This induction of β-1,3-glucanase is not a classical defense response as chitinases and the known acidic II and III β-1,3-glucanases are not induced. So, it has been hypothesized that β-1,3-glucanase weakens the endosperm by digestion of cell wall material which promotes radicle protrusion to facilitate germination (68). Gibberellins which can substitute for light in releasing dormancy in tobacco seeds, induces β-1,3-glucanase in the dark in association with germination. It is well established that 1,3;1,4-β-glucanases, which are structurally related to β-1,3-glucanase but differ in substrate specificity, hydrolyze the cell walls of the starchy endosperm during the germination of cereals (52). Less is known about the function of cereal β-1,3-glucanase, which are present in ungerminated grains and rise markedly in concentration during germination. However it has been proposed that high levels of β-1,3-glucanase are part of a strategy to protect the grain against microbial attack (69). Cordero *et al.* (26) found that one β-1,3-glucanase isoform and three chitinase isoforms are induced in germinating maize kernels infected by *Fusarium moniliforme*.

There is now compelling evidence that β-1,3-glucanase and chitinase, acting alone and particularly in combination, can help defend against fungal infection. It has been proposed that these glucanohydrolases act in at least two different ways: directly, by degrading the cell walls of the pathogen, and, indirectly by promoting the release of cell-wall derived materials that can act as elicitors of defense reactions (9). Several studies have been made in which different β-1,3-glucanase and chitinase isoforms were tested for *in vitro* antifungal activity. In contrast, the class II β-1,3-glucanase PR-2a, PR-2b, and PR-2c did not exhibit antifungal activity alone or in any combinations tested. Similar studies with tomato β-1,3-glucanase and chitinase have shown that the vacuolar class I isoforms, but not the secreted class II isoforms, inhibit growth of *Alternaria solani*, *Trichoderma viride*, and *Phytophthora infestans* (65, 61). Plant β-1,3-glucanase can release oligosaccharides from cell walls of the pathogens, which can then act as elicitors of defense reactions (9). This is well documented for interactions between soybean and the β-glucan elicitor from the pathogenic oomycete *Phytophthora megasperma* f.sp. *glycinea*. Following fungal attack, soybean β-1,3-glucanase releases β-glucans from fungal cell walls, which can then induce accumulation of the phytoalexin glyceollin. The smallest β-glucan released with elicitor activity was a β-1,3-β-1,6-hepta-1,3-glucoside and the structural requirements for the elicitor activity of these oligosaccharides have been investigated. Umemoto *et al.* (98) isolated cDNA for a β-glucan elicitor binding protein (GEPB), which is localized in the plasma membrane of soybean root cells. Expression of soybean GEPB gene has shown to confer β-glucan binding activity to *Escherichia coli* and to tobacco cells cultured in suspension, suggesting that GEPB might be an elicitor receptor. Soybean β-1,3-glucanase have been purified that are able to release active β-glucan elicitors from fungal cell walls. Ham *et al.* (46) presented evidence that fungal pathogens secrete proteins that can selectively inhibit plant β-1,3-glucanase. Fungal pathogenesis appears to involve a complex interplay between host and pathogen β-1,3-glucanase. It is now recognized that species of higher plants produce a broad range of β-1,3-glucanase differing in primary structure, cellular localization, and catalytic activity. The available evidence suggests that different classes of β-1,3-glucanase have different functions in plant microbe interactions. The activity of β-1,3-glucanase and β-D-glucosidases increased in the leaves of potato plants infected by species of the root parasite genus *Globodera* (87). Enzyme assays revealed the appearance of both the extracellular and intracellularly targeted β-1,3-glucanases. The reactions of the cultivars to nematode infection differed, indicating that different pathotypes of the same species or different species of potato cyst nematode elicited the appearance of different classes of β-1,3-glucanase. Kim and Hwang (61) isolated a basic β-1,3-glucanase with inhibitory activity against *Phytophthora capsici* and found a close serological relationship between the pepper 34 kDa β-1,3-glucanase and tomato β-1,3-glucanase. A basic β-1,3-glucanase (34 kDa) induced in pepper (*Capsicum annuum*) stems by mercuric chloride treatment was purified by CM-cellulose cation exchange chromatography. The 34 kDa β-1,3-glucanase inhibited hyphal growth of the chitin-negative fungus *Phytophthora capsici*, but did not show antifungal activity against the chitin-containing fungi *Alternaria mali, Colletotrichum gloeosporioides Magnaporthe grisea,* and *Fusarium oxysporium* f.sp. *cucumerinum.* Treatment with higher amounts of the β-1,3-glucanase caused lysis of the hyphal and zoospore cells of *P. capsici*. Moreover the 34 kDa β-1,3-

glucanase acted synergistically with a pepper basic 32 kDa chitinase in inhibiting hyphal growth of *F. oxysporum* f.sp. *cucumerinum* and *P. capsici*.

## BIOSYNTHESIS OF PR PROTEINS

Biosynthesis of PR-proteins in sugarbeet has been intensively investigated both on plants infected by pathogens and treated with synthetic inducers of SAR: salicylic acid and its function derivates 2,6-dichloroisonicotinic acid (INA) and benzo-1,2,3-thiadiazole-7-carbothioic acid S-methyl ester (BTH). BTH was shown to induce resistance to a number of fungal and viral pathogens, *e.g. Arabidopsis*, wheat and tobacco. Analysis of extracellular fluid isolated from BTH-treated sugarbeet leaves revealed the accumulation of acidic and basic proteins displaying both chitinase and β-1,3-glucanase activities indicating the ability of BTH to activate defence reactions in sugarbeet (18). However, in contrast to this Nielsen *et al.* (78) observed no increase in accumulation of transcripts encoding three chitinase isozymes (including Ch4) and β-1,3-glucanase Glu2 in sugar beet leaves following the treatment with INA, compound similar to BTH, even though the INA pretreatment completely inhibited the development of *Cercospora beticola*. Recently, the BTH capability of inducing SAR to root pathogens was shown on cucumber plants against Pythium damping-off as well as to Phytophthora root rot (2) and even to root-parasitic weed *Orobanche cumana* and root-knot nematodes (81), suggesting that BTH-induced or potentiated defence mechanism might be of more general character. Similarly, the cross-activity of defence responses against diverse pathogens has been demonstrated on rhizomania-diseased sugarbeet and *Heterodera schachtii* root nematodes. Comparison of the determined amino-terminus of a mature acidic PR-1 protein from tobacco and a basic PR-1 protein from tomato with the amino acid sequence derived form their respective cDNA sequences reveals that both proteins, as well as other PR-1-type proteins, are synthesized as preproteins consisting of an amino terminal hydrophobic signal peptide followed by the mature PR-1 domain. Some, but not all, basic PR-1-type proteins feature short (6 to 19 amino acids) carboxy-terminal segment that is absent in most other PR-1-type proteins. This carboxy-terminal extension may function as vacuolar targeting determinant in analogy to the proven role of the carboxy-terminal propeptide of some basic PR-2, PR-3 and PR-5 proteins (75). Complementary DNA clones with sequences similar to tobacco *PR-1* sequences have been isolated from *Brassica napus*, *Solanum tuberosum*, *Medicago truncula*ta and maize and have been purified from tomato, barley and *Arabidopsis* as well as from tobacco (16). PR-1-type proteins have also shown significant homology to some proteins of non-plant origin. PR-8-type chitinases display homology to concanavalin B, an enzymatically inactive protein from jack bean (*Canavalia ensiformis*). More intriguingly, plant PR-8-type proteins show sequence homology with a chitinase from Baker's yeast (*Saccharomyces cerevisiae*). The overall identity between these proteins is 18% including 6 invariant cysteines, with an additional 43% of homologous amino acids in at least five out of the six aligned sequences (13). A single endochitinase from tobacco constitutes the class V chitinases, recently reclassified as PR-11 protein (49, 107). This enzyme is synthesized as a preprotein with an amino-terminal prepeptide and a carboxy-terminal propeptide loosely homologous to the tobacco basic β-1,3-glucanase carboxy-terminal extensions, although an N-glycosylation site is missing. The tobacco PR-11 chitinase shows significant homology to bacterial and nematode chitinases and to a lesser extent also to PR-8-type chitinases. Also remarkable is the fact that some fungal chitinases are more homologous to PR-8 type chitinases (64), while other fungal chitinases are more related to PR-11 type chitinases (47).

## CHARACTERIZATION OF PR PROTEINS

The high resistance of PR-proteins to acidic pH and to proteolytic attack appears to be good adaptation to the conditions of the extracellular environment. Acidic PR-1 proteins were detected in extracellular spaces and xylem elements of TMV-infected tobacco leaves, using biochemical and immuno-localization experiments (35). In addition, (33) described the accumulation of PR-1a, 1b, and 1c after TMV infection of *N. tabacum* Xanthi in the central vacuoles of specialized leaf cells, known as crystal idioblasts. Genomic clones and cDNAs corresponding to basic proteins, which are approximately 65% similar to the acidic PR-1a, 1b and 1c proteins, have also been characterized from the TMV-infected leaves (27, 84). Immunoblot analysis using antibodies against the acidic tobacco PR-1b revealed the existence of the cross-reacting proteins in a number of infected dicots and monocots, including tomato, potato, cowpea, maize, sunflower, barley, *Gomphrena globosa*, *Chenopodium amaranticolor* and *Solanum demissum*. So far there are indications for the two basic PR-1 proteins in tobacco although hybridization studies revealed that the genome contains at least eight (pseudo) genes. (27). Antibodies against the tobacco acidic PR-1 proteins did not react with the basic proteins. β-1,3-glucanases were first purified from leaves of bean and tobacco in the early 1970s (1). Kauffmann *et al.* (58) reported that four pathogenesis related proteins induced in tobacco upon infection were in fact β-1,3-glucanase. Comparison of data from barley β-1,3-glucanases purified from various monocot and dicot plant species indicates that they are all monomers with a molecular weight of about 30 kDa (7, 8). These proteins and their homologs are now known as PR-2-type proteins (107). Phylogenetic analysis of PR-2-type proteins derived from cDNA or genes isolated from several plants yields a dendogram with at least five clusters of related proteins (17). The induction of β-1,3-glucanase as a part of the hypersensitive reaction is a stereotypic response *i.e.*, the pattern of induction is similar for viral, bacterial and fungal pathogens. β-1,3-glucanase show developmental regulation in response to treatment with hormones or infection with pathogens.

More recently, specific β-1,3-glucanase proteins have been measured immunologically and their mRNAs have been measured semi-quantitatively by RNA-blot hybridization. In a limited number of cases, regulation of transcription has been studied using plants transformed with Gus reporter genes under the control of the promoter region of β-1,3-glucanase genes. β-1,3-glucanase accumulates at high concentrations in the roots and in lower leaves of mature, healthy tobacco plants. The β-1,3-glucanase content of leaves decreases toward the top of the plant. Within leaves, they are localized in the vacuole of epidermal cells (59). Biochemical and immunological characterization of tomato PR proteins was elucidated by Fischer *et al.* (39). Acid soluble apoplastic proteins were isolated from leaves of *Lycopersicon esculentum* (Mill) plants infected by *Phytophthora infestans* and separated by two-dimensional gel electrophoresis. Two groups of proteins with either basic (pI $\geq$ 8) or acidic (pI $\leq$ 6) pIs were distinguished. Characterization of defense related proteins that exhibit properties of chitinase and β-1,3-glucanase has also been demonstrated in the wax of "Carnauba" (*Copernica cerifera*), an economically important palm tree, by Cruz *et al.* (29).

## INDUCTION OF PR PROTEINS

Activation of natural weapons before infection, called systemic acquired resistance (SAR) is initiated by pathogens or pathogen-derived elicitors, as well as a number of chemical compounds. Among the main defence genes, which are switched on in response to pathogen infection, belong to encoding PR proteins. The association of PR proteins with SAR, but not with ISR, has led to the hypothesis that accumulation of PR proteins is not a pre-requisite for the induction of resistance, but that PR proteins contribute to the protective state (106). SAR is dependent on the accumulation of SA (salicylic acid), but not JA (jasmonic acid) or ET (ethylene). It appears that only when increases in the levels of any of these signals occur, PR proteins become detectable in the infected plants. The observations indicate that individual PR proteins are induced to various extents by these different signals. Consequently, the mixture of signals released or produced upon microbial stimulation appears to determine the magnitude of the plant's response and its effectiveness to inhibit further infection. In *Arabidopsis* SA-dependent expression of PR-1, PR-2 and PR-5 is required for increased protection against the biotrophic fungus *Perenospora parasitica*, whereas SA-independent but JA-dependent induction of the plant defensin gene *pdf1·2*, as well as of PR-3 and PR-4, is associated with the induced resistance against the necrotrophic fungi *A. brassicicola* (85), *Botrytis cinerea* (96) and *Fusarium oxysporum* f.sp. *matthiolae* (Fom) (36, 37) These results suggest that the SA- and JA-dependent defense pathways in *Arabidopsis* contribute to resistance against distinct microbial pathogens. As a result, PR proteins and similarly induced antimicrobial proteins appear to contribute differentially to the induced resistance against different pathogens. Systemic acquired resistance of plants induced upon infection with a necrotizing pathogen, is characterized by an enhanced capacity to induce cellular defence responses to subsequent pathogen attack. Usually the most effective inducers of PR proteins are pathogens that cause a necrotic reaction; the largest amounts of PR proteins occur in tissues adjacent to the necrotic areas. However, some pathogens can induce large amounts of PR proteins without causing necrosis, for example potato virus Y infection of tobacco induces the systemic production of high concentrations of PR proteins 1a, 1b, 1c and 2 (57) and *Burkholderia solanacearum* induces large amounts of PR-1 proteins in infiltrated tobacco leaf, neither producing any necrotic symptoms. The suppression of mannitol leads to PR-1 accumulation in tobacco following induction by antibiotic (112).

In many plant species, resistance can be induced against a range of pathogens by pre-treating the plant with certain chemicals or compounds of biological origin, ranging from inorganic salts to cell wall fractions from fungi. Pre-inoculation with pathogens or non-pathogens stimulate the same signal transduction pathways. Resistance can be local and restricted to the treated tissue or be induced systemically. The induction of resistance usually coincides with the accumulation of pathogenesis related proteins and their transcripts in resistant tissues (63). Several plant species react to treatment with the stress hormone ethylene with induction of β-1,3-glucanase I activity, protein and mRNA in leaves (4). Although ethylene increases the β-1,3-glucanase I content of epidermal cells slightly, its inductive effect is most pronounced on mesophyll cells of the tobacco leaf. Treatment of plants with ethephon results in no detectable induction, or very weak induction of class II and class III β-1,3-glucanase in leaves of tobacco and tomato. Tobacco cultivars carrying the dominant *N* gene from *N. glutinosa* show a local lesion, HR response to TMV rather than systemic symptoms of mosaic disease. These plants also show decreased disease symptoms in response to secondary infection with certain other viral, bacterial and fungal pathogens. This is an example of SAR the induced long lasting, broad-spectrum resistance phenomenon. Of various chemicals and biological compounds known to induce resistance in different species of plants, only 2,6-dichloroisonicotinic acid (INA) possessed the ability to induce resistance in sugarbeet. Nielsen *et al.* (78) demonstrated that repeated spraying with low concentrations of INA during the week prior to challenge inoculation with the fungus induced complete local and systemic resistance. Eleven acid soluble proteins with apparent molecular masses ranging from 13-82 kD increased in tomato (*Lycopersicon esculentum* Mill.) leaves infected by *Phytophthora infestans* or *Fulvia fulva*. The most prominent changes in the protein pattern were also detectable in the untreated leaves of infected plants indicating systemic effects of the infection. Some of the changes in specific proteins might be useful markers of induced resistance (25). Consequently, the mixture of signals released or produced upon microbial stimulation appears to determine the magnitude of the plant's response and its effectiveness to inhibit further infection. A salicylate-inducible *PR-10* gene subclass from apple, designated *Apa*, was highly activated by wounding and ethephon as well as differentially expressed during infection with a compatible and incompatible race of *Venturia inaequalis* (86). Induction of PR proteins has since been found to be invariably linked in necrotizing infections giving rise to SAR, and has been taken as a marker of the induced state (110, 97, 60). This notion has been reinforced by the characterization in *Arabidopsis* of mutants that either are comprised in both the production of PR proteins and the induction of SAR (*npr1*; 20) (*nim1*; 32), or are constitutive expressors of PR genes as well as SAR (*cpr1*; 11). Systemic accumulation of SA is associated with the HR of tobacco, *Arabidopsis thaliana*, and certain other plants (32). Treatment of mature wild-type tobacco plants with SA strongly induces accumulation of mRNAs of PR related class II and III β-1,3-glucanase and certain other PR proteins. While SA is probably not the long distance systemic signal for SAR activation, it is required for transduction of this signal in leaves distal to the primary infection site. Spraying plants with a solution of salicylic acid mainly induces the acidic PR-1 and acidic β-1,3-glucanase genes (6). Salicylic acid is a naturally occurring compound in tobacco plants and the levels of endogenous salicylic acid rise after TMV infection, not only in the infected leaves but also in the uninfected parts of the plant, which show induced resistance (73). PR proteins induced by BTH, an analogue of SA, in *Carica papaya* also improved resistance responses to *Phytopthora palmivora* (115).

## IMMUNOLOCALIZATION OF PR PROTEINS

Localization of the major, acidic PR proteins in the intercellular space of the leaf seems to guarantee contact with invading fungi or bacteria before these are able to penetrate. In localization studies *in planta*, labelling for β-1,3-glucanases and chitinases was especially pronounced over fungal cell walls confirming their role in plant defence. Most of the investigations were devoted to leaf tissues. In roots, expression of defence genes was studied on infection by pathogens, arbuscular mycorrhizal fungi, antagonistic fungus *Trichoderma harzianum* or non-pathogenic bacterium *Pseudomonas fluorescens* and differences in the expression of distinct classes of chitinases and β-1,3-glucanases were reported in dependence on the particular microbial inducer. Similar results of localization of PR-3 proteins have been reported in potato leaves infected by *Phytophthora infestans* (41). Subcellular localization of PR-1 proteins was studied in roots of resistant *Nicotiana tabacum* cv. *Xanthi* uninfected or infected *in vitro* by the black root rot fungus *Chalara elegans*, using polyclonal or monoclonal antibodies raised against PR-1 protein. In healthy tobacco roots, the PR-1 proteins were found to be present in low amounts in intercellular space material, over cell walls and over secondary thickening of xylem vessels. All these cell compartments were significantly enriched in the PR-1 proteins in infected tobacco root tissues. Their accumulation over the cell walls of inter- and intra- cellular hyphae of *C. elegans* colonizing tobacco roots may reflect an eventual role of these proteins, in association with other PR proteins like β-1,3-glucanases and chitinases in directly hindering hyphal growth of the pathogen (95). Two antisera raised against acidic β-1,3-glucanase and acidic chitinase from tobacco were used to investigate the subcellular localization of the two enzymes in *Fusarium culmorum* infected wheat spike by means of the immunogold labelling technique (56). The studies demonstrated that the distribution of β-1,3-glucanase and chitinase were localized mainly in the cell walls of different tissues including the lemma, ovary and rachis of the wheat spike, while the cytoplasm and the organelles of cells in these tissues showed almost no labelling. However, the accumulation of β-1,3-glucanase and chitinase in the infected wheat spikes differed distinctly between resistant and susceptible wheat cultivars. Immunohistological localization of chitinase and β-1,3-glucanase in rhizomania-diseased and benzothiadiazole treated sugar roots (18) revealed accumulation of plant basic β-1,3-glucanase and basic class IV chitinase in cell walls and extracellular spaces. The deposition of the enzymes was predominantly found in endodermis, cortical cells adjacent to endodermis, and xylem vessels. Similar to the pathogens, benzothiadiazole (BTH) induced deposition of both chitinase and β-1,3-glucanase systemically in cortex and endodermis cells. In *in situ* localization studies of chitinase mRNA and protein in compatible and incompatible interaction triggered by *Phytophthora capsici* in pepper stems. Lee *et al.* (67) showed that induction of chitinase mRNA (*CAChi2*) mRNA started as early as 6 h after inoculation and gradually increased in the incompatible interaction of pepper stems with *P. capsici*. Another interesting finding of the authors was specific immunogold labelling of chitinase on the cell wall of the oomycete as early as 24 h after inoculation. In particular they showed that numerous gold particles were deposited on the cell wall of *P. capsici* with a predominant accumulation over areas showing signs of degradation in the incompatible interaction and that healthy pepper stem tissues were nearly free of labelling. Induction of PR proteins (PR-2 and PR-3) in tea plants using salicylic acid (22, 91), hexaconazole and in suspension-cultured tea cells following inoculation with *E. vexans* or treatment have been characterized biochemically and immunologically using chitinase and β-1,3-glucanase polyclonal antibody (PAb) probes (21, 23). The PR proteins were induced in all interactions where chitinase activity was rapid. Transmission electron microscopic observations of ultrathin sections from induced tea plants embedded in LR-white resin separately treated with PAb-chitinase and PAb-β-1,3-glucanase followed by gold labelling revealed increase in the number of gold particles (24, 92) showing a higher frequency with the PAb-chitinase probe.

## PR PROTEINS AND TRANSGENICS

Plant protection is a major challenge to agriculture worldwide. One of the effective strategies for disease resistance in plants has been the incorporation of disease resistant genes into commercially acceptable cultivars. The most attractive initial candidates for manipulation of the single gene defense mechanisms approach are genes encoding chitinase and β-1,3 glucanase because these two enzymes hydrolyze chitin and β-1,3-glucans which are structural components of the cell walls of several fungi. Chitinase gene from *Rhizopus oligosporus* has been shown to operate as an antifungal system in transgenic tobacco. Transgenic cucumber harboring the rice chitinase genes exhibited enhanced resistance against gray mold, *Botrytis cinerea*. While it is clear that it is possible in several cases to alter the expression of chitinase transgenes to generate plants with increased resistance to the pathogen, it is not clear whether constitutively expressed chitinase alone is responsible for the reduction of disease symptoms as observed in the case of tobacco and canola. Introduction of bacterial chitinase gene from *Serratia marcescens* in transgenic tobacco cells showed up to an eightfold increase in amount of chitinase protein in the plants and conferred resistance to *Rhizoctonia solani*. Expression of β-1,3-glucanase in transgenic tobacco plants was shown to result in enhanced resistance to *Alternaria alternata*. Combined expression of PR-2 (β-1,3-glucanase) and PR-3 (chitinase) gives effective protection against fungal infection as they have been shown to act synergistically. Constitutive expression of individual PR proteins in transgenic plants can lead to reduced pathogen growth and symptom expression, consistent with a role of PR proteins in the expression of acquired resistance (89). Significant suppression of disease symptoms caused by the soil-borne fungus *Rhizoctonia solani* was demonstrated in tobacco or canola expressing vacuolar class I chitinase from bean (15), the basic tobacco chitinase PR-3c (108), tobacco or cucumber PR-8 (66), or the (class II) barley chitinase (53), but enhanced chitinase levels caused no significant protection against *Cercospora nicotianae* (114) or *Fusarium oxysporum* (99). Antifungal activity of chitinases can be synergistically enhanced by β-1,3-glucanses, both *in vitro* and *in vivo*. Thus co-expression of chitinase and glucanase genes in tobacco enhanced resistance against *C. nicotianae* (53). In tobacco simultaneous expression of the basic tobacco chitinase PR-3d and glucanase PR-2e afforded substantial protection against *F. oxysporum* f.sp. *lycopersici*, whereas transgenic plants expressing either one of these genes were not protected (54). An additional 'SAR gene' in tobacco, *SAR 8.2*. when expressed constitutively in transgenic tobacco, was also found to reduce disease caused by *P. parasitica*, but the protein has not been characterized (103). Transgenic plants with chimeric genes containing promoter fragments from a chitinase (class III) gene from *Arabidopsis* fused to the GUS reporter gene were studied by Samac and Shah (90) and was found that the proximal 192 bp from the transcription start site contains elements necessary for salicylate-inducible expression. Naderi and Berger (76) have observed the presence of potato virus Y coat protein (PVY-CP) inside chloroplasts of infected tobacco cells. Induction of symptoms reminiscent of virus infection was observed only in transgenic plants accumulating CP in the chloroplast but not in the cytoplasm. These results suggested that PVY-CP plays a role in symptom development. Plants with high levels of CP in their chloroplasts were more susceptible to PVY, suggesting that there was a host response to the presence of PVY-CP in the chloroplasts. There is strong evidence that expression of β-1,3-glucanase transgenes alone or in combination with chitinase transgenes regulated by the strong CaMV $^{35}$S RNA promoter can reduce the susceptibility of plants to infection by certain fungi (72). Two PR-5-like genes have also been isolated (19) from black nightshade (*Solanum nigrum* L. var. *americanum*) genome. Kruger *et al.* (62) demonstrated that transcription of defense response genes chitinase IIb, PAL and peroxidase is induced by the barley powdery mildew fungus and is only indirectly modulated by R genes.

In plants, SAR is established as a result of *NPR1*-regulated expression of PR genes. Using gene expression profiling in *Arabidopsis*, Wang *et al.* (109) found that in addition to controlling the expression of PR genes, *NPR1* also directly controls the expression of the protein secretory pathway genes. Up-regulation of these genes is essential for SAR, because mutations in some of them diminished the secretion of PR proteins (for example, PR1), resulting in reduced resistance. *NPR1* coordinately regulates these secretion-related genes through a previously undescribed *cis*-element. Activation of this *cis*-element is controlled by a transcription factor that is translocated into the nucleus upon SAR induction. Li *et al.* (70) isolated seven members of the β-1,3-glucanase gene family in barley (*Hordeum vulgare*). It has been also reported that programmed cell death, developmental senescence, and responses to pathogens are linked through complex genetic controls that are influenced by redox regulation.

Expression of green fluorescence protein (GFP) fused to the nonexpressor of PR protein 1 (*GFP-NPR1*) was used by Pavet *et al.* (83) to detect the presence of *NPR1* in the nuclei of transformed plants.

## FUTURE APPROACH

A link between AFPs and resistance is clearly established, although only partial protection is obtained. In the short term it will be possible to select for high levels of characterized AFPs (sormatin, RIP, β-1,3-glucanase, endochitinase) in developing, mature, and infected grain by direct analysis or by using antibodies or molecular markers. In the long term more information is required on the endogenous AFPs of sorghum. There is a need to identify new components, to determine activity and synergism, location, initiation of synthesis and amounts in developing and mature tissues and to relate the above to resistance. The development of molecular markers [*e.g.*, simple sequence repeats (SSRs)] is essential to underpin the selection of resistance and the combination of resistance with yield and end use quality. It will also facilitate dissection of the various aspects of resistance, for example, the separate effects of hardness and AFPs. A more directed approach can also be adopted using markers based on characterized AFPs (*e.g.*, thaumatin) and putative resistance genes (based on other species). Transformation is an essential prerequisite for long-term improvement and must continue to be supported. Despite major advances in the field of plant disease resistance, the precise molecular mechanisms of plant pathogen recognition and the detailed dissection of R-gene mediated signaling networks remain elusive. There is a long-term need to identify and characterize resistance genes in crops and cereals. These genes would be expected to ultimately regulate the manifestation of different phenotypes with resistance to grain mold. In the short term the conserved sequences of the R genes may be used to design primers which can be used to generate polymorphism in marker-assisted selection. In the years ahead, new genomic and proteomic technologies will assist in the identification of signaling components and in the investigation of the biochemical functions of R-proteins and other signaling molecules. Beyond gene identification, proteomic approaches will provide insights into the biochemical properties of proteins involved in disease resistance.

## Literature Cited


1. Abeles FB, Bosshart RP, Forrence LE, Habig WH. 1971. Preparation and purification of glucanase and chitinase from bean leaves. *Plant Physiol.* 47:129–134.
2. Ali Z, Smith I, Guest DI. 2000. Combinations of potassium phosphonate and Bion (acibenzolar-*S*-methyl) reduce root infection and dieback of *Pinus radiata*, *Banksia integrifolia* and *Isopogon cuneatus* caused by *Phytophthora cinnamoni*. *Aust. Plant Pathol.* 29:59-63.
3. Antoniw JF, Kuch JSH, Walkley DGA, White RF. 1981. The presence of pathogenesis-related proteins in callus of Xanthi-nc tobacco. *Phytopathol. Z.* 101:179-184.
4. Beffa R, *et al.* 1995. Cholera toxin elevates pathogen resistance and induces defense reactions in transgenic tobacco plants. *EMBO J.* 14:5753–5761.
5. Bohlmann H. 1994. The role of thionins in plant protection. *Crit. Rev. Plant Sci.* 13:1–16.
6. Bol, JF, Linthorst HJM, Cornelissen BJC. 1990. Plant pathogenesis-related proteins induced by virus infection. *Annu. Rev. Phytopathol.* 28:113-138.
7. Boller T. 1988. Ethylene and the regulation of antifungal hydrolases in plants. In: *Oxford Surveys of Plant Molecular and Cell Biology*. (Ed. BJ Miflin, HF Miflin) Oxford University Press, Oxford. Vol. 5. pp.145–174.
8. Boller T. 1991. Ethylene in pathogenesis and disease resistance. In: *The Plant Hormone Ethylene*. (Ed. AK Matto, JC Suttle) CRC Press, Boca Raton. pp.293–314.
9. Boller T. 1995. Chemoperception of microbial signals in plant cells. *Annu. Rev. Plant Physiol. Mol. Biol.* 46:189-214.
10. Boller T, Gehri A, Mauch F, Vögeli U. 1983. Chitinase in bean leaves: induction by ethylene, purification, properties, and possible function. *Planta* 157:22–31.
11. Bowling SA, *et al.* 1994. A mutation in *Arabidopsis* that leads to constitutive expression of systemic acquired resistance. *Plant Cell* 6:1845–1857.
12. Brederode FT, Linthorst HJM, Bol JF. 1991. Differential induction of acquired resistance and PR gene expression in tobacco by virus infection, ethephon treatment, UV light and wounding. *Plant Mol. Biol.* 17:1117–1125.
13. Broekaert WF, Terras FRG, Cammue BPA. 2000. Induced

and preformed antimicrobial proteins. In: *Mechanisms of Resistance to Plant Diseases*. (Ed. A Slusarenko *et al.*). Kluwer Academic Publishers, The Netherlands. pp.371-477.

14. Broekaert WF, Terras FRG, Cammue BPA, Osborn RW. 1995. Plant defensins: Novel antimicrobial peptides as components of the host defense system. *Plant Physiol.* 108: 1353–1358.
15. Broglie K, *et al.* 1991. Transgenic plants with enhanced resistance to the fungal pathogen *Rhizoctonia solani. Science* 245:1194–1197.
16. Bryngelsson T, *et al.* 1994. Purification, characterization, and molecular cloning of basic PR-1-type pathogenesis-related proteins from barley. *Mol. Plant-Microbe Interact.* 7:267–275.
17. Bucciaglia PA, Smith AG. 1994. Cloning and characterization of *Tag1,* a tobacco anther β-1,3-glucanase expressed during tetrad dissolution. *Plant Mol. Biol.* 24:903–914.
18. Burketova L, Stillerova K, Feltova M. 2003. Immunohistological localization of chitinase and β-1,3-glucanase in rhizomania-diseased and benzothiadiazole treated sugar beet roots. *Physiol. Mol. Plant Pathol.* 63:47-54.
19. Campos MA, *et al.* 2002. Putative pathogenesis-related genes within *Solanum nigrum* L. var. *americanum* genome: isolation of two genes coding for PR5-like proteins, phylogenetic and sequence analysis. *Physiol. Mol. Plant Pathol.* 61:205-216.
20. Cao H, Bowling SA, Gordon AS, Dong X. 1994. Characterization of an *Arabidopsis* mutant that is non-responsive to inducers of systemic acquired resistance. *Plant Cell* 6:1583-1592.
21. Chakraborty BN, Das-Biswas R, Sharma M. 2004. Multi-component coordinated defence strategies in tea plants against *Helopeltis theivora* and *Exobasidium vexans*. *J. Plantn. Crops* 32:289-297.
22. Chakraborty BN, Sharma M, Das-Biswas R. 2004. Antimicrobial proteins of tea. In: *Proc. Int. Conf. O-CHA (tea) Culture and Science* 165-168.
23. Chakraborty BN, Sharma M, Das-Biswas R. 2005. Association of defense enzymes with resistance in tea plants triggered by *Exobasidium vexans* Massee. *Ind. Phytopath.* 58: 298-304.
24. Chakraborty BN, Sharma M, Das-Biswas R, Ghosh A. 2005. Pathogenesis-related proteins of tea: Their induction and immunocytochemical localization. In: *Proc. Asian Conf.*

*Emerging Trends Plant-Microbe Interact.* (Ed. SS Gnanamanickam *et al.*) 194-202.

25. Christ U, Mösinger E. 1989. Pathogenesis-related proteins of tomato: I. Induction by *Phytophthora infestans* and other biotic and abiotic inducers and correlations with resistance *Physiol. Mol. Plant Pathol.* 35:53-65.

26. Cordero MJ, Raventos D, San Segundo B. 1994. Differential expression and induction of chitinases and β-1,3-glucanase in response to fungal infection during germination of maize seeds. *Mol. Plant-Microbe Interact.* 7:23-31.

27. Cornelissen BJC, *et al.* 1987. Structure of tobacco genes encoding pathogenesis-related proteins from the PR-1 group. *Nucleic Acids Res.* 15:6799–6811.

28. Côté F, Cutt JR, Asselin A, Klessig DF. 1991. Pathogenesis-related acidic β-1,3-glucanase genes of tobacco are regulated by both stress and development signals. *Mol. Plant-Microbe Interact.* 4:173-181.

29. Cruz ML, Gomes VM, Fernandes KVS, Machado OLT, Xavier-Filho J. 2002. Indentification and partial characterization of a chitinase and a β-1,3-glucanase from *Copernica cerifera* wax. *Plant Physiol. Biochem.* 40:11-16.

30. Datta SK, Muthukrishnan S. 1999. *Pathogenesis-related proteins in plants.* CRC Press, Boca Raton, LLC, p. 291.

31. Davis BJ. 1964. Disc electrophoresis II. Method and application to human serum proteins. *Ann. NY Acad. Sci.* 121:404-427

32. Delaney T, Friendrich L, Ryals J. 1995. *Arabidopsis* signal transduction mutant defective in chemically and biologically induced diseases resistance. *Proc. Natl. Acad. Sci. USA.* 92:6602-6606.

33. Dixon DC, Cutt JR, Klessig DF. 1991. Differential targeting of the tobacco PR-1 pathogenesis-related proteins to the extracellular space and vacuoles of crystal idioblasts. *EMBO J.* 10:1317-1324.

34. Dong X, Mindrinos M, Davis KR, Ausubel FM. 1991. Induction of *Arabidopsis* defense genes by virulent and avirulent *Pseudomonas syringae* strains and by a cloned avirulence gene. *Plant Cell* 3:61–72.

35. Dumas E, *et al.* 1988. Immunocytochemical location of pathogenesis-related b1 protein induced in tobacco mosaic virus-infected or polyacrylic acid-treated tobacco

plants. *J. Gen. Virol.* 69:2687-2694.

36. Epple P, Apel K, Bohlmann H. 1997. Overexpression of an endogenous thionin enhances resistance of *Arabidopsis* against *Fusarium oxysporum. Plant Cell* 9:509–520.

37. Epple P, Vignutelli A, Apel K, Bohlmann H. 1998. Differential induction of the *Arabidopsis thaliana Thi2.1* gene by *Fusarium oxysporum* f.sp. *matthiolae. Mol. Plant-Microbe Interact.* 11:523-529.

38. Fincher, G.B. (1989). Molecular and cellular biology associated with endosperm mobilization in germinating cereal grains. *Annu. Rev. Plant Physiol. Plant Mol. Biol.* 40:305–346.

39. Fischer W, *et al.* 1989 Pathogenesis-related proteins of tomato: II. Biochemical and immunological characterization. *Physiol. Mol. Plant Pathol.* 35:67-83.

40. Fraser RSS. 1981. Evidence for the occurrence of the pathogenesis-related proteins in leaves of healthy tobacco plants during flowering. *Physiol. Plant Pathol.* 19:69-76.

41. Garcia-Garcia F, Schmelzer E, Hahlbrock K, Roxby R. 1994. Differential expression of chitinase and β-1,3-glucanase genes in various tissues of potato plants. *Z. Naturforsch.* 49c:195–203.

42. Garcia-Olmedo F, Molina A, Segura A, Moreno M. 1995. The defensive role of non-specific lipid transfer proteins in plants. *Trends Microbiol.* 3:72–74.

43. Gianinazzi S, Martin C, Valléé JC. 1970. Hypersensibilité aux virus, température et protéins solubles chez le *Nicotiana* Xanthi nc. Apparition de nouvelles macromolécules lors de la répression de la synthèse virale. *CR Acad. Sci. Paris* 270D: 2883–2886.

44. Gianinazzi S, Pratt HM, Shewry PR, Miflin BJ. 1977. Partial purification and preliminary characterization of soluble leaf proteins specific to virus infected tobacco plants. *J. Gen. Virol.* 34:345-351.

45. Ham KS, Kauffmann S, Albersheim P, Darvill AG. 1991. Host-pathogen interactions XXXIX. A soybean pathogenesis-related protein with β-1,3-glucanse activity releases phytoalexin elicitor-active heat-stable fragments from fungal cell walls. *Mol. Plant-Microbe Interact.* 4:545-552.

46. Ham KS, Wu SC, Darvill AG, Albersheim P. 1997. Fungal pathogens secrete an inhibitor protein that distinguishes isoforms of plant pathogenesis-

related endo-β-1,3-glucanases. *Plant J.* 11:169–179.

47. Hayes CK, *et al.* 1994. Isolation and sequence of an endochitinase-encoding gene from a cDNA library of *Trichoderma harzianum*. *Gene* 138:143 –148.

48. Heitz T, Geoffroy P, Fritig B, Legrand M. 1999. The PR-6 family: Proteinase inhibitors in Plant-Microbe Interactions. In: *Pathogenesis-related proteins in plants* (Eds. SK Dutta and S Muthukrishnan) CRC Press, Boca-Raton, LLC. pp 131-170.

49. Heitz T, *et al.* 1994. Molecular characterization of a novel pathogenesis-related (PR) protein: a new plant chitinase/ lysozyme. *Mol. Gen. Genet.* 245:246–254.

50. Hincha DK, Meins F Jr, Schmitt JM. 1997. β-1,3-glucanase is cryoprotective *in vitro* and is accumulated in leaves during cold acclimation. *Plant Physiol.* 114:1077–1083.

51. Hird DL, *et al.* 1993. The anther-specific protein encoded by the *Brassica napus* and *Arabidopsis thaliana* A6 gene displays similarity to β-1,3-glucanases. *Plant J.* 4:1023–1033.

52. Høj PB, Fincher GB. 1995. Molecular evolution of plant β-glucan endohydrolases. *Plant J.* 7:367–379.

53. Jach G, *et al.* 1995. Enhanced quantitative resitance against fungal diseases by combinatorial expression of different barley antifungal proteins in transgenic tobacco. *Plant J.* 8:97–109.

54. Jongedijk E, Tigelaar H, Van Rockel JSC, Bres-Vloemans SA, Dekker I, Van den Elzen PJM, Cornelissen BJC, Melchers LS. 1995. Synergistic activity of chitinases and β-1,3-glucanases enhances fungal resistance in transgenic tomato plants. *Euphytica* 85:173-180.

55. Joosten MHAJ, Bergmans CJB, Meulenhoff EJS, Cornelissen BJC, De Wit PJGM. 1990. Purification and serological characterization of three basic 15-kilo dalton pathogenesis-related proteins from tomato. *Plant Physiol.* 94:585–591.

56. Kang Z, Buchenauer H. 2002. Immunocytochemical localization of β-1,3-glucanase and chitinase in *Fusarium culmorum*-infected wheat spikes. *Physiol. Mol. Plant Pathol.* 60:141-153.

57. Kassanis B, Gianinazzi S, White RF. 1974. A possible explanation of the resistance of virus-infected tobacco to second infection. *J. Gen. Virol.* 23:11-16.

58. Kauffmann S, Legrand M, Geoffroy P, Fritig B. 1987.

Biological function of 'pathogenesis-related' proteins. Four PR-proteins of tobacco have 1,3-β-glucanase activity. *EMBO J.* 6:3209–3212.

59. Keefe D, Hinz U, Meins F. 1990. The effect of ethylene on the cell-type specific and intracellular localization of β-1, 3-glucanase and chitinase in tobacco leaves. *Planta* 182: 43-51.
60. Kessmann H, Staub T, Ligon J, Oostendorp M, Ryals J. 1994. Activation of systemic acquired disease resistance in plants. *Eur. J. Plant Pathol.* 100:359-369.
61. Kim YJ, Hwang BK. 1997. Isolation of a basic 34 kilo Dalton β-1,3-glucanase with inhibitory activity against *Phytophthora capsici* from pepper stems. *Physiol. Mol. Plant Pathol.* 50:103-115.
62. Kruger WM, Szabo LJ, Zeyen RJ. 2003. Transcription of the defense response genes chitinase IIb, phenylalanine ammonia lyase and peroxidase is induced by the barley powdery mildew fungus and is only indirectly modulated by R genes. *Physiol. Mol. Plant Pathol.* 63:167-178.
63. Kuc J. 1982. Induced immunity to plant disease. *Biosciences* 32:854-859.
64. Kuranda MJ, Robbins PW. 1991. Chitinase is required for cell separation during growth of *Saccharomyces cerevisiae. J. Biol. Chem.* 266:19758–19767.
65. Lawrence CB, Joosten MHAJ, Tuzun S. 1996. Differential induction of pathogenesis-related proteins in tomato by *Alternaria solani* and the association of a basic chitinase isozyme with resistance. *Physiol. Mol. Plant Pathol.* 48:377–383.
66. Lawton K, Uknes S, Friedrich L, Gaffney T, Alexander D, Goodman R, Métraux JP, Kessmann H, Ahl-Goy P, Gut Rella M, Ward E, Ryals J. 1993. The molecular biology of systemic acquired resistance. In: *Mechanisms of Plant Defense Response* (Ed. B. Fritig, M. Legrand). Kluwer Academic Publishers, The Netherlands. pp 422-432.
67. Lee YK, Hippe-Sanwald S, Jung HW, Hong JK, Se BH, Hwang BK. 2000. *In situ* localization of chitinase mRNA and protein in compatible and incompatible interactions of pepper stems with *Phytophthora capsici. Physiol. Mol. Plant Pathol.* 57:111-121.
68. Leubner-Metzger G, Meins F Jr. 1999. Functions and regulation of plant β-1,3-glucanases. In: *Pathogenesis-related pro-*

*teins in plants* (Ed. SK Dutta, S Muthukrishnan) CRC Press, Boca-Raton, LLC. pp 49-76.

69. Leubner-Metzger, G. Fründt, C., Vögeli-Lange, F. and Meins, F. Jr. (1995). Class I β-1,3-glucanases in the endosperm of tobacco during germination. *Plant Physiol.* 109:751-759.

70. Li YF, Zhu R, Xu P. 2005. Activation of the gene promoter of barley beta-1,3-glucanase isoenzyme GIII is salicylic acid (SA)-dependent in transgenic rice plants. *J. Plant Res.* 118 :215-21.

71. Linthorst HJM. 1991. Pathogenesis-related proteins of plants. *Crit. Rev. Plant Sci.* 10:123–150.

72. Lusso M, Kuc J. 1996. The effect of sense and antisense expression of the PR-N gene for beta-1,3-glucanase on disease resistance of tobacco to fungi and viruses. *Physiol. Mol. Plant Pathol.* 49:269-283.

73. Malamy J, Carr JP, Klessig DF. 1990. Salicylic acid – a likely endogenous signal in the resistance response of tobacco to viral infection. *Science* 250:1002-1004.

74. Mauch F, Staehelin LA. 1989. Functional implications of the subecellular localization of ethylene-induced chitinase and β-1,3-glucanase in bean leaves. *Plant Cell* 1:447-457.

75. Melchers LS, *et al.* 1993. Extracellular targeting of the vacuolar tobacco proteins AP24, chitinase and β-1,3-glucanase in transgenic plants. *Plant Mol. Biol.* 21:583–593.

76. Naderi M, Berger, PH. 1997. Pathogenesis-related protein 1a is induced in potato virus Y-infected plants as well as by coat protein targeted to chloroplasts *Physiol. Mol. Plant Pathol.* 51:41-44.

77. Neuhaus J. 1999. Plant chitinases. In: *Pathogenesis-related proteins in plants* (Ed. SK. Dutta and S Muthukrishnan) CRC Press, Boca-Raton, LLC. pp 77-106.

78. Nielsen KK, Bojsen K, Roepstorff P, Mikkelsen JD. 1994. A hydroxyproline-containing class IV chitinase of sugar beet is glycosylated with xylose. *Plant Mol. Biol.* 25:241–257.

79. Ori N, Sessa G, Lotan T, Himmelhoch S, Fluhr R. 1990. A major stylar matrix polypeptide (sp41) is a member of pathogenesis-related proteins superclass. *EMBO J.* 9:3429–3436.

80. Ornstein L. 1964. Disc electrophoresis I. Background and theory. *Ann. NY Acad. Sci.* 121:321-349.

81. Owen KJ, Green CD, Deverall BJ. 2002. A benzothiadiazole applied to foliage reduces

development and egg deposition by *Meloidogyne* spp. in glasshouse grown grape vine roots. *Aust. Plant Pathol.* 31:47-53.

82. Parent JG, Asselin A. 1984. Detection of pathogenesis related proteins (PR or b) and of other proteins in the intercellular fluid of hypersensitive plants infected with tobacco mosaic virus. *Can. J. Bot.* 62:564 – 569.
83. Pavet V, Olmos E, Kiddle G, Mowla S, Kumar S, Antoniw J, Alvarez ME, Foyer CH. 2005. Ascorbic Acid Deficiency Activates Cell Death and Disease Resistance Responses in Arabidopsis. *Plant Physiol.* 139(3):1291-1303.
84. Payne G, Middlesteadt W, Desai N, Williams S, Dincher S, Carnes M, Ryals J. 1989. Isolation and sequence of a genomic clone encoding the basic form of pathogenesis-related protein I from *Nicotiana tabacum. Plant Mol. Biol.*12:595 – 596.
85. Penninckx IAMA, Eggermont K, Terras FRG, Thomma BPHJ, De Samblanx GW, Buchala A, Métraux JP, Manners JM, Broekaert WF. 1996. Pathogen-induced systemic activation of a plant defensin gene in *Arabidopsis* follows a salicylic acid-independent pathway. *Plant Cell* 8:2309–2323.
86. Poupard P, Parisi L, Campion C, Ziadi S, Simoneau P. 2003. A wound- inducible and ethephon-inducible PR-10 gene subclass from apple is differentially expressed during infection with a compatible and incompatible race of *Venturia inaequalis. Physiol. Mol. Plant Pathol.* 62:3-12.
87. Rahimi S, Perry RN, Wright DJ. 1996 Identification of pathogenesis-related proteins induced in leaves of potato plants infected with potato cyst nematodes, *Globodera* species. *Physiol. Mol. Plant Pathol.* 49:49-59.
88. Richard L, Arró M, Hoebeke J, Meeks-Wagner DR, Van KTT. 1992. Immunological evidence of thaumatin-like proteins during tobacco floral different a ion. *Plant Physiol.* 98:337–342.
89. Ryals J, Uknes S, Ward E. 1994. Systemic acquired resistance. *Plant Physiol.* 104:1109-1112.
90. Samac DA, Shah DM. 1991. Development and pathogen-indiced activation of the *Arabidopsis* acidic chitinase promoter. *Plant Cell* 3:1063-1072.
91. Sharma M, Chakraborty BN. 2004. Biochemical and immunological characterization of

defense related proteins of tea plants triggered by *Exobasidium vexans. J. Mycol. Plant Pathol.* 34 :742-760.

92. Sharma M, Chakraborty BN. 2005. Hexaconazole and calixin mediated defense strategies of tea plants against *Exobasidium vexans* Massee. *J. Mycol. Plant Pathol.* 35 : 417-431.

93. Simmons CR. 1994. The physiology and molecular biology of plant 1,3-β-D-glucanase and 1,3,1,4-β-D-glucanase. *Critical Rev. Plant Sci.* 13:325-387.

94. Staehelin C, *et al.* 1994. Perception of *Rhizobium* nodulation factors by tomato cells and inactivation by root chitinases. *Proc. Natl. Acad. Sci. USA* 91:2196-2200.

95. Tahiri-Alaoui A, Dumas-Gadot E, Gianinazzi S. 1993. Immunocytochemical localization of pathogenesis-related PR-1 proteins in tobacco root disease tissues infected *in vitro* by the black root rot fungus *Chalara elegans. Physiol. Mol. Plant Pathol.* 42:69-82.

96. Thomma BPHJ, Eggermont K, Penninckx IAMA, Mauch-Mani B, Vogelsang R, Cammue BPA, Broekaert WF. 1998. Separate jasmonate-dependent and salicylate-dependent defense response pathways in *Arabidopsis* are essential for resistance to distinct microbial pathogen. *Proc. Natl. Acad. Sci. USA* 95:15107–15111.

97. Uknes S, Mauch-Mani B, Moyer M, Potter S, Williams S, Dincher S, Chandler D, Slusarenko A, Ward ER, Ryals JA. 1992. Acquired resistance in *Arabidopsis. Plant Cell* 4:645–656.

98. Umemoto N, Kakitani M, Iwamatsu A, Yoshikawa M, Yamaoka N, Ishida I. 1997. The structure and function of a soybean β-glucan-elicitor-binding protein. *Proc. Natl. Acad. Sci. U.S.A.* 94: 1029-1034.

99. Van den Elzen PJM, Jongedijk E, Melchers LS, Cornelissen BJC. 1993. Virus and fungal resistance: from laboratory to field. *Phil. Trans. R. Soc. Lond.* B 342:271-278.

100. Van Loon LC. 1975. Polyacrylamide disc electrophoresis of the soluble leaf proteins from *Nicotiana tabacum* var. "Samsun" and "Samsun NN" IV. Similarity of qualitative changes of specific proteins after infection with different viruses and their relationship to acquired resistance. *Virology* 67:566-575.

101. Van Loon LC. 1977. Induction by 2-chloroethylphosphonic acid of viral-like lesions, associated proteins, and systemic resistance in tobacco. *Virology* 80:417-420.

102. Van Loon LC. 1985. Pathogenesis-related proteins. *Plant Mol. Biol.* 4:111-116.

103. Van Loon LC. 1997. Induced resistance in plants and the role of pathogenesis-related proteins. *Eur. J. Plant Pathol.* 203:753-756.

104. Van Loon LC. 1999. Occurrence and properties of plant pathogenesis-related proteins; In: *Pathogenesis-related proteins in plants* (Ed. SK Dutta, S Muthukrishnan) CRC Press, Boca-Raton, LLC. pp 1-19.

105. Van Loon LC, Van Kammen A. 1970. Polyacrylamide disc electrophoresis of the soluble leaf proteins from *Nicotiana tabacum* var "Samsum" and "Samsun NN" II. Changes in protein constitution after infection with tobacco mosaic virus. *Virology* 40:199–211.

106. Van Loon LC, Van Strien EA. 1999. Mini Review: The families of pathogenesis-related proteins, their activities, and comparative analysis of PR-1 type proteins. *Physiol. Mol. Plant Pathol.* 55:85-97.

107. Van Loon LC, Pierpoint WS, Boller T, Conejero V. 1994. Recommendations for naming plant pathogenesis-related protein. *Plant Mol. Biol. Rep.* 12: 245–264.

108. Vierheilig H, Alt M, Neuhaus JM, Boller T, Wiemken A. 1993. Colonization of transgenic *Nicotiana sylvestris* plants, expressing different forms of *Nicotiana tabacum* chitinase, by the root pathogen *Rhizoctonia solani* and by the mycorrhizal symbiont *Glomus mossae*. *Mol. Plant-Microbe Interact.* 6: 261-264.

109. Wang D, Weaver ND, Kesarwani M, Dong X. 2005 Induction of protein secretory pathway is required for systemic acquired resistance. *Science* 308 (5724):1036-40.

110. Ward ER, Uknes SJ, Williams SC, Dincher SS, Wiederhold DL, Alexander DC, Ahl-Goy P, Métraux JP, Ryals JA. 1991. Coordinate gene activity in response to agents that induce systemic acquired resistance. *Plant Cell* 3:1085–1094.

111. Wasternack C, Parthier B. 1997. Jasmonate-signalled plant gene expression, *Trends Plant Sci.* 2:302-307.

112. White RF, Reynolds HC, Sugars JM, Antoniw JF. 1996. The suppression of mannitol induced pathogenesis-related protein accumulation in tobacco by antibiotics. *Physiol. Mol. Plant Pathol.* 49:389-393.

113. Yoshikawa M, Tsuda M, Takeuchi Y. 1993. Resistance to fungal disease is transgenic tobacco plants expressing the phytoalexin elicitor-releasing

factor, β-1,3-endoglucanase, from soybean. *Naturwissenschaften* 80:417–420.

114. Zhu Q, Maher EA, Masoud S, Dixon RA, Lamb CJ. 1994. Enhanced protection against fungal attack by constitutive co-expression of chitinase and glucanase genes in transgenic tobacco. *BioTechnology* 12:807-812.

115. Zhu YJ, Moore PH. 2004. Pathogenesis-related (PR) proteins induced by benzothiadiazole (BTH) in *Carica papaya* improved resistance responses to *Phytopthora palmivora. The Joint International Workshop on PR-proteins and Induced Resistance* p. 68.

*Rev. Plant Pathol. Vol. 4, 2008*
*Indian Society of Mycology and Plant Pathology*
*Scientific Publishers (India), Jodhpur pp. 139-159*

# INTEGRATED NEMATODE MANAGEMENT IN GROUNDNUT AND CASTOR CROPS

***D.J. Patel and S.K. Patel***

Department of Nematology, B.A. College of Agriculture, Anand Agricultural University, Anand-388 110, India.

## ABSTRACT

Groundnut production exceeds 28 mt/annum world wide. India is the world's largest groundnut producing country, both from area as well as production view points and occupies 45% oil seeds area producing about 55% of total oil seed production in country. Such an important oil seed crop is dreadly suffered by several biotic and abiotic factors. Of these, plant parasitic nematodes are considered as important biotic factor. More than 90 species of plant parasitic nematodes have been encountered with groundnut and almost 27 species are recorded from groundnut producing regions in India. The key nematodes reported to be associated with groundnut crop are root knot, *Meloidogyne incognita*, *M. javanica*, *M. arenaria* and *M. javanica* pt.2; Stunt, *Tylenchorhynchus brevilineatus*; lesion, *Pratylenchulus* spp.; ring, *Criconemoides* spp.; dagger, *Xiphinema vanderlindei* and yam nematode, *Scutellonema cavenassi*.

India being the largest castor producing country in Asia ranks second in the world after Brazil, occupying about 25% of the worlds total average of castor and contributes about 30% of the total production. The key nematodes reported to be associated with castor are reniform, *Rotylenchulus reniform*, burrowing, *Radopholus similis; lesion*, *Pratylenchus vulnus*, *P. zeae* and root-knot nematodes, *Meloidogyne arenaria*, *M. hapla*, *M. incognita*, *M. acrita* and *M. javanica*.

Soil application of aldicarb, carbofuran and phenamiphos effectively reduced root-knot infestation in groundnut. Seed treated with aldicarb sulfone 75 WP or carbofuran flowable @ 6% checked larval penetration in groundnut. Marshal 25 STD and Oncol 3 G as seed dresses reduced root-knot disease in groundnut effectively. Seed soaking in monocrotophos, phenamiphos, oxamyl and phosalone significantly reduced root-knot nematodes in groundnut and thereby improved plant growth and finally crop yield. Soil application of carbofuran @ 3 kg/ha increased pod yield by 42.8% with 45.2% reduction in root-knot disease in groundnut field. Castor, neem and mustard cakes effectively managed root-knot disease in groundnut and enhanced crop production. Soil application of phenamiphos also increased peanut growth and development. Total pod and fodder yield losses were 50.4%, which resulted in monitory loss of Rs. 6400/ha. Integration of carbofuran with neem oil proved to be best for management of root-knot nematodes giving ICBR of 1:3.47.

Seed yield of castor was reduced by 25% due to infection of reniform nematodes. Application of carbofuran @ 2 kg/ha increased 35.1% yield with 26.3% reduction in nematode population. Nematode played a vital role to break down Fusarium wilt resistance of castor. Root-rot disease was predisposed in the presence of nematode infection.

## INTRODCTION

India ranks second in production of oil seed crops in the world. The most important oilseed crops cultivated in the country are groundnut, castor, mustard, soybean, sesamum, rapeseed, safflower and sunflower which occupy about 25 m ha under cultivation producing about 18 mt oil seeds. Of these, groundnut, castor and mustard are major oilseeds grown in the country. Groundnut (*Arachis hypogea* L.) is a major oil seed crop accounting for 45% of oilseeds area and 55% oilseeds production in the country (3). The oilseed production increased from 5 m t in 1950 to 21.5 m t in 1995. Oilseed scenario in the country has undergone a dramatic change in recent years. The total oilseeds production in 1994-95 is 22.5 mt (40). Consequently edible oil imports have come down sharply. Area from 19 m ha in 1985-86 dramatically increased to 26 m ha in 1994-95 accounting for 36.8% increase over nine years. Product-ivity also linerely increased from 570 kg/ha to 865 kg/ha, a gain of 51.8% over nine years (40).

Among various oilseed crops the productivity of groundnut is dreadly affected by several biotic and abiotic stresses. More than 55 pathogens including nematodes have been reported to attack (Table 1) to hardness quality and quantity of groundnut crop in different regions of the world (45). Among phytonematodes, root-knot nematodes (*Meloidogyne* spp.) attack groundnut crop severely in India.

**Table 1**: Phytonematodes attacking groundnut.

| Nematode | Distribution |
|---|---|
| *Aphelenchoides* sp. | Egypt, India, Cote d lvoire, Burkina Faso, Benin, Nigeria, The Philippines |
| *Aphelenchus* sp. | Burkina Faso Egypt, India, Cote d lvoire, Israel, Nigeria, The Philippines |
| *Criconemoides* sp. | Benin, Burkina Faso Egypt, Gambia, India, Nigeria, Senegal, USA, Venezuela |
| *Helicotylenchus arachisi* Mulk & Jairajpuri | India |
| *H. dihystera* (Cobb) Sher | Burkina Faso, Benin, Gambia, India, Senegal USA |
| *H. retumis* Siddiqui & Brown | India |
| *Helicotylenchus* sp. | India |
| *Helicotylenchus* sp. | Benin, Burkina Faso, Gambia, Egypt, India, Cote d Ivoire, Mali, Malawai Niger, The Philippiness, Sengegal, USA, Venezuela |
| *Heterodera* sp. | Egypt, India |
| *Hirshmannella* sp. | Egypt, India |
| *Hoplolaimus arachidis* Maharaju & Das | India |
| *H. indicus* Sher | India |
| *H. seinhortii* Luc | India |
| *H. seshadri* Mulk & Jairajpuri | India |
| *Hoplolaimus* spp. | Gambia, India, The Philippines, Senegal, USA, Zimbabwe |

| | |
|---|---|
| *M. hapla* Chitwood | Australia, China, India, Israel, Japan, Zimbabwe, S. Korea, USA South Africa |
| *M. incognita* (Kofoid & White) | Egypt, India, Nigeria, Chitwood, Senegal |
| *M. javanica* (Kofoid & White) | Brazil, Egypt, India, Malawi, Nigeria, Senegal, USA, Zimbabwe |
| *Pratylenchulus brachyurus* (Godfreey) Filipijev & Sch. Stekh ) | Australia, Benin, Egypt, Nigeria, Senegal,USA, Zimbabwe |
| *P. coffeae* (Zimmermann) Fipipjev & Sch.Stekh | India |
| *P. delattrei* Luc. | India, Nigeria |
| *P. incidus* Das | India |
| *P. zeae* Graham | India |
| *Radopholus similis* (Cobb) Thorne | India, Malawi, USA |
| *Tylenchorhynchus brevilineatus* Willianms | India |
| *Tylenchorhynchus mashhoodi* Siddiqi & Basir | India |
| *T. phaseoli* Sethi & Swarup | India |
| *T. vulgaris,* Upadhyay, Swarup & Sethi | India |
| *T. zeae* Sethi & Swarup | India |
| *X. insigne* Loos | India |

Castor (*Ricinus communis* L.) is mostly cultivated in Brazil, India, Russia and Argentina. India ranks first in castor production. In India, castor is grown intensively in Andhra Pradesh, Bihar, Gujarat, Maharashtra, Tamil Nadu and Uttar Pradesh covering about 7.14 m ha under cultivation producing about 5.7mt of castor seed (5). Castor is an important commercial crop with respect to its industrial value in various products made from its oil. About 98% of its oil is consumed for industrial purpose like toilet soap making, lubrication, sulphonation, dehydration, preparation of sulpho-recinoleate plasticier, recinolyl alcohol and medicine. Only about 2% is used for non edible domestic purpose. Looking to the importance of castor in agricultural economy of India, it is realised that there is a great need for augmenting its production.

Such an important oil seed crop is attacked by several biotic and abiotic stresses adversely affecting castor production. Phytonematodes are one of them. Beside two important fungal diseases, *viz.* wilt incited by *Fusarium oxysporum* f. sp. *ciceri* and root-rot by *Macrophomina phaseolina*, (2), *M. arenaria thamesi* (9), *M. hapla* (8), *M. incognita* (23), *M. incognita acrita* (18), *M. javanica* (50), *Pratylenchus vulnus* (15), *P. zeae* (22), *Radopholus similis* (6) and *Rotylenchulus reniformis* (43) have been reported to be parasitic to castor crop. Of these, *R. reniformis* is key nematode dreadly affecting castor crop in India.

## GROUNDNUT

### 1. ROOT-KNOT NEMATODES (*MELOIDOGYNE* SPP.)

#### 1. *Meloidogyne arenaria*

Root-knot nematodes are the most destructive soil dwellers limiting groundnut yield. It was originally described on groundnut in Florida, USA. Now-a-days, it is widely distributed throughout the groundnut growing areas of the world. In India, it was reported for the first time from Bihar by Sen in 1960 (43), later on from Delhi (34) and from Punjab (46, 20). Occurrence of *M. arenaria* on groundnut from Saurashtra region of Gujarat State was reported by Vaishnav and Dhruj (51). Because of the conspicuous root 'galls' and 'warts' caused by these nematodes, the seriousness of the damage was recognized early. Yield losses due to root-knot nematode individual infested fields ranged from 14-20%. They can limit development of *Rhizobium* nodules and can aggravate the severity of other soilborne pathogens. Groundnut plants infected with nematodes commonly develop enlarged roots and pegs which develop into galls of various sizes. The galls result from an internal swelling of the root tissue. Thus, they can be distin-guished from *Rhizobium* nodules which usually are appended to the side of root. Pods also become infected and develop knots, protuberances or small warts. Galls on roots, pegs and pods sometimes begin to deteriorate by the time of peanut maturity. Development of the root system is generally much reduced and plants show variable degrees of stunting and chlorosis. Infected plants indicate poor coverage of the ground surface at midseason. The root system is much restricted and vascular elements of infected tissues are disrupted resulting in poor flow of water and nutrients from roots to the shoots. Infected plants show a tendency to wilt during dry periods or when subjected to water stress (38).

**Table 2 :** Effect of *Meloidogyne arenaria* on Groundnut cv. GAUG 10.

| Character | Inocu-lated | Control | % increase(+)/ dec-rease (-) over control |
|---|---|---|---|
| 1. Shoot length, cm | 24.67 | 36.80 | -32.96 |
| 2. Shoot fresh weight, g | 11.03 | 17.80 | -38.03 |
| 3. Shoot dry weight, g | 2.47 | 4.20 | -41.19 |
| 4. Root length, cm | 19.93 | 26.93 | -25.99 |
| 5. Root fresh weight, g | 4.75 | 3.09 | -53.72 |
| 6. Root dry weight, g | 1.12 | 0.70 | +60.00 |
| 7. Leaf lamina size (10$^{th}$ leaf) cm$^2$ | 11.56 | 17.47 | -33.83 |
| 8. Petiole length (10$^{th}$ leaf), cm$^2$ | 5.29 | 6.01 | -11.98 |
| 9. No.of internodes on main shoot | 12.00 | 14.00 | -14.29 |
| 10. Fifth internode length, cm | 1.92 | 2.81 | -31.67 |
| 11. Chlorophyll content, mg/g | | | |
| Chlorophyll a | 2.11 | 0.68 | -75.69 |
| Chlorophyll b | 0.70 | 3.39 | -79.35 |
| Total chlorophyll | 2.81 | 12.07 | -76.72 |
| 12. Number of galls/plant | 2305 | -- | -- |
| 13. Final population | 136432 | -- | -- |

Groundnut plant infected by *M. arenaria* remains stunted because of reduction in various plant growth parameters due to increased nematode population (52). The losses in groundnut yield due to *M. arenaria* was estimated to the tune of 59.46%. Similarly *M. arenaria* induced 22.1% loss in pod yield of groundnut cv. GAUG 10 in Saurashtra area of Gujarat State. Total chlorophyll content of leaves and leaf lamina was also reduced by 76.72 and 33.83% respectively over normal plants (Table 2). As a result, infected plants look stunted with smaller chlorotic leaves. In field, the infected rows appear thinner compared to healthy rows and give the patchy appearance. The root system of such infected plants shows severe galling with numberous egg-masses. The nematode completes its cycle within 28 days at a temperature range of 27-33°C.

**Table 3.** Effect of *Meloidogyne arenaria* on formation of *Rhizobium* nodules on groundnut cv. GAUG 10.

| Treat-ment | Plant ht., Cm | Fresh wt, g | | Dry wt., g | | No. of Nodules/ Plant |
|---|---|---|---|---|---|---|
| | | Shoot | Root | Shoot | Root | |
| R | 41.00 | 19.33 | 3.42 | 4.84 | 0.90 | 145 |
| $N_1$ | 26.00 (36.59) | 12.00 (32.92) | 4.55 (33.04) | 3.00 (37.89) | 1.20 (33.33) | 0 |
| $N_2$ | 24.50 (40.24) | 11.25 (41.80) | 4.75 (38.89) | 2.82 (41.61) | 1.25 (38.89) | 0 |
| $RN_1$ | 33.42 (18.49) | 15.45 (20.07) | 4.65 (35.96) | 3.87 (19.88) | 1.23 (36.67) | 97 (33.48) |
| $RN_2$ | 31.00 (24.39) | 14.50 (24.99) | 4.83 (41.23) | 3.63 (24.84) | 1.28 (42.22) | 87 (40.02) |

R = 25 ml *Rhizobium* bacteria (13 x $10^9$ cells/ml)

$N_1$= 1000 *M. arenaria* $J_2$ $N_2$ *M. arenaria* $J_2$

Figures in parentheses indicate per cent decrease over R.

The pathogenicity study of root-knot nematodes on groundnut conducted by Patel (31) indicated that initial inoculum of 100 J2/kg soil was detrimental to the growth and development of groundnut cv. GAUG 1 while it was 1,000 J2/kg soil each for GAUG 10 and JL 24 varieties. Regarding interaction studies between *M. arenaria* and *Fusarium solani,* concomitant inoculation of both the organisms induced wilt symptoms earlier on groundnut and suppressed plant growth than fungus alone, whereas interaction effect of *M. arenaria* and *Aspergillus niger* had neutral type for plant growth and nematode multiplication. However, synergistic interaction effect between *M. arenaria* and *A. flavus* in reducing shoot growth of groundnut cv. GAUG 10 was observed with the concomitant inoculation of 4 g mycelium of *A. flavus* and either 1,000 or 2,000 *M. arenaria* J2/plant. Adverse effect of concomitant inoculation of *M. arenaria* with either *A. niger* or *A. flavus* was observed on number of galls/plant as well as final nema population (32). The presence of *M. arenaria* significantly reduced numbers of *Rhizobium* nodules/plant on groundnut (Table 3) (31). *Pythium myriotylum* interacted synergistically with *Fusarium solani* and *M. arenaria* in causing pre-emergence damping-off of peanut seedlings. Pod rot was also more in presence of these organisms (7).

### *b. Meloidogyne javanica*

Root-knot nematodes (*Meloidogyne javanica*) was first time recorded on groundnut cvs. JL 24 and J 11 in Kapadwanj areas of Kheda district in middle Gujarat (28). It has been also reported on groundnut from Uttar Pradesh, Punjab, Orissa, Maharashtra and other states of India. Infected plants exhibit similar symptoms as induced by *M. arenaria*. However, root galling pattern of *M. arenaria* is quite conspicuous. Root galling by nematode is observed on all the roots of plant including tertiary roots like root galling pattern on vegetable crops. Whereas in case of *M. arenaria*, profuse galling (comparatively smaller in size) is seen mainly on tertiary roots and very few galls are produced on main root portion. Host differential study conducted by Patel *et al,* (29) indicated that generally *M. incognita* and *M. javanica* do not infect groundnut crop. But *M. javanica* reported on groundnut cvs. JL 24 and J 11 produced heavy root galling and nematode was reported as *M. javanica* pathotype 2 to cause discernible damage because of very conspicuous root galls produced first time from India. This *M. javanica* pt.2 infects cotton, mungbean, urdbean, fennel, and pigeonpea crops severely besides groundnut in the area. Nematode infection drastically reduces plant growth and thereby crop production is poor (25).

Pathogenicity study conducted by Joshi (16) revealed that an inoculum level of 1000 *M. javanica* J2/kg soil/plant drastically reduced plant growth of groundnut cvs. GG 2 and JL 24. Nematode infection as groundnut cv. GG 2 significantly reduced plant height, fresh as well as dry shoot and root weights, leaf area, no. of internodes and branches and produced more galls with higher nematode population (Table 4). Groundnut plants infected by *M. javanica* had increased leaf and cavette temperatures and transpiration rate and decreased diffusion resistance over healthy plants. Similarly nematode infected plants recorded less peroxidase and polyphenol oxidase activities and higher total phenol contents than that of healthy plants.

**Table 4.** Effect of *Meloidogyne javanica* on plant growth and nematode multiplication on groundnut cv. GG 2.

| Character | Inocu-lated | Control | % increase (+) / decrease (-) over control |
|---|---|---|---|
| 1. Shoot length, cm | 13.6 | 17.10 | -20.46 |
| 2. Shoot fresh weight, g | 8.50 | 12.00 | -29.16 |
| 3. Shoot dry weight, g | 2.60 | 4.10 | -36.58 |
| 4. Root length, cm | 7.10 | 9.60 | -26.04 |
| 5. Root fresh weight, g | 5.10 | 3.60 | +41.66 |
| 6. Root dry weight, g | 1.08 | 0.76 | -42.10 |
| 7. Leaf area, $cm^2$ | 3.00 | 5.25 | -42.85 |
| 8. No. of internodes in main shoot | 9.00 | 11.0 | -18.18 |
| 9. Fifth internodes length, cm | 1.40 | 2.25 | -37.33 |
| 10. Number of branches | 2.00 | 3.00 | -33.33 |
| 11. Number of galls/plant | 4.80 | -- | -- |
| 12. Root-knot index (0-5)* | 4.05 | -- | -- |
| 13. Final population | 70.285 | -- | -- |

* 0= Free, 5= Maximum Disease Intensity.

Significant differences in enzymatic activities *viz.* peroxidase, polyphonel oxidase and total phenol contents are observed in groundnut roots due to *M. javanica* at varying inoculum levels. Higher peroxidase activity is recorded in 10,000 J2/plant followed by 1,000 J2/plant. Uninoculated control plants had significantly less peroxidase activity. In case of polyphenol oxidase activity and total phenol contents, similar trends are noticed (30). Significant reduction is recorded in chlorophyll a,b and total chlorophyll contents of leaves due to nematode inoculum levels of 100, 1,000 and 10,000 J2/plant (16). Reduction in chlorophyll content of groundnut leaves due to infection by *M. arenaria* was also observed by Patel (31). An inoculum level of 100 J2 and above/ plant significantly reduced plant growth parameters *viz.* plant height, fresh and dry shoot weights. Fresh and dry root weights progressively increased with an increase in nematode inoculum level from 10 to 10,000 J2/plant due to heavy root galling and profuse egg sac formation on roots and pods (37). Nematode inoculum level of 10,000 J2/plant had significantly minimum plant height, fresh and dry shoot weights. Total nematode population increased with an increase in nematode inoculum levels and it was maximum of 4782.2 times at the level of 10 J2/plant and minimum of 10.5 times at the highest level of 10,000 J2/plant.

### c. Other nematodes

Kalahasti malady disease caused by *Tylenchorhynchus brevilineatus* is characterised by small brownish yellow lesion appearance on the pegs, pods, stalks as well as on young developing pods. The margins of the lesions are slightly elevated because of the proliferation of host cells around the lesions. Pod stalks are much reduced in length and pod surface becomes discoloured on advanced stages of the disease. Root discolouration is also observed. Affected plants remained stunted and greener than normal foliage.

The root lesion nematodes (*Pratylenchus* spp.) are migratory endoparasites that attack peanut roots, pegs as well as pods and feed within the parenchymatous tissues. The nematode can enter or live in the roots, pods or pegs during all of its developmental stages. Both mechanical and chemical damage resulted from the nematode feeding within the tissues. Root lesions developed and on large nematode population, these lesions coaleased causing extensive discolouration and damage that resulted in reduction in growth and pod production. Both degree of damage and pod loss at harvest are positively correlated with nematode numbers which can increase to over 3,000 /g tissue in the roots, pegs and pods. In addition nematode feeding within the pegs weakened them resulting in pod loss at harvest. Lesion nematode numbers were positively correlated with percentage of damaged seeds and negatively correlated with yields of pods, percentage of sound mature seeds, percentage of total seeds and value. Other microorganisms may colonize the damaged tissues and penetrate through the pod, causing seed damage. The combined effect of nematodes and associated micro-organisms can result in significant reduction in yield, quality and value of the crop. *Pratylenchus* spp. also caused disfiguration of the roots and pods of groundnut in USA (11,49). *Criconemoides* spp. causing stunting and loss in green weight of tops and roots in groundnut crop are also reported by Graham (12). While studying host-parasite relationship and pathogenicity of *P. brachyurus* on groundnut, it was observed that nematodes can cause dark coloured necrotic lesions on the shell and cause peg rot condition in the crop (10). *Xiphinema vanderlindei* is reported from the rhizosphere of groundnut in South Africa by Heyhs (13). In Senegal, *Scutellonema cavenassi* on peanut affected all plant parts of the plant and symbiotic nitrogen fixation apperatus (9).

## MANAGEMENT

Soil application of aldicarb, carbofuran and phenamiphos effectively reduced root-knot infestation in groundnut (1, 41). Seeds treated with either aldicarb sulfone 75 WP or carbofuran flowable each @ 6% had minimum larval penetration to the roots which ultimately resulted in significantly more fresh and dry shoot weights of groundnut plants, finally giving significantly less root galls and final nema population (1). Marshal 25 STD and Oncol 3 G used as seed dressers also effectively reduced root-knot nematodes in groundnut at lower concentrations (1 and 2%) as compared to higher concentrations (3 and 4%) (2).

Seed soaking in monocrotophos, phenamiphos, oxamyl and phosalon significantly reduced penetration of root-knot nematodes in groundnut roots and thereby improved plant growth with reduced root-knot disease. Monocrotophos and phenamiphos were more effective. Higher concentrations (500 ppm) of each of these chemicals effectively reduced nematode infection as compared to lower ones. Organic amendments *viz.*, castor, mustard and neem cakes also significantly reduced *M. arenaria* population and thereby increased groundnut yield (3). Varietal screening (227 genotypes) against *M. arenaria* did not yield any genotype resistant to nematode. Simultaneous inoculation of either *Rhizoctonia bataticola* or *Fusarium solani* with *M. javanica* on groundnut indicated that both the fungi reduced multiplication of *M. javanica* subsequently giving less galling on groundnut roots. *R. bataticola* proved more antagonistic to *M. javanica* than *F. solani* (42).

Soil application of carbofuran @ 3 kg/ha increased dry pod yield by 42.8% and dry fodder yield by 28.8% with reduction in root-knot disease by 45.2% over control. There was 36.3% increase in pod yield and 36.5% reduction in RKI due to phorate @ 3 kg/ha, lower dose of 2 kg/ha of the nematicides was less effective (30).

**Table 5.** Management of *M. javanica* in groundnut field through orgnic amendments.

| Organic amendment | Yield, q/ha | RKI (1-5)* |
|---|---|---|
| Castor cake | 67 (+28.85) | 2.00 (-38.46) |
| Neem cake | 73 (+40.38) | 2.00 (-38.46) |
| Mustard cake | 62 (+19.23) | 2.75 (-15.38) |
| Carbofuran | 75 (+44.23) | 1.75 (-46.15) |
| Control | 52 | 3.25 |
| S.Em | 5.4 | 0.25 |
| C.D. 0.05 | 16.2 | 0.71 |
| C.V.% | 13.2 | 13.9 |

1= Free, 5= Maximum Disease Intensity.
Figures in parentheses indicate increase(+)/decrease (-) over control.
Each Cakes @ 1 t/ha and Carbofuran @ 2 kg/ha were applied

Castor, neem and mustard cakes effectively managed *M. javanica* on groundnut and enhanced crop production (Table 5). Higher doses (3 kg/ha) of carbofuran, phenamiphos and phorate proved effective in reducing *M. javanica* infection and helped to improve plant growth (16). Soil application of phenamiphos @ 2 kg/ha significantly increased peanut growth and development over control in field. Dry pod and fodder yield losses were 30.0 and 20.4% respectively over control. Total pod and fodder yield losses were estimated to the extent of 50.4% which resulted in monitory loss of Rs. 6400/ha (35). Application of nematicides significantly reduced root-knot disease. Sebufos was most effective closely followed by phenamiphos in reducing the disease. While application of neem oil gave highest yield closely followed by neemcake (Table 6). From economic view point combination of carbofuran with neem oil (1381 kg/ha) gave highest ICBR of 1:3.47 over control and next treatment combination is carbofuran with castor cake (1260 kg/ha) having ICBR 1:1.58.

**Table 6.** Effect of Soil application of nematicides and organic amendments on root-knot disease control and yield of groundnut with economics.

(Pooled for 5 years)

| Treatment | Yield kg/ha Dry Pod | Dry Fodder | RKI (1-5)* | Total Income (pod fod-der), Rs/ha | Additional income, Rs/ha | Additional Expendi-ture, Rs/ha | ICBR |
|---|---|---|---|---|---|---|---|
| CAR + CC | 1260 | 3498 | 2.00 | 20997 | 6887 | 4345 | 1:1.58 |
| CAR + NC | 1138 | 3365 | 1.77 | 19273 | 5163 | 5345 | 1:0.45 |
| CAR + NOI | 1381 | 3671 | 1.81 | 22770 | 8660 | 2495 | 1:3.47 |
| PHEN + CC | 1082 | 3484 | 2.21 | 18751 | -- | -- | -- |
| PHEN + NC | 1364 | 3650 | 2.02 | 22525 | -- | -- | -- |
| PHEN + NOI | 1387 | 3597 | 2.12 | 22734 | -- | -- | -- |
| SEB + NC | 1310 | 3571 | 1.66 | 21731 | -- | -- | -- |
| SEB + CC | 1409 | 3214 | 1.55 | 22433 | -- | -- | -- |
| SEB + NOI | 1336 | 3577 | 1.71 | 22065 | -- | -- | -- |
| CON | 818 | 2590 | 3.70 | 14110 | -- | -- | -- |
| S.Em | 154 | 307 | 0.25 | -- | -- | | |
| C.D.0.05 | 436 | 869 | 0.71 | -- | -- | -- | -- |
| C.V.% | 13.6 | 15.6 | 13.9 | -- | -- | -- | -- |

1= Free, 5= Maximum Disease Intensity.

1. Cost of materials Carbofuran @ Rs. 50/ kg, Neem cake @ Rs. 3.00/kg Castor cake @ Rs. 2 /kg; Neem oil Rs. 30/litre.

2. Selling rate : G'nut pods @ Rs. 12.50/kg ; G'nut dry fodder Rs. 1.5/ha

1= Free, 5= Maximum Disease Intensity. CAR = Carbofuran, Phen = Phenamiphos, SEB = Sebufos, CC = Castor cake, NC = Neem cake

## CASTOR

### Reniform nematode (*Rotylenchulus reniformis*)

Reniform nematode *R. reniformis* was highly pathogenic to castor cv. Anand 6-4, reducing plant height as well as root and shoot weights by 30.72, 69.99 and 43.32% respectively (Patel, 1981). Castor growth was drastically reduced at 100 J4/plant and above. Due to nematode infection, uptake of nutrients viz. N, P, K, Ca and Mg by castor plant was hampered significantly, the reduction increased subsequently with an increase in initial nematode population (26). Seed yield and test weight of castor seed reduced significantly at 1,000 and 500 *R. reniformis* J4/kg soil/plant over control. Nematode inoculated plants (both levels) had 20.08% reduced yields over unioculated plants (Table 7) also affected oil contents of castor seeds. There were 3.14 and 6.60% reduction in oil contents of castor seeds due to *R. reniformis* infection at 500 and 1,000 J4/kg soil, respectively (Table 7). Leaf chlorophyll content was also significantly reduced due to nematode inoculation (27). Application of carbofuran @ 2 kg/ha increased 11.53% plant height and 35.07% yield with 26.0% loss in grain yield over control with 26.31% reduction in nematode population in castor cv. GAUCH 1 (Table 8), while in GCH 4, plant height and yield were increased by 9.8 and 33.27% with 24.96% loss in yield with 47.8% reduction in nematode population (34) (Table 8). Castor is also reported to be highly susceptible to *R. reniformis* with reproduction rate of 22.40% at inoculum level of 1000 larve/kg soil. An increase in inoculum level corresponding increased total population with reduced nematode multiplication rate (17).

**Table 7.** Effect of *R. reniformis* on yield and quality of castor cv. GCH 4.

| Nematode | Seed yield, g | | | Test weight, g | | | Per cent |
|---|---|---|---|---|---|---|---|
| level/kg soil | 1992-93 | 1993- 94 | Mean | 1992-93 | 1993- 94 | Mean | oil content |
| 1000 | 15.34[b] | 23.52[ab] | 19.43[b] (-20.08) | 15.52[b] | 15.96[b] | 15.74[b] | 40.71[a] (-6.60) |
| 500 | 16.53[b] | 22.32[b] | 19.43[b] (-20.08) | 15.80[b] | 15.99[b] | 15.90[b] | 42.22[a] (-3.14) |
| Control | 20.13[a] | 28.45[a] | 24.29[a] | 18.58[a] | 18.66[a] | 18.62[a] | 43.59[a] |
| S.Em | 0.52 | 0.97 | 0.68 | 0.40 | 0.18 | 0.26 | NS |
| Year | - | -- | NS | -- | -- | NS | -- |
| Y X T | -- | -- | NS | -- | -- | NS | -- |
| C.V.% | 7.4 | 9.6 | 9.0 | 5.9 | 2.6 | 4.5 | 6.3 |

**Table 8.** Estimation of avoidable yield losses due to reniform nematode in castor.

INP : 170 J4/200 g soil

| Treatment | GAUCH 1 | | | | GCH 4 | | | |
|---|---|---|---|---|---|---|---|---|
| | Plant height, cm | Yield, kg/ha | Final nema. popn./ 200 g soil | % loss in yield | Plant height, cm | Yield, kg/ha | Final nema. popn./ 200 g soil | % loss in yield |
| CAR | 141.2 | 2484 | 560 | 25.96 | 141.0 | 2700 | 480 | 24.96 |
| CON | 126.6 | 1839 | 760 | | 128.2 | 2026 | 920 | |
| S.Em. | 7.3 | 72 | | | 9.4 | 67 | | |
| | NS | 208 | | | NS | 197 | | |
| | S | NS | | | S | NS | | |
| | 7.0 | 18.4 | | | 5.8 | 16.7 | | |
| Percent Increase(+)/ decrease(-) over control | +11.53 | +35.07 | -26.31 | | +9.98 | +33.27 | -47.82 | |

## Host- Parasite Relationship

Histopathology of castor roots infected with *R. reniformis* showed damage to phloem and cortical parenchyma, killing feeder roots by nematode feeding which hindered uptake of nutrients and water from the soil. This lead to poor development of plant. The infected tissues at the site of nematode infection exhibited hypertrophy, hyperplasia, cell walls thickening, granular protoplasm and enlarged nuclei and nucleoi (26). Giant cells were formed as many as 150-200 on castor roots. The infected cells at feeding site in castor roots contained numerous starch granules than that of normal cells (48). Castor crop infested with *R. reniformis* had symptoms of stunting and necrosis resulting in yield reduction (53). The nematode multiplicated more rapidly on 45 days old seedlings of *Ricinus* sp. than that of 30 or 15 days old seedlings. Pathogenic effect of nematode on castor was observed with respect to growth reduction, leaf shedding, early flowering, malformation, seed discolouration with inferior quality seeds containing lower oil content (47). Water absorption capacity of castor roots infected by *R. reniformis* on account of poor root development was noticed (14). Browning and necrosis of cells surrounding nematode head as well as away from the head due to feeding of *R. reniformis* in castor were also reported (24).

## Interaction with other pathogenic organisms

*Fusarium oxysporum* f. sp. *ciceri* (Foc), a highly pathogenic wilt causing fungus, inoculated either alone or alongwith reniform nematode(N) retarded the growth and development of castor plants over control. Fungus inoculation four weeks after nematode inoculation induced wilt symptoms earlier than fungus inoculation alone. Plant height was significantly reduced in N-Foc treatment, whereas fresh plant shoot and root weights were minimum in N + Foc treatment followed by N-Foc and Foc-N treatments. The treatment of N alone had maximum (469 times) reproduction rate as compared to other treatments. Wilt disease appeared three weeks earlier in different combinations of N and Foc than Foc alone. Among the combinations, N inoculated 4 wks prior to Foc inoculation proved highly detrimental inducing 66.6% wilt disease in castor cv. GAUCH 1(27). In case of GCH 4 variety, plant height, fresh shoot and root weights were significantly less in N alone treatment followed by N+Foc and N-Foc treatments. This treatment also produced maximum nematode population build up/ plant and had higher reproduction rate of 523 times over other treatments. Castor wilt disease incited by Foc was not observed in the treatment of Foc alone because of resistance against wilt disease in the castor variety. But wilt disease appeared in the plants inoculating with nematodes 4 wks prior to fungus inoculation and inoculation of both the pathogens together. This clearly proved that nematode played a vital role to break down the wilt resistance nature of castor cv. GCH 4 (36). Interaction between N and *Macrophomina phaseolina* (Mp) on castor cv. GCH 4 gave reduced plant height and fresh root weight. But fresh shoot weight was drastically reduced in N alone. The N alone produced maximum nematode population build up/plant and had higher reproduction rate of 528 times over other treatments. Root-rot disease appeared earlier in different combinations of N and Mp over Mp alone. The maximum plant mortality of 41.7% was recorded in the treatment of N+Mp (34).

## Management

Castor varieties RG 5, 450 (VI 2-1), 460 (VI 18); SHB 110, 118, SKI 50; SPS 43-2 & 3; VHB 131; JI 5, 59; RC 879/1, 13676, 1203, 55212 and AS 7 possess resistance to nematode (33). For management aspects, the research work so far done is very limited. Carbofuran, phorate and phenamiphos each @ 0.5 and 1.0 kg/ha reduced *R. reniformis* enhancing castor yield under microplot conditions, phenamiphos @ 1.0 kg/ha being more effective (Anon., 1989-91). Applications of carbofuran @ 2 kg/ha increased 35.1% yield with 26.3% reduction in reniform nematode population in Castor.

## Future Research

- Intensive and rigorous screening of germplasm should be done in order to locate resistance source(s) in crops.
- Resistant genotypes should be used for transferring resistant factors into susceptible high yielding varieties.
- Screening of large germplasm of crops yielded very few entires/lines to be resistant to nematode and hence work on mutation breeding through mutagenic agents like Sodium azide, EMS (Ethyl Methane Sulfonate), irradiation, etc should be initiated.
- Effects of nematode infection on quantitative and qualitative aspects of crops should be studied.
- Use of soil solarization (slow soil pasteurization) should be explored for management of nematode diseases, wherever feasible.
- Emphasis should be given on use of organic amendments and crop rotations.
- Integrated Management Practices adopting cultural practices, biological agents, resistant varieties, low rate of nematicides, organic amendments, etc. should be planned for economic management of nematode diseases in crops.

- Intensive survey work should be carried out to get awarness about the damage by different plant parasitic nematodes beside root-knot especially in groundnut crop.
- Role of nematodes in aggravating fungal diseases like stem/collar rot, root-rot, pod-rot, etc. needs to be investigated and management practices be evolved in groundnut crop.
- Detailed studies on complex involving *R. reniformis* wilt and root-rot disease including integrated management need to be initiated in castor crop.

### *Literature Cited*


1. Anonymous. 1985. Biennial Report, AICRP (Nematodes), Anand Centre, Gujarat p.81-82.
2. Anonymous. 1986. Annual Report, AICRP (Nematodes), IARI, New Delhi.
3. Anonymous. 1987. Tech. Bult, Plant Protection in Groundnut, Junagadh, Gujarat.
4. Anonymous. 1991. Biennial Report. AICRP (Nematodes), Anand Centre, Gujarat.
5. Anonymous. 1994. Statistics of crop production. The Hindu Survey of Indian Agriculture, P. 212.
6. Birchfield W. 1956. New and suspected host plants of the burrowing nematode, *Radopholus similes* (Cobb). Thorne. *Pl. Dis. Reptr.* 40: 866-868.
7. Garcia R, Mitchell DJ. 1975. Interaction of *Pythium myriotylum* with *Fusarium solani, Rhizoctonia solani* and *Meloidogyne arenaria* in pre-emergence damping off of peanut. *Pl. Dis. Reptr.* 59: 665-669.
8. Gaskin TA, Crittenden HW. 1956. Studies of the host range of Meloidogyne hepla. *Pl. Dis. Reptr.* 40: 265-270.
9. Germani, G. 1981. Pathogenicity of the nematode *Scutellonema cavenessi* on peanut and soybean. *Revue de nematologie.* 4: 203-208.
10. Good JM, Boyle LW, Hammons RO. 1958. Studies of *Pratylechulus brachyurus* on Peanuts. *Phytopathology.* 48: 530-535.
11. Graham TW. 1949. Nematode root-rot of tobacco and other crops. *Phytopathlogy* 39:8.
12. Graham TW. 1955. Pathogenicity of *Rotylenchus brachyurus* on tobacco and of *Criconemoides* sp. on tobacco and peanuts. *Phytopathology* 45: 347.
13. Heyhs J. 1962. A report on South African nematodes of the families Longidoridae, Belondi-

ridae and Alaimidae (Nemata : Dorylaimoidea) with descriptions of three new species. *Nematologica* 8:15-20.

14. Ismail W, Alam MM. 1975. Inhibition in the Water absorption Capability of castor and tomato roots by the infection of reniform Nematode. *Geobios* 2: 23-24.
15. Jensen HJ. 1953. Experimental greenhouse host range studies of two Root-lesion nematodes, *Pratylenchus vulnus* and *Pratylenchus penetrans*. *Pl. Dis. Reptr.* 37: 384-387.
16. Joshi PR. 1989. Root-knot nematode, *Meloidogyne javanica* on groundnut. M.Sc. (Agri.) Thesis, Gujarat Agricultural Univ., Sardar Krushinagar (G).
17. Khan FA, Khan AM. 1973. Studies on reniform nematode, *Rotylenchulus renif-ormis,* host range and popul-ation changes. *Indian J. Nematol* 3: 24-30.
18. Linde WJ, Van Der. 1956. The *Meloidogyne* Problem in South Africa, *Nematologica* 1: 177-183.
19. Linde WJ, Van Der; Clemitson JG, Crous ME. 1959. Host parasite Relationships of South Africa root-knot eelworms *Meloidogyne* spp. *Sci. Bull. Dep. Agric. Tech. Seru., Un. S. Afc.* 385: 1-16.
20. Mahajan R, Chhabra MK 1979. Occurrence of plant parasitic nematodes in Punjab, India. *PANS*: 25: 46-49.
21. Martin GC. 1958. Root-knot nematodes *Meloidogyne* spp. in the Federation of Rhodesis and Nyasaland. *Nematologica* 3: 332-349.
22. McBridge JM, Jons DM, Canter CR. 1961. Relative host response of interplanted weeds and corn to *Pratylenchus zeae* and *P. brachyurus* (Nematoda: Tylenchida). *Phytopathology* 51:644.
23. Minz G. 1956. The root-kbnot nematode, *Meloidogyne* spp., in Israel. *Pl. Dis. Reptr.* 40: 798-801.
24. Nath RP, Swarup G, Rama Rao GVSV 1969. Studies on the Reniform nematode, *Rotylenchulus reniformis* Linford and Oliveira, 1940. *Indian Phytopath.* 22: 99-104.
25. Patel BA, Patel DJ, Sharma SB, Patel HV, Patel SK. 1996. Nematode problems of groundnut and their management in Gujarat. India. *International Arachis Newsl.* 16: 38-39.
26. Patel DB. 1981. Effect of *Rotylenchus reniformis* Linford and Olivera, 1940, on castor. M.Sc. Thesis submitted to Guj. Agricultural Univ., Anand, Gujarat.
27. Patel DB. 1995. Studies on Reniform Nematode, *Rotylenchulus Renifornmis* Linford and Oliveira, 1940 on castor, Ph.D. Thesis Submitted to Gujarat

Agricultural Univ., Anand, Gujarat.

28. Patel DJ, Patel BA, Chavda, JC. 1988. Record of *Meloidogyne javanica* on groundnut in Gujarat, India. *International Arachis Newsl.* 3: 16-17.

29. Patel DJ, Patel BA, Patel HV. 1993. Pathotypes of *Meloidogyne javanica* in India. *Nematol. Medit.* 21: 207-208.

30. Patel HV, Patel DJ, Patel SK. 1998 b. Management of root-knot Nematodes with seed treatment and soil application of granular nematicides in groundnut field. Nematological society of India, National Symposium held at BACA, GAU, Anand. P. 56-57.

31. Patel HR. 1983. Studies on root-knot nematode of groundnut, Ph.D. Thesis, Gujarat Agricultural University, Sardar Krushinagar, Gujarat.

32. Patel HR, Vaishnav MU, Dhruj IU. 1985. Interaction of *Meloidogyne arenaria* and *Fusarium solani* on groundnut *Indian J. Nematol.* 15: 98-99.

33. Patel SK, Patel DJ, Patel HV, Patil RK. 1995a. Host status of Castor (*Ricinus communis* L.) varieties/lines to *Rotylenchulus Reniformis*, paper presented in "National Symposium on nematode Problem of India. An Appraisal of the Nematode Management with Eco-friendly Approaches and Biocomponents," held on March 24-26, 1995 at IARI, New Delhi.

34. Patel SK, Patil RK, Patel HV, Patel DJ. 1995b. Determination of avoidable yield losses due to *Rotylenchulus reniformis* Linford and Oliveira, 1940, in Castor. *Indian J. Mycol. & Pl. Path.* 25:73.

35. Patel SK. 1998. Studies on root-knot nematode *Meloidogyne javanica* (Pathotype 2) on groundnut cv. GG 20. Ph.D. Thesis Guj. Agricultural Univ., Sardar Krushinagar, Gujarat.

36. Patel SK, Patel DJ, Talati JG. 1998 Biochemical alterations induced by *Meloidogyne javanica* pt.2 infection in groundnut cv. GG 20. Nematological Society of India, National Symposium held At BACA, GAU, Anand. P. 76-78.

37. Patel SK, Patel, DJ. 1998. Determination of threshold level of *Meloidogyne javanica* pt.2 attacking groundnut cv. GG 20. Nematological Society of India, National Symposium held at BACA, GAU, Anand PP 82-85.

38. Porter DM, Smith DH, Rodriguez Kabana R. 1984. Compedium of Peanut Disease Americal Phytopathological Society P 38-41.

39. Prasad SK, Dasgupta DR and Mukhopadhyaya NC. 1964. Nematodes associated with commercial crops in northern India and host range of

*Meloidogyne javanica* (Treub, 1885) chitwood 1949. *Indian J. Ento.* 26: 438-446.

40. Reddy PS. 1995. The Hindu survey of Indian Agriculture, M/s. Kasturi & Sons Ltd., Madras P. 72-79.

41. Sakhuja PK, Sethi CL. 1986a. Effect of some systemic nematicides on groundnut and multiplication of *Meloidogyne javanica. Indian J. Nematol.*16:23-26.

42. Sakhuja PK, Sethi CL. 1986b. Multiplication of *Meloidogyne javanica* as affected by *Fusarium solani* and *Rhizoctonia bataticola* on groundnut. *Indian J. Nematol.* 16: 1-3.

43. Sen AC. 1960. Some notes on biology and control of the parasitic nematodes on economic crops in Bihar. *Indian J. Ent.* 2: 61-64.

44. Seshadri AR, Sivakumar CV. 1963. A preliminary note on the Occurrence of the reniform nematode (*Rotylenchulus reniformis)* on A numer of cultivated crops in South India. *Madras Agric. J.* 50:134-137.

45. Sharma SB. 1985. A world list of nematodes pathogens associated with chickpea, groundnut, pearl millet, pigeonpea and sorghum, ICRICAT, Hyderabad, (AP).

46. Sharma NK, Sharma SK, Midha SK. 1978. Root-knot disease on groundnut in Punjab. *Indian Farming* 28: 19-21.

47. Sivakumar CV, Seshadri AR. 1971. Pathogenicity of the reniform nematode, *Rotylenchulus reniformis*, to castor. *Indian J. Nematol.* 1 :227-236.

48. Sivakumar CV, Seshadri AR. 1972. Histopathology of infection by the reniform nematode, *Rotylenchulus reniformis* to castor. *Indian J. Nematol.* 2: 173-181.

49. Steiner G. 1919. Nematodes the growers should know. *Soil Sci. Soc. Pl. Proc.* 4/B: 72-117.

50. Tarjan AC. 1953. Geographical distribution of some *Meloidogyne* spp. In Israel. *Pl. Dis. Reptr.* 37: 315-316.

51. Vaishnav MU, Dhruj IU. 1981. Occurrence of root-knot nematode on groundnut in Gujarat. Nematological society of India, Symp. held at TNAU, Coimbatore on Feb. 13-15, P. 33.

52. Vaishnav MU, Dhruj IU. 1985. Pathogenicity of *Meloidogyne arenaria* on *Arachis hypogaea. Indian J. Nematol.* 15: 235-236.

53. Verma SK, Prasad SK. 1969. The reniform nematode, *Rotylenchulus reniformis* Linford & Oliveira, 1940, Biological studies. *Indian. J. Ento.* 31: 36-47.

*Rev. Plant Pathol. Vol. 4, 2008*
*Indian Society of Mycology and Plant Pathology*
*Scientific Publishers (India), Jodhpur pp. 161-191*

# ZEARALENONE A POTENT MYCOTOXIN OF GREAT CONCERN

***S.M. Reddy, S. Girisham and K. Narasimha Rao***

Department of Microbiology, Kakatiya University, Warangal - 506 009.

Key words: Mycotoxin, zearalenone, zearalenol, *Fusarium,* hyperester-ogenism, pharmacokinetics, management

## ABSTRACT

Zearalenone, mycotoxin of diverse nature, proved to be elaborated in various foods and feeds by different species of *Fusarium.* Inspite of non-steroidic nature of ZEN, it exerts most potent oestrogenic effect. Different domestic animals are reported to be affected by this toxin. Swine is reported to be highly susceptible. Dietary levels of Zearalenone as low as 1 μg/g cause signs of hyperester-ogenism in pigs. Abortion and still births, reduced litter size and weight of piglets, redness and swelling of the vulva, slight swelling of the mamma and numerous follicles and some cystic follicles on the ovaria are some of the disturbances caused by ZEN. Zearalenone binds to estrogen receptors (ER) and get transformed to elicit toxicity. ZEN is partially decomposed by heat and dry milling of maize. Reduction by 80-90% in other fractions including flour and grits reveals its elaboration in surface layers of food grains. Mechanism of action and pharmacokinetics of ZEN are discussed. Finally management of ZEN and future areas of research on ZEN were also discussed.

## INTRODUCTION

Zearalenone (ZEN), a phenolic resorcylic acid lactone, was first isolated and characterized in 1962 from cultures of *F. graminearum* obtained from corn and implicated in a field outbreak of hyperesterogenism in pigs in the United States by Stob *et al.* (139). Zearalenone is a nearly universal corn contaminant and often occurs in the same sample with trichothecenes (75). Despite its dissimilarity to steroid compounds, ZEN produces potent hyperesterogenic responses in susceptable animals which is unique among known mycotoxins (90). Pitt (104) and Gaumy *et al.* (36) and Agag (2) have excellently reviewed different aspects of incidence and toxicity of Zearalenone

## STRUCTURE AND PROPERTIES

The chemical name of Zearalenone is 6–(10-hydroxy-6-oxo-trans-1-undecenyl)-β-resorcylic acid lactone, and empirical formula is $C_{18}H_{22}O_{54}$ with molecular weight of 318 (31). ZEN structure was first elucidated by Urry *et al.* (156). The name Zearalenone is derived partly from the generic name of the host plant infected by *Fusarium* (Zea) and partly from its chemistry (resorcylic acid lactone, en=double bond at C-12 and 1 one=ketone (156). It is a white crystalline compound that exhibits blue-green fluorescence when excited by long wave length UV light (360 nm) and a more intense green fluorescence when excited with short wave UV (260 nm). Zearalenol (α-Zearalenol) is an hydroxylated derivative of ZEN due to ZEN reductases present in animal tissues (27).

In methanol, UV absorption maxima occur at 236 nm, 274 nm and 316 nm. Maximum fluorescence in ethanol occurs with irradiation at 450 nm. Solubility in water is about 20 μg/ml. Zearalenone is slightly soluble in hexane and progressively more so in benzene, acetonitrile, methylene chloride, methanol, ethanol and acetone. It is also soluble in aqueous alkali. Structure of ZEN and its derivatives are presented in Fig. 1 and some synthetic analogues of ZEN are depicted in Fig. 2.

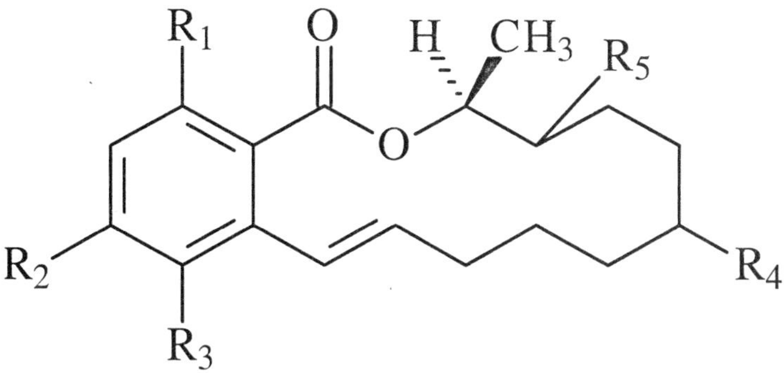

| $R_1$ | $R_2$ | $R_3$ | $R_4$ | $R_5$ | Cl' et $C_2$ | Nom |
|---|---|---|---|---|---|---|
| OH | OH | H | O | $H_2$ | CH=CH | Zearalenone |
| OH | OH | H | OH | $H_2$ | $CH_2=CH_2$ | α et β-zearalanol |
| OH | OH | H | OH | $H_2$ | CH=CH | α et β-zearalenol |
| H | H | H | $H_2$ | $H_2$ | $CH_2=CH_2$ | dideoxyzearalane |
| OH | OH | H | $H_2$ | O | $CH_2=CH_2$ | tetrahydrozearalane |
| O-$CH_3$ | O-$CH_3$ | H | O | O | CH=CH | dimethylzearalenone |
| O-$CH_3$ | OH | H | O | O | CH=CH | o-methykzearalenone |
| OH | O-$CH_3$ | H | O | O | CH=CH | p-methylzearalanone |
| H | H | H | O | O | $CH_2=CH_2$ | dideoxyxzearalanone |
| H | OH | H | O | O | $CH_2=CH_2$ | 2-deoxyzearalanone |
| OH | H | H | O | O | $CH_2=CH_2$ | 4-deoxyzearalanone |
| OH | OH | OH | O | O | $CH_2=CH_2$ | 5-hydroxyzearalanone |
| OH | OH | H | O | O | $CH_2=CH_2$ | Zearalanone |
| OH | OH | H | $H_2$ | $H_2$ | $CH_2=CH_2$ | Zearalane |
| OH | OH | H | O | OH | CH=CH | 8=hydrozearalenone |
| OH | OH | H | O | H | CH=CH | 7-dehydrozearalenone |
| OH | OH | HC=O | O | $H_2$ | CH=CH | 5-formylzearalenone |
| OH | OH | H | OH | OH | CH=CH | 6,8-dihydroxyzearalene |

**Fig 1.** Structure of zearalenone and its derivatives

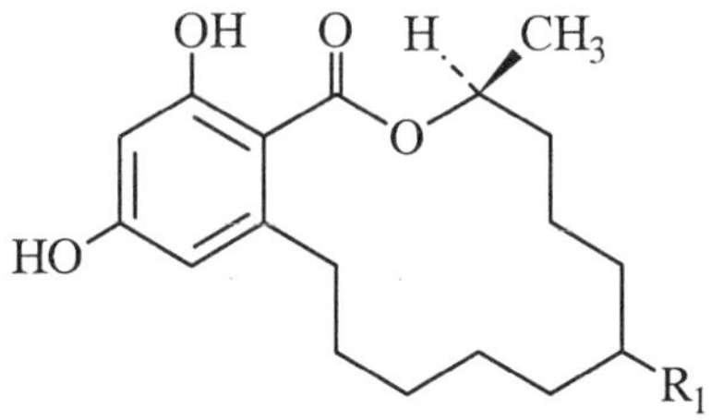


Zearalanone $R_1$ = O; Zearalane $R_1$ = H; Zearalanol $R_1$ = OH

**Fig. 2 :** Synthetic zearalenone analogs

## ORGANISMS

The most important producer of ZEN is *F. graminearum*, which frequently is referred to incorrectly in the literature as *F. roseum* (76). Isolates of *F. graminearum* that produce large amount of ZEN in culture have been obtained from mouldy corn implicated in porcine hyperestrogenism in the United States and elsewhere (158). Isolates of *F. graminearum* from scabby barley and wheat in Japan (48) have shown to produce ZEN in culture have also been obtained from many other agricultural commodities (160, 169). Some other species of *Fusarium* that produce ZEN are *F. culmorum* (111, 133) and *F. crookwellense* (141, 159). *F. sporotrichioides* and *F. equiseti* (76) and *F. semetectum* (164). *F. moniliforme*, *F. tricinctum* and *F. oxysporum* (2). These fungal species occurred naturally in corn, barley, animal feeds, hay, oats, sesame and sorghum. The natural occurrence of ZEN is favored by high humidity and low (10-15°C) temperatures (20,164)

## ANALYTICAL METHODS

ZEN and some of its derivatives fluorescence blue-green when excited by UV radiation (360nm) and fluorescence even more intensively when irradiated at 260 nm. This property has been exploited to detect ZEN on TLC plates (38).The current methods for the detection of Zearalenone include techniques like thin layer chromatography (5, 134, 168) and gas chromatography, Mass spectrometry (99) and HPLC (65). Some of the detection methods of TLC are précised in Table 1.

**Table 1:** Detection of Zearalenone produced by *Fusarium* with TLC

| Solvent system | Rf value | Spray reagent | Detection | | Reference |
|---|---|---|---|---|---|
| | | | UV | Visible | |
| Toluene : Ethyl acetate : Formic acid (6:3:1) | 0.78 | $Ce(SO_4)_2$ 1% in 6N.$H_2SO_4$ | -- | Brown | Grost-Allman and Steyn (40) |
| | | 2,4-DNP | -- | Dark orange | Mirocha *et al.* (87) |
| | | $FeCl_3$ 3% in ethanol | -- | Pink | Pathre *et al.* (102) |
| Benzene : Methanol : Acetic acid (24:2:1) | 0.42 | 50% $H_2SO_4$ in $CH_3OH$ | Yellowish brown | -- | Ishili *et al.* (49) |
| | | $H_2SO_4$ | Charring | -- | Kamimura *et al.* (59) |
| | | 20% $AlCl_3$ | Blue green | -- | Kamimura *et al.* (59) |
| | | β-Anisaldehyde Methanol 85 ml $H_2SO_4$ 5 ml Glacial acetic acid (10 ml) | -- | Faint brown | Scott *et al.* (126) |

HPLC with direct fluorescence detection is considered to be the usual choice for determination of ZEN due to its precission, specificity and reliability (6, 65, 161, 162).

Enzyme linked immunosorbent assays of Zearalenone using polyclonal, monoclonal and recombinant antibodies suggested by Park *et al.* (100) proved to be good for detection of ZEN in cereals. Liquid-liquid partition (5, 109, 134), open chromatographic columns (62), chromatographic cartridges (99, 162), multifunct-ional (113) and immunoaffinity columns (125) are some of the clean up steps employed in the detection of ZEN.

Krska (65) considers the use of multifunctional and immunoaffinity clean up columns for purification of ZEN to be the best as it is a time saving. Use of immunochemical techniques such as Enzyme-Linked Immunosorbant Assays (ELISA) (11) and immunoaffinity columns with flourimetry have also been suggested for determination of ZEN by Scott and Trucksess (125). Liquid and gas chromatographic methods which are commercially employed for ZEN analysis are as follows.

**(i) Sample preparation:** A mixture of organic solvent (acetonitrile, ethyl acetate, methanol, chloroform and acetone) with water or aqueous acidic solutions is commonly used for extracting ZEN from cereals (120). For purification of the extract, SPE cartridges (128, 174), liquid-liquid partition (58, 124), MycoSep column (130) and recently specific IA columns (58, 65, 66 161, 174) have been used.

**(ii) Separation and detection techniques:** HPLC has been the most often used technique for ZEN analysis in cereals (58, 140). Reversed phase separation with fluorescence detection, due to native fluorescence of the ZEN molecule, has frequently been applied. In addition to fluorescence detection, electrochemical, UV and diode array detection have been used (120, 140). GC techniques, mainly with MS detection, have also been utilized. With reports of simultaneous detection of different *Fusarium* toxins, and LC-MS and LC-MS/MS also appear to be a promising techniques (91, 99, 114, 115, 121,140, 148, 174). Some of the common HPLC methods currently used for ZEN analysis in cereals are perecied in Table 2.

**Table 2.** Some common HPLC methods

| Toxin analyzed | Extraction | Clean-up | Mobile phase | Separation, detection | Reference |
|---|---|---|---|---|---|
| ZEN | ACN-water | MycoSep 224 | MeOH-water | HPLC-Flu | 130 |
| ZEN | CAN | IA column | ACN-water | HPLC-Flu | 119 |
| ZEN | CAN-water | IA column | ACN-water-MeOH | HPLC-Flu | 161 |
| ZEN (OA) | Chloroform-water-phosphoric acid | Silica-column | MeOH-water phosphoric acid | HPLC-Flu | 72 |

ACN = acetonitrile, MeOH = methanol, HPLC-Flu = high performance liquid chromatography with fluorescence detector.

## NATURAL INCIDENCE

The natural occurrence of ZEN is favored by high humidity (2.5) and low temperature(10-15°C) (20). Weidenborner (164) feels that a temperature between 12 to 14°C are more favorable for ZEN production but production may also take place at temperature below 10°C and even below freezing. However, it requires a minimum of 22 to 25% moisture to grow on cereal grains (52). The fungus survives best between crops of corn and small grains and in residues on the soil surface (160). The spores of this fungus spreads by rain splash or air movement to corn silk where they germinate and colonize on the silks during wet weather. By growing down the silk channel, the fungus is able to attack developing ear and cause ear rot. A reddish - brown discoloration of the kernels progresses from the tip downwards and effect ears. A whitish red mold growth may present on the highly affected ears (160). High levels of ZEN in cereals specially maize accumulate during storage. Mature *Fusarium* infected grains that have not significantly dried because of wet weather at harvest or in grains that were stored wet (moisture >22%). Variety of food grains and other foods including corn products (167), corn beer (74), wheat flour (144) and walnuts (55) are reported to be contaminated by ZEN. The presence of ZEN in whole plants and parts of maize used for silage making in Germany (97) were contaminated by ZEN concentrations up to 300 μg/kg which mainly accumulated at the end of the ripening process with subsequent contamination of the silage. *Fusarium* species may produce some ZEN and other toxins (21). Abnormally high concentrations of both nivalenol and ZEN have been recorded in Japanese barley samples and in maize in New Zealand (106).

World wide occurrence of ZEN in cereals and its co-occurrence along with trichothecenes particularly deoxynivalenol has been reported by Tanaka and Ueno (143) from all over the world. Natural incidence of ZEN in different agricultural commodities has been reported from different parts of the world (55, 144, 150). Large number of reports are also available about the natural occurrence of ZEN in corn and corn based foods in Brazil (96, 116), Argentina (112, 135), Uruguay (103), Korea (62, 100), Canada (123), UK (127), Italy (161) and Indonesia (4). Sydenham and Shepherd (140) recorded ZEN in 1-69% of samples with a contamination range of 4-8036 μg/kg. FAO (30) considers contaminated cereals as the main source of ZEN in Nordic countries (29). Trigo-Stockli *et al.* (155) revealed that the level of ZEN was more in the bran and low in flour indicating the ZEN exists on the surface of the grains. ZEN production has been reported on grains in the field, during harvest, commercial grain processing and / or subsequently during storage of any food or feed stuff containing the grain (64). ZEN has also been detected in barley and malt (101), pecans (118), tomatoes (136) and bananas (56)

## BIOLOGICAL ACTIVITY

The toxicity of Zearalenone is unique among known mycotoxins inspite of its dissimilarity to steroid compounds. It produces potent hyperesterogenic responses in susceptible animals (23). Zearalenones were classified as a weak estrogen having anabolic properties (31). Swines are the most commonly affected animals, cattle, poultry and laboratory rodents are also affected but to a lesser degree (20). Swine appears to be most susceptible to ZEN (27). The biological activity of ZEN appears to have hormonal activity in regulating perithecia formation during the sexual stage of the life cycle of *F. graminearum* (170). Low concentrations of ZEN (0.1 to 1.0 μg) increased the number of perithecia by nearly 100%; where as higher levels (> 10 μg) almost totally blocked the formation of perithecia in 4 day old cultures (89). ZEN shows predominantly oestrogenic properties that are manifested in the form of hyperesterogenism in female swine, cattle and sheep causing severe reproductive problems (54). Esterogenicity includes atrophy of seminal vesicles and testis, squamous, metaplasmia of prostrate gland, osteropersis, myelofibrosis of bone marrow, cytoplasmic vacuolization of adrenal glands, hyperkeratosis of vagina and endometrial hyperplasia. ZEN caused androgenic response in male turkey poults (98). Young male pig exposed to ZEN undergoes symptoms of feminization (94). Transmission of ZEN or its metabolites into milk appears to be minimal (107).

Estrogenic disturbances are also suspected in sheep. Abortion mild vulvo vaginitis and hypertrophy of the uterus are recorded (110). Reduction in the incidence of ovulation and infertilization were seen in ewes dosed with ZEN for 10 days before bating (132). In the outbreak of ZEN mycotoxicosis on two farms in Northern Queensland most prepubertal gilts in the herds displayed enlarged testes and signs of oestrus such as red, swollen vulvas. In several cases both rectal and vaginal prolapses occurred. In one of the farms, 25 pigs died.

Smith *et al.* (132) observed exposure of ewes to ZEN caused decline in ovulation rate, cycle length and an increase in duration of estrus but did not affect pregnancy rate or embryonic loss. Bailey *et al.* (7) recorded decreased growth rate, food intake, fertility, resorptions, still births, abortion and bone malformations in fueturesres in rats due to intake of ZEN at 10 mg/kg. Egg production decreased in hen due to consumption of ZEN (119).

According to Marasas and Nelson (77) dietary levels of ZEN as low as 1 µg/g cause clinical signs of hyperestrogenism in pigs. Several alterations of immunological parameters were recorded at high ZEN concentration, *in vitro* (29). Lang *et al.* (71) recorded degeneration of embryos due to progressive disorganization of the embryonic disk. Marin *et al.* (29) recorded increase in interleukin 1L-2 and IL-5 in murine thymosa cell line EL-4 in the presence of ZEN suggests its immunotoxicity. In Newzealand ZEN in pasture is a recognized cause of infertility in sheep (152). Its acute toxicity is low and LD50 (mg/kg bw) for oral and intraperitonial administration are 2000 and >500 respectively in the mouse, >4000 > 10,000 > 5500 in the rat and >15,000 for oral administration in chicks. Coulambe and Roher (25) and Hussain and Brasel (47) feel that swine highly sensitive to ZEN. Mirocha *et al.* (88) and Mirocha and Christensen (85) gave detailed account of ZEN toxicity. Several closely related metabolites of ZEN produced by *Fusarium* spp also possess similar properties, although a few have been proven to occur naturally. Garies (35) reported that swine fed by a diet containing 50 mg/kg of pure ZEN suffered abortion and stillbirths, while levels above 10 mg/kg reduced the litter size and reduced the weight of piglets. Trial feeding of female pigs revealed that a concentration of 0.25 mg/kg or even less, produced distinct redness and swelling of the vulva, slight swelling of the mamma and numerous vesicular follicles and some cystic follicles on the ovarian (8). ZEN decreased the serum level of oestrogen in animals (50).

Coulombe and Roher (25) feel that sexually mature boars are relatively insensitive to ZEN contamination. They also discussed at length biological action of ZEN in different animal systems. In mice ZEN affects ovary DNA and exhibited mitotoxicity (3).

Kuiper-Goodman (68) recorded increased incidence of pituitary adenomas in both male and female animals with progression to malignancy with the consumption of ZEN. Gimeno and Quni-tanilla (39) reported the natural outbreak of ZEN mycotoxicosis in horse. Symptoms in exposed mares were enlarged edematous vulva, prolapsed vagina, oversized uterus and internal haemorrhage

ZEN at concentration of 2.0 to 5.6 and 1.5 ppm respectively was reported in a feed given to swine showing hyperesterogenism (85). Pigs showing clinical signs of hyperesterogenism was reported to be fed by maize containing 8 ppm ZEN (14). α-Zearalenone is three times more estrogenic than β-Zearalenone based on a rat uteropic assay (43). Elghor (28) and Hassnane (44) have reported that the increase in cytotoxic effect was usually associated with decrease in number of cells with chromosomal aberrations.

Prelusky *et al.* (108) feel that ZEN may cause immuno-toxicity. ZEN has also been considered as a possible agent that caused pubertal changes among young children of Puerto Rico (117). Coppock *et al.* (24) observed that ZEN about 750 ppb and 500 ppb of DON caused poor consumption of feed, depressed milk production, diarrhea and total reproductive failure. Towers *et al.* (153) and Sporsen and Towers (137) and Smith *et al.* (131) feel that measurement of urinary zinc can successfully give the ZEN intake. In sheep ZEN causes lower conception, reduced ovulation, increased twinning rates and a 10-20% decline in fertility of ewes. In dairy cattle reproductive problems arise due to ZEN at 400 ppb. Kuiper-Goodman *et al.* (69) concluded that pure ZEN is less toxic than naturally occurring contaminated cereals indicating the presence of additional toxic substances in matrix.

Presence of ZEN in red meat and poultry may cause pre mature teleroche in Italy and premature teleroche pubersche, gynocomastia and precocious pseudopuberty in Puerto Rico (45). Carcinogenic activity of ZEN still remains controversial (46, 151). Resistance of ruminants to ZEN is attributed to its degradation by resident microorganisms.

Infertility, reduced milk production, and hyperesterogenism in cows have been reported to be associated with ZEN (27). Reduced feed intake, vaginitis, vaginal secretions, poor reproductive performance and mammary gland enlargement in virgin heifer (57) are some of the toxic syndromes of ZEN. Schoental (129) described 'cold cow syndrome' characterized by decreased body temperature, diarrea and swelling of the vulvae and other symptoms due to excess intake of ZEN through feed.

## BIOSYNTHESIS

Gafoor and Trail (33) have reviewed the biosynthetic pathways of Zearalenone. From studies on biosynthesis revealed the entire ZEA molecule would be synthesized from a single acetyl-CoA molecule and eight malonyl-CoA molecules (Fig. 3) as previously determined biochemically (86). The first 10 carbon additions are catalyzed by ZEA2p, which has the reductive domains KR, DH, and ER and could in variably reduce the keto groups on the polyketide chain. The keto group can be partially reduced to yield an alkene or can be fully reduced to an alkane depending on the length of the carbon chain growing on the ACP domain. The remaining three rounds of $C_2$ additions are generated by ZEA1p, which is a nonreducing PKS and which lacks KR, ER, and DH domains. The unreduced ketones added by ZEA1p are highly reactive and form a ring that is rapidly aromatized due to enolization of the ketone groups via a non enzymatic mechanism is for $C_{11}$ and $C_{12}$ to be added by ZEA1p, not ZEA2p, an alternative that remains to be tested experimentally. The two ZEA proteins may be dimerized or may act independently. Dimerization would facilitate the transfer of the product of ZEA2p to ZEA1p (Fig. 3). In *in vitro* studies, PKSs can be promiscuous in terms of starter unit specificity (163). If ZEA1p and ZEA2p are associated, then ZEA1p should use the correct starter unit and there by maximize the efficiency of ZEA biosynthesis.

## MECHANISAM OF ACTION

ZEN and some of its metabolites have shown to competitively bind to estrogen receptors (ER). Binding to specific receptor has been demonstrated in uterus, mammary glands, liver and hypothalamus from different species. Effects of Zearalenone are similar to Zearalenol. Zearalenone is generally considered to produce estrogenic effects 5-10 times greater that those of ZEN (52). α-Zearalenol is 10 times more estrogenic than ZEN, where as β-isomer is considered less than that of ZEN (164). α-Zearalenol showed comparatively stronger binding than β-zearalenol. The relative binding affinities to rat uterine cytoplasmic receptor for ZEN and derivatives were α zearalenol > zearalenol > β-zearalenol > ZEA > β-zearalenol (29). ZEN was shown to bind and activate both ERα and ERβ in cells transected with ERα and ERβ. For ERα, ZEA was found to be a full antagonist and for ERβ it is a mixed antagonist (67). In blood ZEN and zearalenol bind to human sex hormone-binding globulin to some extent (29).

Hayes (45) feels that ZEN interacts with estrogen receptors translocation of receptor–ZEN compete to the nucleus, combination with chromatin receptors, selective RNA transcription leading to biochemical effects including increased water and lowered liquid content in muscle and increased permeability of the uterus to glucose, RNA, RNA protein precursors (Fig. 4). It also induces biphasic changes in the luteinzing hormone but not the follicle stimulating hormone in sperm (16). Estrogenic activity of ZEN and several of its derivatives is due to depression of goanadotrophin levels (110).

ZEN competes with estuarial 11B (E2) in binding to cytosolic estrogen receptors, activates the expression of initial protein and increases RNA polymerase I and II activities in nuclei (17, 60).

A number of factors including sex, species of animal, environmental factors, nutritional status and mycotoxin synergism etc can either directly or indirectly modulate toxic effects.

## PHARMACOKINETICS

Studies on pharmacokinetics and metabolism indicate that ZEN is fairly, rapidly adsorbed following oral administration and can be metabolized by intestinal tissue in pigs and possibly in humans during its absorption with the formation of α and β-zearalenol, which are subsequently conjugate with glucouronic acid. Quantitatively glucouranide conjugates with the parental compound and α-zearalenol predominates in these species (165). Differences between species in the metabolism of ZEN were recorded. In the rat, most of the ZEN is found free Zearalenone or glucouranide conjugate there of while only small amounts of zearalenol and their conjugates were formed (53, 69).

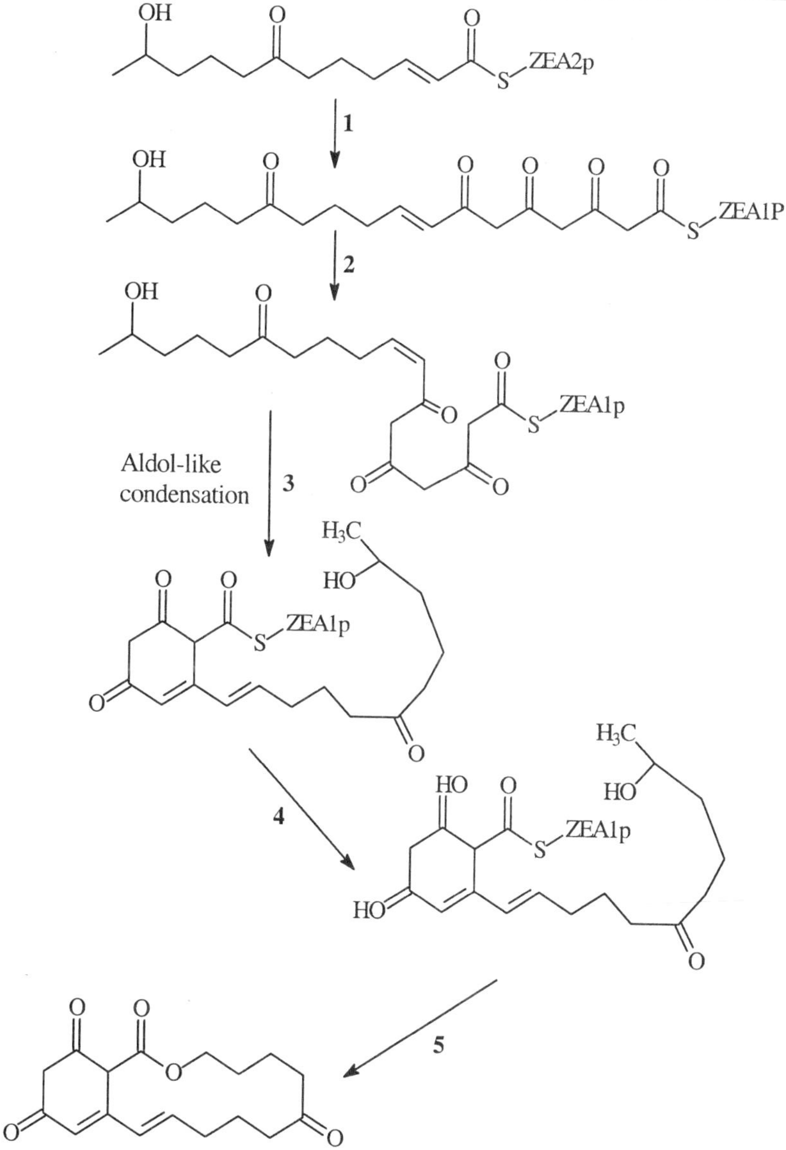


Zearalenone

**Fig - 3 : Biosynthetic pathway of Zearalenone**

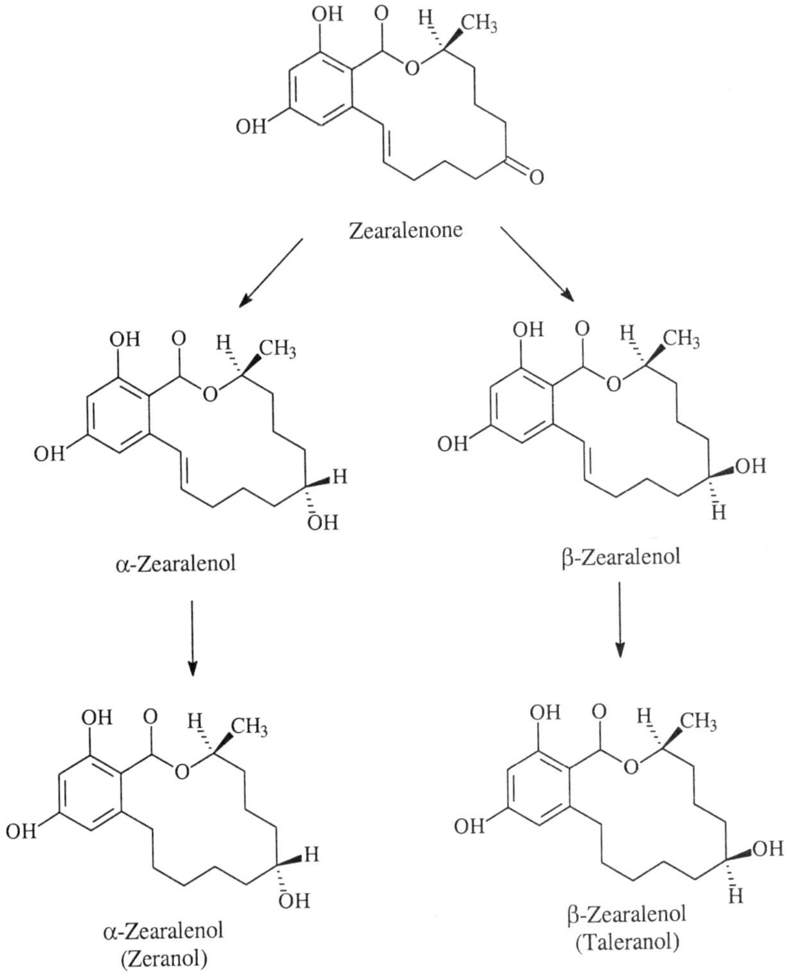


**Fig-4 : Metabolic derivatives of Zearalenone**

Broiler chicks and a laying hens are affected very little by dietary ZEN even when massive doses are given (52,73,113). Feed containing 300 ppm of ZEN in chicks resulted in increased weight of bursa of Fabricius, increased comb weight and increased number of cysts in the genital tract (82). ZEN is rapidly adsorbed after oral administration is extensively absorbed by tissues in rats, rabbits and humans (165). Beihl *et al.* (12) recorded about 80-85%

of ZEN absorbed by pig tissues. Kuiper–Goodman *et al.* (69) could provide evidence for deposition of ZEN in adipose tissues. ZEN and its metabolites are excreted mainly in the bile in most animal species except rabbits in which urine is the main route. in pigs 45% ZEN is recorded in urine during first 48 hrs, 22% was recorded in feces (12). In chickens ZEN is distributed chiefly to the liver and gall bladder (84) and excreted in the form of ZEN or the derivatives. In laying hens most is excreted in faeces but residues may occur in yolk (26)

Biliary excretion which enters hepatic circulation occurs in rats and mice, while urinary excretion predominates in rabbits. Urinary excretion is also the main route of excretion in pigs inspite of the demonstrated enter-hepatic circulation of ZEN owing to high degree of reabsorption in the gut. Urinary excretion of ZEN is also significant in human (69).

In humans and pigs ZEN was found mainly as glucouranide conjugates of ZEN and Zearalenone in urine with 24 hrs of uptake. Turkeys metabolize ZEN to Zearalenone, while chickens tends to produce equal amounts of ZEN to highly active Zearalenone (171).The amount of detectable ZEN in animal tissues depends on the contamination of feed, treatment of animals with ZEN or α-Zearalenol, duration of exposure to the toxin, the persistence of ZEN in the animal and species variation (165). Chicken fed with 100 mg/kg Zen for 8 days, contained 59-103 mg/kg ZEN in muscle and up to 681 mg/kg in liver (84). A dose of 544.5 mg/day given to cow for 21 days yielded maximum concentration of Zearalenone of only 2.5mg ZEN, 3 mg/L ZEN in mice (107). Eggs accumulated metabolites of ZEN in the yolk even after 94% of the dose had been eliminated in excreta (26). Both bovine and ovine rumen fluid rapidly reduce ZEN to the more readily excreted α and β-Zearalenone (61)

## MANAGEMENT OF FUSARIAL MYCOTOXINS

Besides usual strategies both at pre-harvest and post-harvest stage in the management of mycotoxins are also apply in true spirit in the management of ZEN.

Factors associated with the inhibition of ZEN production in the field are complex and genetics associated with such properties would be equally more complex. Increased understanding of the nature, origin, formation of ZEN may contribute to its control of contamination of foods and feeds.

Methods aimed at preventing or reducing level of ZEN contamination can be broadly classified as preventive or curative. Under this, there have been five basic approaches (i) prevention of the initial growth of the *F. graminearum* and subsequent production of ZEN (ii) detection of ZEN in food and selective removal of contaminated portion which, however, reduces the toxin only to a negligible level, (iii) inactivation or destruction of the ZEN by chemical, physical and biological means, (iv) development of effective and economic methods for controlling the mould growth and subsequent production of ZEN and (v) utilization of genetic resource.

Wet milling of corn reveals that ZEN concentrates in the gluten (48-56%) none is found in the starch. On the other hand, dry milling results in high concentration of toxin in the germ and feed fraction. During bread making 34-40% of ZEN in wheat flour is lost and in making instant noodles 48-62% is lost and in preparing biscuits 16-27% in lost (122). Fermentation of corn results as ZEN free ethanol, found in solids used for animal feed (9).

Addition of non-nutritive sorbents or bacteria, yeast and modified yeast cells capable of reducing the bioavailability of ZEN are assuming increasing interest and are critically reviewed by Galvano *et al.* (34).

A lactone hydrolase was found to be responsible for detoxification of ZEN (142). Bauer *et al.* (8) detoxified ZEN by employing calcium hydroxide monomethylamine. Grosse *et al.* (42) reported that vitamins A, C and E reduced DNA- ZEN adduct formation in kidney and liver of mice. Vitamin E was able to prevent genotoxicity of ZEN (37). Modified mannanoligosaccharide derived from the cell wall of *S. cerevisiae* was reported to bind ZEN. Trenholm *et al.* (154) reported that cholestyramine was capable of binding ZEN efficiently.

Gut flora of pig is reported to degrade ZEN (64). Similarly Megharaj *et al.* (81) reported that mixed culture of bacteria can remove ZEN from culture medium. Nezami *et al.* (95) reported that *Lactobacillus rhamnosus* GG and *Lactobacillus rhamnosus*

LC705 removed the levels of ZEN and its main derivative of ά-zearalenol in liquid media to a significant level.

## FUTURE STUDIES

Inspite of intensive studies, a large gap still exists in the understanding of ZEN and its biological effects and needs concentrative efforts under the following areas

1. Additional studies to determine the hormonal effect level prepubertal pigs should be performed.
2. Need for more detailed studies on potential genotoxicity of ZEN, to clarify its ability to bind covalently to DNA.
3. Need for comparative studies on species difference including use of human cells in the metabolism of ZEN to elucidate different findings in long term carcinogenicity.
4. Blood level of ZEN in human with known dietary ZEN intake could help to clarify the toxic kinetic behavior of Zearalenone.
5. More research efforts are necessary to verify the *in vivo* detoxification ability of the proposed agents, their mode of action possible long-term draw backs of these detoxification and decontamination procedures and their economic feasibility.

## *Literature Cited*


1. Abbas HK, Mirocha CJ. 1988. Production of Fusarenone-X, nivalenol and zearalenone by *Gibberella zeae* isolates and their toxicity in fibro blasts and rats. *Mycotoxin Res.* 4 : 67-75.
2. Agag BI. 2004. Mycotoxin in foods and feeds 3-zearalenone. *Ass. Univ. Bull. Environ. Res.* 7 : 159-176.
3. Aidia EL-Makawy S, Mohamed, Hassanane El-Sayed AM, Alla ABD. 2001. Genotoxic evaluation for the estrogenic mycotoxin zearalenone. *Reprod. Nutr. Dev.* 41 : 79-89.
4. Ali N, Sarjono AY Yashizawa J. 1998. Natural co-occurrence of aflatoxins and *Fusarium* mycotoxins (Fumonisins, deoxynivalenol, nivalenol and zearalenone) in corn from Indonesia. *Food additives and Contaminants* 15 : 377-384.
5. AOAC. 1995. Zearalenone in corn, thin layer chromatographic method, No. 976-22, Official Methods of Analysis of

AOAC. International, 16th ed. (Gaithersberg, M.D., AOAC International), pp. 44-45.

6. Bagneris RW, Gaul JA Ware G.M 1986. Liquid chromatographic determination of zearalenone and zearalenol in animal feeds and grains using fluorescence detection. *J. Assoc. Off. Anal. Chem.* 69 : 894-898.

7. Bailey DE, Cox GE, Morgreidge K, Paylor J. 1976. Acute and subacute toxicity of zearalenone in the rat. *Toxicol. Appl. Pharmacol.*, 37 : 144.

8. Bauer J, Heinritzi K, Garies M, Gedek B. 1987. Veranderungea an Genitaltrakt des Weiblichen schweines nach Verfutterung praxisrelevanter Zearalenon mengen. *Tierdrztl-iche Praxis* 15 : 33-35.

9. Bennett GA., Lagoda AA, Shotwell OL, Hesseltine CW. 1981. Utilization of zearalenone contaminated corn for ethanol production. *J. Amer. Oil Chem. Soc.* 58 : 974-976.

10. Bennett GA, Shotwell OL, Kwalek WF. 1985. Liquid chromatographic determination of α-zearalenol and zearalenone in corn : collaborative study. *J. AOAC.* 68 : 958-961.

11. Bennett GA, Nelson TC, Miller BM, 1994. Enzyme linked immuno sorbent assay for detection of zearalenone in corn, wheat and pig feed Collaborative Study. *J. AOAC* 77 : 1500-1508.

12. Biehl ML, Prelusky DB, Koritz GD, Hartin KE, Buck WB, Treholm HL. 1993. Biliary excretion and enterohepatic cycling of zearalenone in immature pigs. *Toxicol. Appl. Pharmacol.* 121 : 152-159.

13. Blaney BJ, Dodman RC. 1988. Production of the mycotoxins zearalenone, 4-deoxynivalenol and nivalenol by isolates of *F. graminearum* groups 1 and 2 from cereals in queensland. *Aust. J. Agric. Res.* 39 : 21-29.

14. Blaney BJ, Bloomfield RC, Moore CJ. 1984. Zearale-none intoxication of pigs. *Austr. Vet. J.* 61(1) : 24-27.

15. Bolliger G, Tamm C. 1972. Vierneue metabolite von zibberella zeal : 5-Formyl-zearalenon, 7'-dehydro zearalenon, 8'-hydroxy und 8'-epihydroxy-zearalenon. *Helv. Chim. Acta.* 55 : 702-709.

16. Bongiovanni AM. 1983. An epidemic of premature thelarche in Puerto Rico. *J. Pediatr.*, 103: 245-246.

17. Boyd PA, Wittliff JL. 1978. Mechanism of Fusarium mycotoxin action in mammary gland. *J. Toxicol. Environ. Health* 4 : 1.

18. Branten SL, Deaton JW, Hagler WM, Mashin WR, Hardin JM. 1989. Decreased

egg production in commercial laying hens, fed zearalenone and deoxynivalenol contaminated grain sorghum. *Avians Dis.* 33 : 804.

19. Branton SL, Deaton JW, Hagler WM, Maslin JrWR, Hardin JM. 1989. Decreased egg production in commercial laying hens fed Zearalenone and deoxynivalenol contaminated grain sorghum. *Avian Dis.* 33 : 804.
20. Cast, Council for Agric. Sci. and Technol, 1989. Mycotoxins economic and health risks. Task free report No. 116, No. 1989.
21. Christensen CM, Mirocha CJ, Meronuck RA. 1988. Molds and mycotoxins in feeds. *Minn. Ext. Serv. Bull.* AG-FO-3538, p. 14.
22. Cole RJ, Kirksey JW, Cutler HG, Doupnik BC, Peckham JC. 1973. Toxin from *Fusarium moniliforme* effects on plants and animals, science, 179 : 1324-1326.
23. Columbe RA Jr.1993. Biological action of mycotoxins. *J. Dairy Sci.* 76 : 880-891.
24. Coppeck RW, Mostrom MS, Sparling CG, Jocobsen B, Ross SC. 1990. Apparent zearalenone intoxication in a dairy herd from feeding spoiled acid-treated corn. *Vet. Hum. Toxicol.* 32 : 246.
25. Coulombe, Roher A. 1993. Biological action of mycotoxins. *J. Dairy Sci.* 880-891.
26. Daily, RE, Reese RE, Brouwer A, 1980. Metabolism of [$^{14}$C] zearalenone in laying hens. *J. Agric. Food Chem.* 28 : 286-291.
27. Diekman MA, Green ML. 1992 Mycotoxins and reproduction in domestic livestock. *J. Anim. Sci.* 70 : 1615-1627.
28. Elghor AA 1994. *In vivo* sensitivity of male, maternal and foetal cells of mice to the clastogenic action of dimethoate. *Proc. Zool. Soc. A.R. Egypt.* 25 : 183-193.
29. Eriksen GS, Alexander J. 1998. *Fusarium* toxins in cereals – a risk assessment. Nordic Council of Ministers, Tema Nord, 1998, 502, pp. 7-27 and 45-58.
30. FAO 1998. Animal Feeding and Food safety, Report of an FAO Expert Consultation, Rome 0-14 March 997, FAO, Food and Nutrition Paper 69, Food and Agriculture Organi-zation of the United Nations, FAO, Rome.
31. FDA, Food and Drug Administration, 1979. Conference on mycotoxins in animal feeds and grains related to animal health. June, 8, 1979, Rockville, Maryland.
32. Fitzpatrick DW, Picken CA, Nurphy LC, Buhr MM. 1989.

Measurement of the relative binding affinity of Zearalenone α-zearalenol and β-zearalenol for uterine and oviduct estrogens receptors in swine, rats and chickens : An indicator of estrogenic potencies. *Comp. Biochem. Physiol. Comp. Pharmacol. Toxicol.* 94 : 691.

33. Gaffore, Iffa and France Trails 2006. Characterization of two polyketide synthase genes involved in Zearalenone biosynthesis in *Gibberella zeae. Appl. Environ. Microbiol.* 72(3): 1793-1799.

34. Galvano Fabio; Andrea Piva; Alberto Ritieni and Giacomo-Galvano, 2001. Dietary strategies to counter act the effects of mycotoxins : A review. *Journal of Food Production,* 64:120-131.

35. Garies M. 1993. *Fusarium* mycotoxins an animal feeds and effects on live stock, in proceedings of the UK Workshop on occurrence and significance of mycotoxins, Slough, Central Science Laboratory, MAFF, 21-23 April, 1993 (Ed. K.A. Scudamore), M.A.F.F. London, pp. 7-15.

36. Gaumy JL, Bailly JD, Burgatet V, Gurre P. 2001. Zearalenone properties toxicite experiments. *Revue. Med. Vet.* 152 : 219-234.

37. Ghedria-Chekir L, Maaroufi K, Zakhama A, Ellouz F, Dhouib S, Creppy EE, Bacha H. 1998. Induction of a SOS repair system in bacteria by zearalenone and its prevention by vitamin E. *Chem. Biol. Interact.* 113 : 15-25.

38. Gillepsie AM, Schenk GM, 1977. Fluorescence and phosphorescence of ochratoxins, sterigmatocystins, patulin and zearalenone : quantitation of ochratoxins. *Analytical Litters* 10(2) : 161-172.

39. Gimeno A, Quintailla JA, 1983. Analytical and mycolog-ical study of a natural outbreak of zearalenone mycotoxicosis in horses. *Proc. Int. Symp. Mycotoxins,* pp. 387-392.

40. Gorst-Allman ChP, Steyn PS, 1979. *J. Chromatogr.* 175 : 325-331.

41. Grigo-Stockli DM, Deyoe CW, Satumbaga RF, Pedersen JR. 1996. Distribution of deoxynivalenol and zearalenone in milled fractions of wheat. *Cereal Chemistry*, 73 : 388-391.

42. Grosse Y, Chekir-Ghedria L, Huc A, Obrecht-Pflumino S, Dirheimer G, Bacha H, Pfhl-Leszkowicz A, 1997. Retinol, ascorbic acid and a tocophenol prevent DNA adduct formation in mice treated with the mycotoxins ochratoxin A and zearalenone. *Cancer Letters,* 114 : 225-229.

43. Hagler WM, Mirocha CJ, Pathre SV, Behrens JC, 1979.

Identification of the naturally occurring isomer of zearalenol produced by *Fusarium roseum* Gibbosum in rice culture. *Appl. Environ. Microbiol.* 37: 849.

44. Hassanane, MS, El-Sayed A, Abdalla El-Fiky MA, Amer S, Handy A, 2000. Mutagenecity of the mycotoxin, diacetoxyscirpenol on somatic and germ cells of mice. *Mycotoxin Res.* 16 : 53-63.

45. Hayes AW. 1994. Principles and methods of toxicology, 3rd ed., p. 343, Raven Press, New York.

46. Hilakivi-Clarke, L, Onojafe I, Raygada M, Cho E, Skaar T, Russo I, Clarke R. 1999. Prepubertal exposure to zearalenone or genistein reduce mammary tumorigenesis. *Br. J. Cancer,* 80 : 1682-1688.

47. Hussain, H.S., Brasel, J.M. 2001. Toxicity, metabolism and impact of mycotoxins on human and animals. *Toxicol.,* 167 : 101-134.

48. Ichnoe, M., Kurata, H., Sugiura, Y., Ueno, Y. 1983. Chemotaxonomy of gibberalle zeae with special reference to production of trichothecenes and zearalenone. *Appl. Environ. Microbiol.,* 46 : 1364-1369.

49. Ishili, K., Sawanot, M., Ueno, Y., Tsunada, T. 1974. *Appl. Microbiol.*, 27 : 625.

50. Ito, Y., Ohtsubo, K. 1994. Effects of neonatal administration of zearalenone on the reproductive physiology of female mice. *J. Vet. Med. Sci.*, 56 : 1155-1159.

51. Jackson, R.A., Fenton, S.W., Mirocha, C.J., Davis, G. 1974. Characterization of two isomers of 8'-hydroxy zearale-none and other derivatives of zearalenone. *J. Agric. Food. Chem.*, 22 : 1015.

52. Jacobsen, B.J., Bowen, K.L. Shelby, R.A., Diener, U.L., Kemppainen, B.W., Floyd, J. 1993. Mycotoxins and mycotoxicosis circular ANR-767. Aubern Univ., Aubern, Alabama.

53. JECFA 2000. Joint FAO / WHO Expert committee on food additives, 53rd Report. Safety Evaluation of Certain Food Additives. WHO Food Additives Series, 44.

54. Jemmali, M. 1987. Trade and economic implications of mycotoxins need for greater uniformity Mye 87/6. Joint FAO/ WHO UNEP second international conference on mycotoxins Bangkok, Thailand, 26 Sept. to 3 October.

55. Jemmali, M., Mazerand, C. 1980. Presence of dezearalenone on $F_2$ danslesnoixdecommerce. *Annules de Microbiologie*, (Paris), B 31 : 319-321.

56. Jimenez, M., Huerta, T., Mateo, R. 1997. Mycotoxin production by *Fusarium* species isolated from Bananas. *Applied and Environmental Microbiology,* 63 : 364-369.

57. Jones, F., *et al.* 1994. Understanding and coping with effects of mycotoxins in livestock feed and forage. Electronic publication No. dRO-29, NCCES, North Carolina State Univ., Raleigh.

58. Josephs, R.D., Schumacher, R. Krishna, R. 2001. International interlaboratory study for the determination of the *Fusarium* mycotoxins zearalenone and deoxynivalenol in agricultural commodities. *J. Food Additives and Contaminants*, 18 : 417-430.

59. Kamimura H, *et al.* 1981. Simultaneous detection of fusarial toxins. *J. Assoc. Off. Anal. Chem.* 64 : 1067-1073.

60. Kawabata Y. Tashiro F, Ueno Y. 1982. Synthesis of a specific protein induced by zearalenone and its derivatives in rat uterus. *J. Biochem.* 91 : 801.

61. Kiessling KH, Patterson H, Sandholm K, Olsen M. 1984. Metabolism of aflatoxin, ochratoxin, zearalenone and three trichothecenes by intact rumen fluid, rumen protozoa and rumen bacteria. *Appl. Environ. Microbiol.* 47 : 1070.

62. Kim JC, Kang HJ, Lee DH, Lee YN, Yoshizawa T, 1993. Natural occurrence of *Fusarium* mycotoxins in barley and corn in Korea. *Appl. Environ. Microbiol.* 59 : 3798-3802.

63. Klang DT, Kennedy BJ, Pathre SV, Mirocha CJ. 1978. Binding characteristics of zearalenone analogs to estrogen receptors. *Cancer Res* . 38 : 3611.

64. Kollarczik, B., Gareis, M, Hamett, M. 1994. *In vitro* transformation of the *Fusarium* mycotoxins deoxynivalenol and zearalenone of pigs. *Nat. Toxins* 2 : 105-110.

65. Krska R. 1998. Performance of modern sample preparation techniques in the analysis of *Fusarium* mycotoxins in cereals. *J. Chromnatography* 815 : 49-57.

66. Kruger SC, Kohn B, Ramsey CS, Prioli R. 1999. Rapid immunoaffinity-based method for determination of zearalenone in corn by fluorometry and liquid chromatography. *J. AOAC Int.* 82 : 1364-1368.

67. Kuiper GG, *et al.* 1998. Interaction of estrogenic chemicals and phytoestrogens with estrogen receptor beta. *Endrocrinol.* 1139 : 4252-4263.

68. Kuiper-Goodman T, 1991. Risk assessment of ochratoxin A residues in food. In : mycotoxins, Endemic Nephropathy

and Urinary tract tumors, IARC Scientifici Publications, No. 115, International Agency for Research on Cancer, World Health Organizatiions, Casteg-haro, M., Piestina, R., Dirhe-imer, G., Cherno-Zemsky, I.N. and Bartsch, H. (ed.), p. 307-320, Oxford University Press, New York.

69. Kuiper-Goodman T, Scott, P, Watanabe H. 1987. Risk assessment of the mycotoxin zearalenone. *Regul. Toxicol. Pharmacol.* 7 : 253-306.

70. Kupper-Goodman T, 1998. Food safety: Mycotoxins and phyco-toxins in perspective in Mycotoxins and Phycotoxins-developments in Chemistry, Toxicology and Food safety (Miraghi M. Van Egmond, H., Bera, C. and Gilbert, J. (ed.)), Allaken, Fortcollin, p. 25-48.

71. Lang GG, Turek J, Diekman MA, Scheidt AB. 1992. Effect of zearalenone on days 7 to 10 post mating on blastocyst development and endometerial morphology in sows. *Veternary Pathology* 29 : 60-67.

72. Langseth W, Ellingsen Y, Nymoen U, Okland EM, 1989. High performance liquid chro-matographic determination of zearalenone and ochratoxin A in cereals and feed. *J. Chromatogr.* 478 : 269-274.

73. Lesson S, Diaz GJ, Summers JD. 1995. Poultry metabolic disorders and mycotoxins, p. 326-332, University Books, Guelph, Ontario, Canada.

74. Lovelace CEA, Nyathi CB. 1977. Estimation of the fungal toxins zearalenone and aflatoxin contaminating opaque maize beer in Zambia. *J. Sci. Food Agric*. 28 : 282-292.

75. Lud Y, Yoshizawa T, Katayama T. 1990. Comparative study on the natural occurrence of *Fusarium* mycotoxins (Trich-othecenes and Zearalenone) in corn and wheat from high and low risk areas for human esophageal cancer in China. *Environ. Microbial.* 56 : 3723.

76. Marasas WFO. 1991. Toxi-genic Fusaria. In mycotoxins and animal foods (J.E. Smith and R.S. Hederson Eds.), CRC Press, Boca Raton, FL, p. 119-139.

77. Marasas WFO, Nelson PE. 1987. Mycotoxicology : Introdu-ction to the mycology plant pathology, chemistry, toxicology and pathology of natural occurring mycotoxicosis in animals and man, Pennsy-lvania State, University Press, University Park, 1987.

78. Marasas WFO, Nelson PE, Tousson TA. 1984. Toxigenic *Fusarium* species. Identify and mycotoxicology, Pennsylvania

State University Press, University Park.

79. Marin S, Sanchis V, Magan N. 1996. Water activity, temperature and pH, effect on growth of *F. moniliforme* and *F. proliferatum* isolates from maize. *Can. J. Microbiol.* 41 : 1063-1070.

80. Matsura Y, Yoshizawa T. 1981. Effect of food additives and heating on the decomposition of zearalenone in wheat flour. *J. Food Hygie. Soc. Jap.*, 22 : 293-298.

81. Megharaj M, Garth Waite I, Thiele JH. 1997. Total biodegradation of the oestrogenic mycotoxin of zearalenone by bacterial culture. *Lett. Appl. Microbiol.* 24 : 329-333.

82. Meronuck RA, Garren KH, Christensen CM, Nelson GH Fern Bates, 1970. Effects on turkey poults and chicks of rations containing corn invaded by *Penicillium* and *Fusarium* species. *Am. J. Vet. Res.* 31 : 551-555.

83. Miller JD, Taylor A, Greenhalgh R. 1983. Production of deoxynivalenol and related compounds in liquid culture of *F. graminearum. Can. J. Microbiol.* 29 : 1171-1178.

84. Mirocha C, Robison TS, Pawlosky RJ, Allan NK. 1982. Distribution and residue determination of [$^{3}$H] zearalenone in broilers. *Toxicol. Appl. Pharmacol.* 66 : 77-87.

85. Mirocha CJ, Christensen CM. 1974. Oestrogenic mycotoxins synthesized by *Fusarium*. In Mycotoxins, I.F.H. Purchase (Ed.) Elsevier, Amsterdam, pp. 129-148.

86. Mirocha CJ, Pathre SV. 1979. Mycotoxins – their biosy-nthsis in fungi – zearalenone biosynthesis. *J. Food Prot.* 42:821-824.

87. Mirocha CJ, Schauerames B, Pathre SV. 1974. Isolation, detection and quantitation of zearalenone in maize and barley. *J. Assoc. Off. Anal. Chem.* 57 : 1104-1110.

88. Mirocha CJ, Christensen CM, Nelson GH. 1971. Microbial toxins, Chapter 4, Academic Press, New York.

89. Mirocha CJ, Pathre SV, Christensen CM. 1977. Zearalenone, in mycotoxins in Human and animal health (J.V. Rodricks, C.W. Hesseltine and M.A. Mehlmah eds.), Pathorox, Park forest sourth, III, 345.

90. Mitterbauer R, *et al.* 2003. A sensitive and inexpensive yeast bioassay for the mycotoxin zearalenone and other compounds with estroge- nic activity. *Appl. and Environ. Microbiol.* 69 : 805-811.

91. Muller HM, Reimann J, Schumacher U, Schwadorf K.

1998. Natural occurrence of *Fusarium* toxins in oats harvested during five years in an area of southwest Germany. *Food Addit. Contam.* 15 : 801-806.

92. Neish GA, Cohen H. 1981. Vomitoxin and zearalenone production by *F. gramnearum* from winter wheat and barley in Ontario. *Can. J. Plant. Sci.* 61 : 811-815.

93. Neish GA, Farnwarth ER, Greenlaugh R, Young JC. 1983. Observations on the occurrence of *Fusarium* species and their toxins in corn in eastern Ontario. *Can. J. Plant Pathol.* 5 : 11-16.

94. Newberne PM. 1987. Interaction of nutrients and other factors wilty mycotoxins. In P. Krogh (ed.), Mycotoxins in Food, pp. 177-216, Academic Press, New York.

95. Nezami, El, Hani, Nektaria Polychronaki, Seppo Salminen, Hannu Mykkar, 2002. Binding rather than metabolism may explain the Interaction of two food-grade *Lacto bacillus* strains with zearalenone and its derivative α-zearalenol.

96. Nordin N, Luchese RH. 1998. Deteccaode aflatoxina e zearalenona em mitho (*Zea may* L) destinado a alimentacao animal. *Boletim. SBCTA* 32:35-39.

97. Oldenberg E. 1993. Occurrence of zearalenone in maize. *Mycotoxin Res.* 9 : 72-78.

98. Olsen M, Mirocha CJ, Abbas HK, Johansson B. 1986. Metabolism of high concentrations of dietary zearalenone by young male turkey poults. *Poult. Sci.* 65 : 1905.

99. Onji Y, Aoki Y, Jani N, Umebayashi K, Kitada Y, Dohi Y, 1998. Direct analysis of several *Fusarium* mycotoxins in cereals by capillary gas chromatography–Mass spectrometry. *J. Chromato.* A 815 : 59-65.

100. Park JW, Kim EK, Shan DH, Kim YB. 2002. Occurrence of zearalenone is Korena barley and corn foods. *Food Additives and Contaminants* 19: 158-162.

101. Park KJ, Park AR, Lee XW. 1992. Natural occurrence of *Fusarium* mycotoxins of the 1990 barley crop in korea. *Food Additive and Contaminant* 9 : 639-645.

102. Pathre SV, Mirocha CJ. 1979. Trichothecenes natural occurrence and potential hazard. *J. Amer. Oil Chem. Soc.* 56 : 820-823.

103. Pincire MS, Silva GE. 1997. *Fusarium* toxins in Uruguay : Head blight, toxin levels and grain quality, cereal. *Res. Commun.* 25 : 805-806.

104.Pitt SI. 2000. Toxigenic fungi : which are important ? *Med. Mycol.* 38 : 17-22.

105.Pittet A. 1998. Natural occurrence of mycotoxins in foods and feeds an update review. *Rev. Med. Vet.* 149 : 479-492.

106.Placenta CM, D'Melo JP, McDonald AM. 1999. A review of worldwide contamination of cereal grains and animal feeds with *Fusarium* mycotoxins. *Animal Feed Science and Technology* 78(1-2) : 21-37.

107.Preiusky DB, Scott PM, Trenholm HL, Lawrence GA, 1990. Minimal transmission of a zearalenone to milk of dairy cons. *J. Environ. Sci. Health Part B Pest, Food Contam. Agric. Wastes* 25 : 87.

108.Prelusky DB, Rotter BA, Rotter RG. 1994. Toxicology of mycotoxins, In : Mycotoxins in Grain : Compounds other than Aflatoxin. J.D. Muller and H.L. Trenhorm (eds.), Egan Press, S.E. Paul M.N., Ch. 9, pp. 359-403.

109.Quiraga NM, Sola SL, Varsavsky E. 1994. Selection of a simple and sensitive method for detecting zearalenone in corn. *J. AOAC Int.* 77:939-941.

110.Radostits OM, Blood DC, Gay CC 1994. Veterinary medicine, 8th ed., pp. 1604-1605, W.b. Saunders Co., Ltd., London.

111.Rama Krishna Y, Bhat RV, Ravindranath V. 1989. Production of deoxynivalenol by *Fusarium* isolates from samples of wheat associated with a human mycotoxicosis out break and from sorghum cultivars. *Appl. Environ. Microbiol.* 55 : 2619-2620.

112.Reshik S, Neare A, Pacin A, Martinez E, Apro N, Latreite S, 1996. A survey of the natural occurrence of aflatoxins and zearalenone in Argentina field maize, 1984-94. *Food Add. And Contam.* 13 : 115-120.

113.Romer, Labs Inc. Mycotoxins control consultants, 1994. Quantitative method for DON, aflatoxin, $B_1$ and zearalenone. 14: 333-339, Method # MY8402, Version 94.2.

114.Rosenberg E, Krska R, Wissiack R, Kmetov V, Josephs R, Razzaazi E, Grasserbauer M. 1998. High performance liquid chromato-graphy atmospheric-pressure chemical ionization mass-spectrometry as a new tool for the determination of the mycotoxin zearalenone in food and fodder. *J. Chromatogr.* A819 : 277-288.

115.Ryu JC, Yang JS, Song YS, Kwon OS, Park J, Chang JM. 1996. Survey of natural occurrence of trichothecene mycotoxins and zearalenone in Korea cereals harvested in 1992 using gas chromatography / Mass spectrometry. *Food Add and Contam.* 13 : 333-341.

116.Sabino M, Rodriguez-Amaya DB, 1993. Review mycotoxin research in Brazil, *Cienciae culture* 45 : 359-371.

117.Saenz de Rodrigues CA 1984. Environmental hormone contamination in Puero Rico. *The New England Journal of Medicine* 310 : 1741-1742.

118.Schroeder HW, Starey JB, 1976. Development of aflatoxin in stuart decans as affected by shell integrity. *Horticultural Science* 11 : 53-54.

119.Schumacher R, Kriska, Grasserbauer M, Edinger W, Lew H. 1998. Immunoaffinity columns versus conventional clean up a method–comparison study for the determination of zearalenone in corn. *Fresenius. J. Anal. Chem.* 360 : 241-246.

120.Schumacher R, Krska R, Weingaertner J, Grasser-bauer M. 1997. Inter laboratory comparison study for the determination of the *Fusarium*, mycotoxins deoxynivalenol in wheat and zearalenone in maize using different methods. *Fresenius. J. Anal. Chem.* 359 : 510-515.

121.Schwadorf K, Muller HM. 1992. Determination of α- and β-zearalenol and zearalenone in cereals by gas chromatography with ion-trap detection. *J. Chromatogr.* 595 : 259-267.

122.Scott GE, King SB, 1984. Site of action of factors for resistance to *Fusarium moniliforme* in maize. *Plant Dis.* 68 : 804-806.

123.Scott PM. 1977. Aflatoxins and sterigmatocystin. *Fortschr. Chem. Org. Naturst.* 31: 119-51.

124.Scott PM 1999. Natural toxins. In L Official methods of analysis of AOAC International, Cuniff, P (ed.), p. 45-46, Gaithersburg, AOAC International.

125.Scott PM Trucksess MW. 1997. Application of immuno-affinity columns to mycotoxin analysis. *J. AOAC Int.* 80 : 941-949.

126.Scott PM, Lawrence JW, Van Walbeek W. 1970. Detection of mycotoxins thinlayer chromatography. Application to Screening of Fungal Extracts, *Appl. Microbiol.* 20 : 839-842.

127.Scudamore, KA, Nawaz S, Hetmonski MJ. 1998. Mycotoxins in ingredients of animal feeding stuffs II. Determination of mycotoxins in maize and maize products. *Food. Add. Contam.*

128.Scudamore KA, Patel S. 2000. Survey for aflatoxins, ochratoxins A, zearalenone and fumanisins in maize imported into the United Kingdom. *Food Addt. and Contam.* 17 : 407-416.

129.Shoental R. 1983. Cold cow syndrome. *Vet. Rec.,* 112 : 594.

130.Silva CMG, Vargas EA. 2001. A survey of zearalenone in corn using Romer mycosep[TH] 224 column and high perfor-mance liquid chromatography. *Food. Addit. Contam*. 18 : 39-45.

131.Smith J, Wesselink C, Parr J, Sprosen JM, Fowke EA, Towers NR, Laboyrie D. 1995. Effect of zearalenone on ewes pregnancy rates. In : "Taxinology and Food Safety", pp. 41-43.

132.Smith JF, Di Menna ME, McGowan LT, DiMenna ME, 1990. Reproductive performance of coopworth ewes following oral doses of zearalenone before and after mating. *J. Reprod. Fert.*, 89 : 99-106.

133.Snijders CHA, Perkowski J. 1990. Effects of head blight caused by *Fusarium culmorum* on toxin content and weight of wheat kernels. *Phytopathology*, 80 : 566-570.

134.Soares LMV, Rodrigues – Amaya DB. 1989. Survey of aflatoxins, ochratoxin A, zearalenone and sterigma-tocystin in some Brazillian foods by using multitoxin thin layer chromatographic method. *J. AOAC* 72 : 22-26.

135.Solovey MM, Somoza C, Cano G, Pacin A, Resnik S. 1999. A survey of fumonisins, deoxynivalenol zearalenone and aflatoxins contamination in corn based food products in Argentina. *Food Addit. Contam.* 16 : 325-329.

136.Sphan JA, Dreifuss PA, Schulten HR. 1997. Field desorption mass spectrometry of mycotoxins and mycotoxin mixture and its application as screening technique for food stuffs. *J. AOAC* 60 : 73-82.

137.Sprosen JM, Towers NR. 1995. Urinary zearalenone metabolite concentrations in herds with fertility problems. In: "Toxinology & Food Safety". Toxicology and food safety, research group. Ruakura Research Centre, Hamilton, New Zealand, pp. 45-46.

138.Steek JA, Mirocha CJ, Pathre SV. Metabolism of zearalenone by *Fusarium roseum*. *J. Toxicol. Environ. Health* 3 : 35.

139.Stob J, Bardwin RS, Tuite J, Andrews FN, Gillette KG. 1962. Isolation of an ana-bolic uterotrophic compound from corn infected with *Gibberellic zea*. *Nature* 196 : 1318.

140.Sydenham EW, Shephard GS. 1996. Chromatographic and allied methods of analysis for selected mycotoxins. In: Progress in food contaminant analysis, Gilbert, J. (ed.), 65-146.

141.Syndenham EW, Marassas WFO, Thiel PG, Shephard GS, Nieuwenhusis JJ. 1991. Production of mycotoxins by selected *Fusarium grami-*

nearum and *F. crookwellense* isolates. *Food Addi. Contam.* 8 : 31-41.

142. Takaliahi-Ando N, Ohsato S, Takehito S, Hiroshi H, Yamaguchi I, Kimura M. 2004. Metabolism of zearalenone by genetically modified organisms expressing the detoxification gene from *Clonostachys rosea. Appl. and Envi. Microbiol.* 70: 3239-3245.

143. Tanaka T, Ueno Y. 1988. World wide natural occurrence of *Fusarium* mycotoxins, nivalenol, deoxynivalenol and zearalenone. In : mycotoxins and phycotoxins (eds. S. Natori, K. Hashmoto and Y. Neho), Elsevier Science Publishers, Amsterdam, P. 88.

144. Tanaka T, Hasegawa A, Matsuki Y. 1985. Rapid and sensitive determination of zearalenone in cereals by high-performance liquid chromatography with fluorescence detection. *J. Chromatogr.* 328 : 271-278.

145. Tanaka T, Hasegawa A, Yamamoto S, Lee US, Sugiura Y, Ueno Y. 1988. World wide contamination of cereals by the *Fusarium* mycotoxins nivalenol, deoxynivalenol and zearalenone. *J. Agric. Food. Chem.* 36 : 979-983.

146. Tanaka T, Hasegawa A, Matsuki Y, Ueno Y. 1985. *Food. Add. Contamin.* 2 : 259-265.

147. Tanaka T, Hasegawa A, Matsuki Y, Matsui Y, Lee US, Ueno Y. 1985. Co-contamination of the *Fusarium* mycotoxins nivalenol, deoxynivalenol and zearalenone in scabby wheat grains harvested in Hokkaido, Japan. *J. Food Hygi. Soc. Jap.* 26 : 979-983.

148. Tanaka T, Yoneda A, Inove S, Sugiura Y, Ueno Y, 2000. Simultaneous determin-ation of trichothecene mycoto-xins and zearalenone in cereals by gas chromatography–mass spectrometry. *J. Chromatogr.* A882 : 23-28.

149. Temmali M, Ueno YM, Ishili K, Fransineti C, Ethine M, 1987. Natural occurrence of trichothecene (Nivalenol, deoxynivalenol, T-2) and zearalenone on cron. *Experieatia* 34 : 1333.

150. Thiel, PG, Meyer CJ, Marasas WFO. 1982. Natural occurrence of moniliforme together with deoxynivalenol and zearalenone in Transkeian corn. *J. Agric. Food. Chem.* 30 : 309-312.

151. Tomaszewski J, Miturski R, Semczuk A, Kotarski J, Jakowicki J. 1998. Tissue zearalenone concentration in normal hyperplastic and neoplastic endometrium. *Ginekol. Pol.* 69 : 363-366.

152. Towers NR, Sprosen JM. 1992. *Fusarium* mycotoxins in pastoral forming zearalenone induced infertility in ewes. *In Recent Advances in Toxicology Research,* 272-284.

153. Towers NR, Sprosen JM, Webber W. 1995. Zearalenone metabolites in cycling and non-cycling cows. In: "Toxicology and Food Safety" Toxicology and Food Safety Research Group, Ruakura Research Centre, Hamilton, New Zealand.

154. Trenholm HL, Charmley LL, Underhill KL, Prelusky DB. 1996. Reducing toxicity of mycotoxins in feed. In Proceedings of the 4th International Feed Production Converence. Catholic University of Sacred Heart, Piacenza, Italy, pp. 69-86.

155. Trigo-Stockli DM, Deyoe CW, Salimbaga RF, Pedersen JR. 1996. Distribution of deoxynivalenol and zearalenol in milled fractions of wheat. *Cereal Chemistry* 73 : 388-391.

156. Urry WH, Wehrmeister HL, Bodge EB, Hidy PH. 1966. The structure of zearalenone. *Tetrahedron Lett.* 27 : 3109-3114.

157. Valvano Fabio, Andrea Piva, Alberto Ritieni, Glacono Gaivano, 2001. Dietary strategies to counteract the effects of mycotoxins : A Review. *Journal of Food Production* 64: 120-131.

158. Vesonder RF, Ellis JJ, Rohwedder WK. 1981. Elaboration of vomitoxin and zearalenone by *Fusarium* isolates and the biological activity of *Fusarium* produced toxins. *Appl. Environ. Microbiol.* 42 : 1132-1134.

159. Vesonder, R.F., Golinski, P., Plattner, R., Zietkiewiez, D.L. 1991. Mycotoxin formation by different geollgraphic isolates of *Fusarium crookwellense. Mycopathologia.*, 113 : 11-14.

160. Vincelli, P., Paker, G. 1999. Mycotoxins in corn produced by *Fusarium* fungi. ID-121-Electronic file created in 1999. Corn Growers Association, Kentucky, Frankfort.

161. Vissconti, A., Pascale, M. 1998. Determination of zearalenone in corn by means of immunoaffinity clean up and high performance liquid chromatography white fluorescence detection. *J. Chromato. A.*, 815 : 133-140.

162. Ware, G.M., Zhao, Y., Kuan, S.S., Carman, A.S. 1999. Preparative method for isolating α-zearalenol and zearalenone using extracting disk. *J. AOAC,* 82 : 90-94.

163. Watanabe, C.M.H., Wilson, D., Linz, J.E., Townsend, C.A. 1996. Demonstration of the catalytic roles and evidence for the physical association of type

I fatty acid synthases and a polyketide synthase in the biosynthesis of aflatoxin $B_1$. *Chem. Biol.,* 3 : 463-469.

164. Weidenborner, M. 2001. Encyclopedia of food mycotoxins. Springer Publisher, Berlin, New York and London.

165. WHO, World Health Organization, 2000. "Safety evaluation of certain food additives and contaminants, WHO Food additives series, 44.

166. Williams, B.A., Mills, K.T., Burroughs, C.D. and Bern, H.A. 1989. Reproductive alterations to female C57BL/Org mice exposed neonatally to zearalenone, an estrogenic mycotoxin. *Cancer Lett.,* 46 : 225.

167. Williams, B.C. 1985. Mycotoxins in foods and feed stuffs, in mycotoxins : A Canadian perspective (eds. P.M. Scott, H.L. Trenholm and M.D. Sutton), National Research Council Canada, Ottawa, pp. 49-53.

168. Williams, S.S. 1984. Official methods of analysic of the association of official analytical chemists, AOAC, Arlington, VA.

169. Windels, C.E. Mirochas, C.J., Abbas, H.K., Xie, W. 1989. Perithecium production in *Fusarium gramnearum* populations and lack of correlation with zearalenone production. *Mycologia,* 81 : 272-277.

170. Wolf, J.C., Mirocha, C.J. 1973. Regulation of sexual reproduction in *Zibberellazeae* (*Fusarium roseum*) by F-2 (Zearalenone). *Can. J. Microbiol.,* 19:725.

171. Wyatt, R.D. 1991. Poultry, pp. 553-578, In: Smith, J.E. and R.S. Henderson, 1991.

172. Xan, H.H. *et al.* Effect of zearalenone and tamoxifen on swine and mink reproduction. *J. Appl. Toxicology,* 15 : 223-232.

173. Zollner, P. *et al.* 2002. Concentration levels of zearalenone and its metabolites in urine, muscle tissue and liver samples of pigs fed with mycotoxin contaminated oats. *J. Agric. Food Chem.*, 50 : 2494-2501.

174. Zollner, P., Jodlbeuer, J., Linder, W. 1999. Determination of zearalenone in grains by high-performance liquid chromatography-tandem mass spectrometry after solid phase extraction with RP-18 Leoni, LAB, Valente Soares, L.M. and Oliveira.

*Rev. Plant Pathol. Vol. 4, 2008*
*Indian Society of Mycology and Plant Pathology*
*Scientific Publishers (India), Jodhpur pp. 193-216*

# TAXONOMY OF GLOMEROMYCOTA : A REVIEW

***B.F. Rodrigues***

Department of Botany, Goa University Goa 403 206, India

KEY WORDS : Glomeromycota, Arbuscular mycorrhizae, molecular markers

## ABSTRACT

Arbuscular mycorrhizal (AM) fungi, a major component of soil microbial community, forms symbiotic association with the roots of more than 90% of terrestrial plants. AM colonization has also been reported in hepatics and hornworts. Fossil records indicate that the AM fungi may have played an important role in the success of early terrestrial plants. AM fungi play a very important role in the improvement of plant growth. They are reported to be vital for the uptake and accumulation of ions from the soil and their translocation to the hosts because of their high metabolic rate and strategically diffuse distribution in the upper layers.

Since the time of use of molecular methods to elucidate the phylogenetic relationships among these fungi, their classification has been in a rapid transition. Molecular field studies have also revealed a large number of putative new species. Presently, the AM fungi are placed in the phylum Glomeromycota, which currently comprises of approximately 150 described species distributed among 10 genera. With the exception of genus *Geosiphon*, remaining all are exclusively mycorrhizal.

## INTRODUCTION

Arbuscular mycorrhizae (AM) are the most common types of mycorrhizae present in nearly 90% of all plant families. Only in the last few decades botanists in general and mycologists in particular have realized that most terrestrial plants live in symbiosis with fungi (54). Though AM may have been described as early as 1842 (50), they only remotely resemble the AM.

Frank (23) was fairly certain that these symbiotic plant-fungus associations were required for the nutrition of both partners. Studies carried out by different workers (73, 21, 34, 61, 26, 62, 35, 39) provided detailed description of AM fungi other than those found in Ericales and orchids (now known as arbuscular mycorrhizal fungi). As early as 1889, Schlicht had already observed the basic anatomical relationships between host and fungal tissues. Janse (34) called the intramatrical spores "vesicules" and "arbuscules" by Gallaud (26), were located in the inner cortex. He made very accurate observations of the arbuscle and concluded that it is entirely surrounded by a host membrane. He also noted that partial digestion of the arbuscule resulted in a structure called the "sporangiole" by Janse (34). These observations were later confirmed (20). Gallaud (26) further distinguished between *Arum* and *Paris* types of arbuscules. Many herbaceous plants exhibit the *Arum* type of colonization, which involves extensive intercellular growth of the fungus as it penetrates the root cortex followed later in the colonization by formation of arbuscules. In *Paris* type of colonization growth into the root is slow, being primarily intra-cellular, and the fungus coils inside each cell with rare or minimally structured arbuscules (26). Jones (35) described the appresorium, which are hyphal swellings between two adjacent root epidermal cells. These are sites where hyphae first penetrate root cells by exerting pressure and/or enzymatic activity.

More recently, mycorrhizas have been defined as associations between fungal hyphae and organs of higher plants concerned with absorption of substances from the soil (31). Most mycorrhizal associations occur in roots, which evolved to house fungi (15), but they also occur in the subterranean stems of certain plants and the thallus of bryophytes (68, 69, 80, 65). A new, broader definition of mycorrhizas that embraces the full diversity of mycorrhizas (while excluding all other plant-fungus associations) has been put forth by Brundrett (16). He defines mycorrhizas as a symbiotic association essential for one or both partners, between a fungus (specialized for life in soils and plants) and a root (or other substrate-contacting organ) of a living plant that is primarily responsible for nutrient transfer. Mycorrhizas occur in a specialized plant organ where intimate contact results from synchronized plant-fungus development.

It is now evident that the AM fungi originated more than 460 million years ago (66). These fungi may have played a crucial role in facilitating the colonization of land by plants. These fossilized fungal hyphae and spores strongly resembled modern AM fungi. This evidence indicates that AM fungi represent an ancient phylogenetic clade within the fungi. Until recently there was confusion about their monophyly (48). However, the later studies have clearly indicated that they belong to a distinct monophyletic group quite separate from other fungi (74).

The AM fungal taxonomy has advanced greatly during the last three decades. This paper gives an overview of the taxonomic advances in Arbuscular mycorrhizal (AM) fungi. It also gives an account of the various characters, which aid in the identification of these fungi.

## CHARACTERS USED FOR IDENTIFICATION OF ARBUSCULAR MYCORRHIZAL FUNGI

Various morphological characters play an important role in establishing the taxonomic identity and relationships of AM species and thus aids in the construction of a workable system of classification. Hence, undoubtedly, certain characters will be of greater importance than others. Various morphological characters used for the identification of AM fungi are listed (Table 1) and described below:

### *Sporocarp Morphology*

The sporocarpic species produce spores in a loose arrangement or in a highly ordered arrangement around a hyphal plexus (29). The sporocarps may be formed in soil, root, empty seed coats, insect or rhizomes. *Glomus* species form single spores or spores in sporocarps where the spores are arranged randomly in the matrix hyphae. Peridium may be present around the sporocarps in the form of loosely or compact interwoven hyphae, a patchy covering over the sporocarps or as hyphal network covering single or small clusters of spores. The presence or absence of peridium accounts for much of the variation observed in size of sporocarps. Sporocarps are not known in *Entrophospora, Gigaspora* and *Scutellospora* (69, 93). Sporocarps are also typically absent in *Acaulospora* with the exception of *Acaulospora myriocarpa* wherein the spores are in a cluster (71) and *A. sporocarpa* that has an aggregation of spores in a network of hyphae (9). External sporocarp colour range from white to brown, while the internal sporocarp colour range from white to black and brown.

**Table 1**. Morphological characters of spores used for identification of AM fungi.

| | Morphological characters |
|---|---|
| Sporocarp morphology | Size, Shape, Peridium. |
| Spore morphology | Colour, Shape, Size, Content. |
| Subtending hyphae | Shape, Width, Pore Occlusion. |
| Auxiliary Cells | Ornamentation. |
| Mycorrhizal Anatomy | Hyphal characters, Intraradical spores. |
| Spore wall structure | Colour, Dimension, Number, Type, Ornamentation, Reaction. |
| Spore Germination | Direct, Indirect. |

*Spore Morphology*

Spores in the soil may be produced terminally, laterally on subtending hyphae or on a single suspensor, like cell. Characters such as spore colour, shape and size may vary considerably depending on the developmental stage and environmental conditions. Spore colour varies from hyaline to white to yellow, red, brown and black with all intermediate shades. The difference in colour may be due to pigmentation in spore wall or in the spore content (46).

Morton (46) suggested that variation in spore shape might be due to the result of environmental stress. Shape of spores is mainly governed by the genotype of the fungus and the substrate in which the spores are formed. Intraradical spores are mainly globose, subglobose to ellipsoidal, while the extraradical spores may be globose, sub-globose, ellipsoidal, oblong ovate to highly irregular shaped.

Of all the known species of AM fungi, *Glomus tenue* is the smallest with an average diameter of 10-12 μm. On other hand, *Gigaspora gigantea* is the largest spore with dimensions ranging from 183-500 x 291-812μm. Spore size varies considerably within the same species and hence both immature and mature spores are taken into account while describing the species.

*Subtending hyphae*

At generic level of classification, the shape of the subtending hyphae or the sporophore assumes great importance. The subtending hyphae may be simple to recurved or sometimes swollen in *Glomus* species. The sporophore in *Gigaspora* and *Scutellospora* is bulbous and bears one or more peg like hypha. During extractions from the soil, sporophore may get detached from the spore, but the bulbous structure continues with the wall of the spore (91) and usually remains attached to the spore. In *Entrophospora*, the sporophore is swollen and straight but at times may be totally absent due to a detachment close to the spore. Sometimes, the spores are sessile or may bear small pedicel as in *Acaulospora*. The width of the hyphae varies considerably within different genera and species of AM fungi.

The mechanism of pore occlusion at the point of attachment of the subtending hypha to the spore has some taxonomic significance. Walker (91) suggested three distinct lines with regard to the occlusion of the spore content in *Glomus viz.,* spores possessing a complete endospore formed by more or less flexible inner wall group, spores sealed by the ingrowths and thickening of the wall layer of the subtending hypha, and occlusion by the septum usually somewhat distal to the spore base.

*Auxiliary cells*

The size and the shape of the auxiliary cells have been found to be of little importance in differentiating species of *Gigaspora* or *Scutellospora*. In *Gigaspora*, the auxiliary cells are echinulate with spines that are forked dichotomously (8), whereas, in

*Scutellospora*, the projections on the surface of the auxiliary cells are highly variable in shape and size (47).

### *Mycorrhizal anatomy*

Generally, fungal anatomy in roots is not used in taxonomic descriptions to separate taxa below the generic level. Colonization of the root with AM fungi initiates a series of developmental processes culminating in a morpholo-gically and functionally unique symbiosis. Arbuscular mycorrhizal fungal hyphae penetrate host roots through the epidermis. The hyphae grow inter- and intra-cellularly in the cortical cells (56). The hyphae from intercellular lateral branches penetrate the cell walls and form a branching, tree-like structure termed as 'arbuscule' inside the cell lumen, and the plant invaginates the plasma membrane, matching the branching pattern of the fungus. The resulting dendritic structure has a large membrane surface area, and these membrane interfaces are where nutrient transfer between plant and fungus occurs. It is possible to differentiate among certain AM fungi, using visual differences in morphology of fungal hyphae and vesicles within roots (4). Harley and Smith (31) suggested that vesicles might perform the function of storage because lipids and glycolipids are the most abundant substances in them. Biermann and Linderman (12) thought that intraradical vesicles in some species of AM fungi act as propagules and contribute significantly to the colonization of other roots.

It has been suggested that certain hyphal characters such as long infection units with 'H' connections between parallel strands of hyphae in *Glomus* (3), pale staining of intraradical hyphae by trypan blue in *Acaulospora* (8), constrictions near branch points in hyphae of *Acaulospora* and *Entrophospora*, and coiled, irregularly swollen hyphae with lateral projections or knots in *Gigaspora* or *Scutellospora* may be utilized as diagnostic features to identify genera in mycorrhizal roots (50). Intraradical spores in Glomaceae usually are globose, subglobose to elliptical, whereas those in *Acaulospora* are pleomorphic, knobby and stain lightly in trypan blue. Morton (46) hypothesized that either the host or environmental factors may cause variation in morphological structures. Abbott and Robson (2) concluded that the anatomy of AM fungi formed by a particular endophyte species grown under a range of conditions may not be as variable as has been generally assumed.

Abbott (1) developed a key for 10 AM fungal species using 20 characteristics such as hyphal diameter, mode of branching, vesicles, arbuscules, staining reactions, *etc.* and concluded that these characteristics are stable in different hosts and soil environments.

*Spore wall structure*

Spore wall characteristics have been universally accepted as more stable and reliable criteria than other spore features (41). A spore wall has been defined as the first individual structure formed, originating from the wall of sporogenous hypha and differentiating into phenotypically distinctive layers (53). Spore wall characteristics became the important morphological characteristics. Walker (88) developed the concept of "wall groups" He defined a wall group as "an aggregation of walls that are either adherent, or that remain close together when a spore is crushed" and suggested the use of standard terminology and wall murographs. This concept proved to be a subject of much variation and interpretation because the degree of separation was often influenced greatly by condition of the spores (fresh, fixed, parasitized, aged), amount of pressure applied to a spore when it is crushed on a slide, and the type of mounting medium used. In all, eight-wall layer types *viz.,* evanescent, laminated, membranous, unit (80), expanding (11), coriaceous (89), amorphous (45) and germinal (82) have been described so far. They are distinguished mainly on the basis of their morphological features and their reaction to certain chemicals such as lactophenol and Melzer's reagent.

The number, width and position of wall layers differ among species and they have been increasingly relied upon for identification purpose. Differentiation of sub-cellular morphological characters in spores of *Gigaspora* (8) and *Scutellospora* species (47) are used for identification. Ornamentation on the spore wall layer appears to be an important taxonomic criterion in identification of species, especially when other morphological characters are overlapping.

*Spore germination*

Ultrastructural studies of spore germination processes may play a role in the identification of AM fungal species. Spores of glomalean fungi have all the necessary metabolic constituents and genetic information to germinate and produce new hyphae (77), although they cannot continue to grow without a host. However, spore germination in AM fungi has been studied in only a few species. Two methods of spore germination is known to exist in the Glomales *viz.,* 1. Direct germination takes place when the inner wall layers protrude though a weakened area of the outer wall layer as a germ tube initially, later elongating into a typical hypha. This type of germination has been observed in *Glomus* and *Gigaspora*. 2. Indirect germination takes place by the development of "germination shield" prior to emergence of germ tube. This type of germination is known to occur in *Acaulospora, Entrophospora* and *Scutellospora*. In *Acaulospora leavis* the germination shields are difficult to see and many a times are not observed because of the kinds of mountants used to prepare permanent slides (Spain, 1992). Even though in *Acaulospora* and *Scutellospora,* the germination shields are formed in the same way, the wall types associated with germination differ. In Acaulosporaceae, the germination shield is formed on an innermost flexible wall that has beaded layer, while in *Scutellospora*, it is formed on the innermost flexible wall with a coriaceous layer (81, 24).

## TAXONOMY OF ARBUSCULAR MYCORRHIZAL (AM) FUNGI

The term mycorrhiza, is the combination of two words, one Greek "mikes" (fungus) and the other Latin "rhiza" (roots). Traditionally, Albert Bernhard Frank (22), a German botanist introduced the word "Mycorhiza". Later Frank (23) distinguished between ectomycorrhiza and endomycorrhiza on the basis of his studies carried out on Cupuliferae for the former on Ericales and orchids for the later. It was Kelley (36, 37) who incorporated the second 'r' to the word "mycorrhiza".

The arbuscular mycorrhizae (AM) are also known as vesicular arbuscular mycorrhizae (VAM) or endomycorrhizae.

### *i. History of the group*

The history of members of Glomales is long and convoluted. German mycologist, Link (38) established the genus *Endogone*. Tulasne and Tulasne (87) were the first to describe the genus *Glomus* then comprising of two species *viz., Glomus microcarpum* and *G. macrocarpum* known only from spore clusters found in the soil. They considered *Glomus* to be closely related to *Endogone*.

Dangeard (21) for the first time described arbuscular mycorrhiza from popular roots. He regarded this as a disease and named the fungus as *Rhizophagus populinus* and placed it in Chytridales.

Fries (25) established the family Endogonaceae and placed it in the Tuberales, which later transferred to the Mucorales by Bucholtz (17) due to the affinities of *Endogone* with members of the Mortierellaceae. The genus *Sclerocystis* was first described by Berkeley and Broome (13). Thaxter (84) revised the family Endogonaceae and placed all members of *Glomus* in the genus *Endogone*, while maintaining the genus *Sclerocystis*. The family Endogonaceae was placed in order, Endogonales by Moreau (43), which was later, validated to class Zygomycetes (7). The characteristic features of AM fungi that prompted mycorrhizo-logists to place them in the class Zygomycetes include:

a. The presence of chitin in the cell wall (14).
b. Presence of nonseptate and coenocytic mycelium.
c. Formation of non-motile spores, the chlamydospores.
d. Formation of putative zygospores in *Gigaspora decipiens* (85).
e. Features of nuclei in spores similar to the spores of other Zygomycetous fungi (40).

Mosse and Bowen (57) did not use the production of sporocarps as a major characteristic, instead considered the developmental stages of spores as major characteristics. Nicolson and Gerdeman (59) divided the fungi into two groups of *Endogone viz.*, one forming extrametrical azygospores/zygospores arising from the tip of a swollen hyphal suspensor but producing number of intramatrical vesicles, and the other forming extrametrical chlamydospores and intramatrical vesicles.

Gerdemann and Trappe (29) suggested a division of *Endogone*, the largest and most heterogenous genus in the family Endogonaceae, into seven genera (Fig. 1). Of these, three genera do not form mycorrhizal associations. These include the genus *Endogone* that is known to form putative ecto-mycorrhizal associations (18). It produces zygospores and have saprobic mode of nutrition and is retained in the order Endogonales with the sole family Endogonaceae. Mehrotra and Baijal (42) suggested that Endogonales need not be retained as a separate order and the only genus *Endogone* be transferred to the order Mucorales as the formation of zygospores in *Endogone* is somewhat similar to that formed by the members of the order Mucorales. The phylogenetic analysis of the nuclear small subunit ribosomal RNA strongly suggests that *Endogone* (Endogonales) and the Glomeromycota do not form a clade (27). The genus *Modicella* produces thin-walled sporangia and, the genus *Glaziella*. Trappe and Schenck (86) transferred the genus *Modicella* to the family Mortierellaceae in the order Mucorales of the Zygomycetes. Similarly, the teleomorphic stage of *Glaziella* was found to be an Ascomycete and hence the entire genus *Glaziella* was transferred to the Ascomycetes in a new family, Glaziellaceae and a new order, Glaziellales (30). The remaining four genera form mycorrhizal associations. These include *Glomus, Gigaspora, Acaulospora* and *Entrophospora*. The genus *Glomus* produces sporocarpic and non-sporocarpic chlamydospores. They also described the genera *Acaulospora* and *Gigaspora*. The genus *Gigaspora*, is nonsporocarpic, and produces azygospores while *Acaulospora* is characterized by non-sporocarpic producing spores that were borne singly and laterally on a hypha, which terminates in a large thin-walled vesicles and formed mycorrhizal associations. Ames and Schneider (6) described the genus *Entrophospora*. They transferred *Glomus infrequens* into the new genus *Entrophospora* and called it *E. infrequens*, as the fungal spores were unlike any other described species in the Endogonaceae.

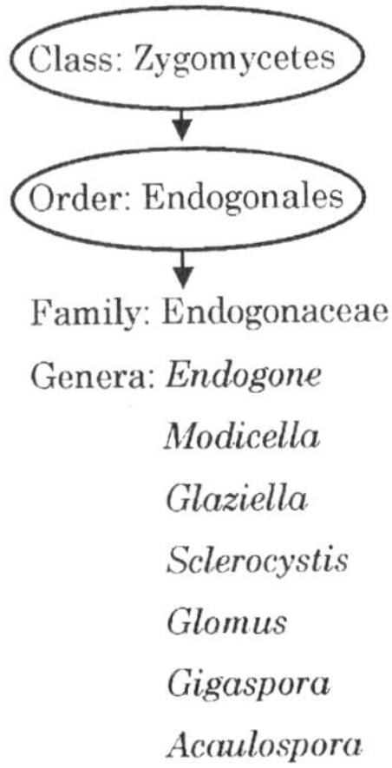


**Fig. 1.** Classification of endomycorrhizal fungi (29).

The genus *Gigaspora* was split into two genera *viz.*, *Gigaspora* and *Scutellospora* (93). The creation of new genus *Scutellospora* was based on details of spore germination, spore wall structures and morphology of auxiliary cells. Morton and Benny (51) erected the order Glomales that is characterized by the unique ability of its members to form arbuscular mycorrhizae in mutalistic symbiosis with living plants. The order Glomales, includes two suborders *viz.*, the Glomineae, which is characterized by the presence of arbuscules and vesicles in the roots and formation of chlamydospores (thick wall, asexual spores) borne from subtending hyphae and, Gigasporineae, which is characterized by the absence of vesicles in the roots and formation of auxiliary cells and azygospores (spores resembling a zygospore but developing asexually from a subtending hypha resulting in a distinct bulbous attachment) in the soil. The suborder Glomineae was separated into two families *viz.*, the Glomaceae (63) and Acaulosporaceae (51), while the suborder Gigasporineae contains the sole family Gigasporaceae (Fig. 2). All these families were characterized by the mode of spore formation and were initially supported by molecular data (78).

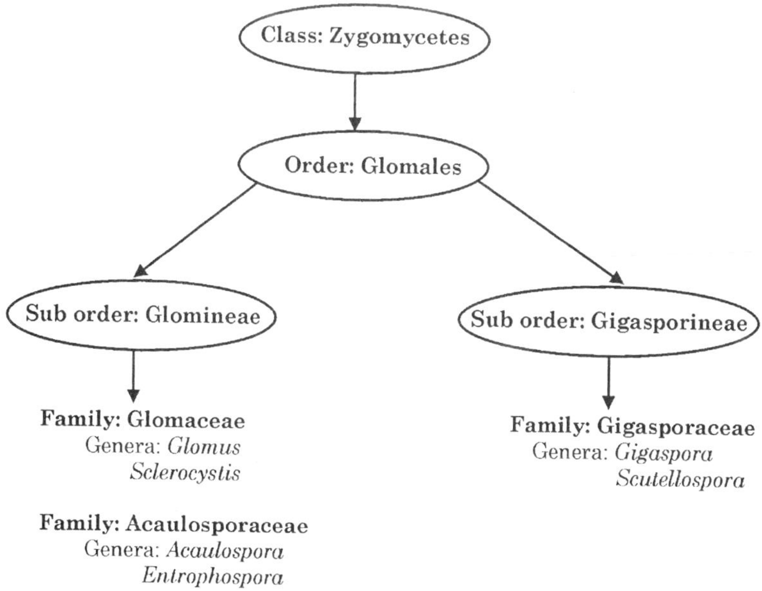


**Fig. 2** : Classification of AM fungi (51).

The family Glomaceae consists of two genera *viz., Glomus* and *Sclerocystis*. In *Glomus*, spores are globose to ellipsoidal, thick walled hyaline to black, usually attached to single subtending hypha. Spores germinate by germ tube through subtending hypha. While in *Sclerocystis*, spores are similar to *Glomus* except they are clavate in shape and are arranged in a sprorcarp.

The members of family Acaulosporaceae develop spores from or within neck of sporiferous saccule while the members of the family Gigasporaceae produce large azygospores with bulbous base. Germination of the spores is either through the production of germ tube through the cell wall or from the germination shield. They do not produce vesicles but produce auxiliary cells in the soil.

The family Acaulosporaceae consisted of two genera *viz., Acaulospora* and *Entrophospora.* In both the genera spores are large (up to 600 $\mu$m) with a typical bulbous base with characteristic lateral projections. The genus *Acaulospora* is characterized by spores that are formed laterally on the neck of a small thin walled saccule, whereas the genus *Entrophospora* differs from *Acaulospora* as in the former the spores are formed inside the parent hypha just below the sporiferous saccule. The family Gigasporaceae consisted of two genera *viz., Gigaspora* and *Scutellospora.* Species of *Gigaspora* and *Scutellospora* form large spores in the soil. Spores are produced terminally on a single, bulbous-shaped suntending hypha, termed as bulbous suspensor (46). The subtending hypha has one or more small lateral projections pointing towards and sometimes in contact with the outer surface of the spore wall. The function of these small lateral projections is not known. Gerdemann and Trappe (28) used the term azygospore to describe spores in *Gigaspora* because of the similarity of the single subtending hypha to a gametangium in *Endogone*. There is no evidence that spore development in *Gigaspora* or *Scutellospora* is parthenogenic or that development evolved from zygospores via the reduction of one of a pair of gametangia (64). In *Gigaspora*, germ tube develops directly from the spores, while in *Scutellospora* the germ tube emerges from germination shields formed in between outer and inner cell walls (51).

*Acaulospora* and *Entrophospora* tend to form spores associated with a small hyphal chamber. Spores in *Acaulospora* are formed laterally on the stalk in a large terminal and thin walled hyphal chamber (9). The spores are hyaline, yellow to reddish-brown, globose to ellipsoid with spore wall ornamentation. However, in *Entrophospora*, spores are produced completely within the neck of the hyphal chamber (6). The small chamber in *Acaulospora* and *Entrophospora* has been variedly named such as a mother spore (55), a vesicle (29), a hyphal terminus (72), a sporiferous saccule (94), a sporogenous saccule (9) or a swollen sac (46). Microscopic examination indicates that spores in *Acaulospora* species may not develop from the side of the chamber neck. Spores develop within a lateral swelling of the chamber neck since the outer wall of the spore may be continuous with the wall of the chamber itself (92). Berch (10) first used the term "sporogenous vesicle" to describe the hyphal swelling associated with spores of *Entrophospora*. Walker (90) suggested that the so-called azygospores in *Acaulospora* are sporangiospores and the saccule is a sporangium.

*ii. Recent Advances in Taxonomy*

The genus, *Glomites* was erected by Taylor *et al.* (83) to describe fossil fungi that closely resemble modern day *Glomus* species. The genus *Gigaspora* was re-described by Bentivenga and Morton (8) incorporating developmental patterns of morphological characters.

The separation between *Glomus* and *Sclerocystis* became controversial in the early 1990's. Almeida and Schenck (5) placed all *Sclerocystis* species in *Glomus* with the exception of *S. coremioides*. They were of the opinion that an unbroken continuum of morphological characters existed between sporocarpic *Glomus* species and all the *Sclerocystis* species except one (*S. coremioides*). They considered *S. coremioides* unique and, therefore, separated from the *Glomus* clade based on following four morphological traits:

a. Spore formation on separate subtending hyphae rather than from branching sporophores.

b. A well-defined septum at the same position near the spore base.

c. Arrangement of spores in hemispherical layer.

d. New sporocarps are formed from older sporocarps often fuse into columns.

Wu (97) resisted this change on the basis of comparative studies of spore ontogeny and sporocarps morphology of the *Sclerocystis* species and showed that the above mentioned traits were shared to a varying degrees by other *Sclerocystis* species. He further hypothesized a model of a smooth evolutionary transition between relatively unorganized, *Glomus*-like sprorcarps of *S. rubiformis* and intermediate forms like *S. clavispora, S. liquidambaris* and *S. sinuosa* to *S. coremioides*. He concluded that *S. coremioides* was not unique and reverted to Gerdemann and Trappe's (29) classification scheme. However, Wu's rationale for genus-level separation, based on sporogenesis and sporocarp formation, was not convincing. Later Redecker *et al.* (67) carried out phylogenetic analysis of the 18S ribosomal subunit of *Glomus sinuosum* (= *S. sinuosa*) and *S. coremioides* which revealed both species are close relatives and fall within a monophyletic clade comprising the well characterized species, *Glomus mosseae, G. intraradices* and *G. versiculiferum*, to the exclusion of several other *Glomus* species. Their study indicated that formation of complex sporocarps is an advanced character of some *Glomus* species, but the sporocarpic trait is not sufficiently unique to group these species into a separate genus *Sclerocystis*.

Later it was clear that the mode of spore formation is not a useful diagnostic feature for some genera. A number of AM fungal species in *Glomus* (Glomaceae) and *Acaulospora* (Acaulosporceae) have not fit easily into family or genus definitions (51). Spores of *Glomus* and *Acaulospora* types were reported to be produced by several distinct, deeply divergent lineages (67). Subsequently, Morton and Redecker (49) erected two new families *viz.*, Archaeosporaceae and Paraglomaceae. This work was based on combination of 18S rDNA sequences, fatty acid profiles, immunological reactions against specific antibodies, and mycorrhizal morphology. On the basis of type specimens, culture manipulations (52) and rDNA sequence data (67, 70), the two AM species *viz., Acaulospora gerdemannii* and *Glomus leptotichum* were found to be synanmorphs of the same organism. Similarly, *Glomus gerdemannii* also proved to be dimorphic in monospecific pot cultures and both spore morphotypes bore striking resemblance to synanamorph of *Acaulospora gerdemannii*/ *Glomus leptotichum* (49). Root colonization by *Glomus occultum* is so faint that roots often appear to be nonmycorrhizal, despite of abundant sporulation (44). Similarly, root colonization by *Glomus brasilianum* and *Acaulospora trappei* also show faint staining, but both show abundant sporulation. *Glomus brasilianum* is almost indistinguishable from *Glomus occultum* when observed under a dissecting microscope, while *Acaulospora trappei* forms spores with a subcellular structure more like that of *Acaulospora gerdemannii* than other species of Acaulosporaceae. Nucleotide sequence data sets from ITS, 5.8S and 18S rDNA regions established that these species are more closely related to the dimorphic species than species in either Acaulosporaceae or Glomaceae and are grouped in two distinct clades (67). One clade of species forming acaulosporoid spores *viz., Acaulospora gerdemannii* (=*Glomus leptoticum*)/*Glomus gerdemannii* along with *Acaulospora trappei* defines the genus *Archaeospora* of the new family Archaeosporaceae, while the other clade of species forming only glomoid spores *viz., Glomus occultum* and *Glomus brasilianum* is placed in the genus, *Paraglomus*, of the new family Paraglomaceae (49).

Schüßler *et al.,* (74) using molecular data elevated the group of arbuscular mycorrhizal fungi to the level of phylum (Glomeromycota), which was shown to have monophyletic origin. The endocytobiotic fungus *Geosiphon pyriformis* analysed phylogenetically by their small subunit (SSU) rRNA gene sequences was added to the Glomeromycota. They reported that Glomeromycota probably diverged from the same common ancestor as the Ascomycota and Basidiomycota. They also erected new orders *viz.,* Archaeosporales (Archaesporeaceae, Geosiphonaceae), Paraglomerales (Paraglomaceae) and Diversisporales (Acaulosporaceae, Diversisporaceae, Gigasporaceae) and Glomerales (Glomeraceae) with their respective families given in parenthesis. The most recent classification of AM fungi is given in Table 2.

*Geosiphon pyriforme* inhabiting the surface of humid soils of Germany represents the only known example of endocytobiosis between a fungus and cyanobacteria (*Nostoc* sp). Except for the genus *Geosiphon*, all the species included in the phylum Glomeromycota are exclusively mycorrhizal (49, 66). The cyanobacterium provides carbon and nitrogen to the fungus, as in a lichen (27). Interestingly, current molecular phylogenetic studies place *Geosiphon* inside the Glomalean clade, not in the outlying position that its wildly divergent anatomy and physiology would seem to indicate (66).

**Table 2:** Classification of AM fungi (76)

- Phylum: Glomeromycota
- Class: Glomeromycetes
  - Order: Glomerales
    - Family: Glomeraceae
      - Genus: *Glomus* (group A and B)
  - Order: Diversisporales
    - Family: Gigasporaceae
      - Genus: *Gigaspora*
        *Scutellospora*
    - Family: Acaulosporaceae
      - Genus: *Acaulospora*
        *Entrophospora*
    - Family: Pacisporaceae
      - Genus: *Pacispora*
    - Family: Diversisporaceae
      - Genus: *Diversispora*
        *Glomus* (group C)
  - Order: Paraglomales
    - Family: Paraglomaceae
      - Genus: *Paraglomus*
  - Order: Archaeosporales
    - Family: Geosiphonaceae
      - Genus: *Geosiphon*
    - Family: Archaeosporaceae
      - Genus: *Archaeospora*

Schüßler *et al.* (75) corrected the formerly orthographically incorrect term Glomales to Glomerales, which represents one of the four orders within the phylum Glomeromycota. *Glomus* groups A and B (76) are exemplified by the well-known species of *Glomus mosseae* and *Glomus claroideum*, respectively. These two groups are genetically relatively distant but still form a monophyletic group in rDNA phylogenetic trees.

rDNA phylogenies have shown that the genus *Glomus* is several times polyphyletic (67, 76). Oehl and Siverding (60), erected a new genus *Pacispora*. According to them, it comprised of *Pacispora scintillans* (the type species) and *P. dominikii* (as a separate species) and *P. chimonobambusae*, all formerly placed in the genus *Glomus* of the Glomeraceae along with four newly identified species *viz., P. franciscana*, *P. robigina*, *P. coralloidea* and *P. boliviana*. However, the morphological and molecular data presented by Walker *et al.* (94) show that the two names *viz., Pacispora scintillans* and *P. dominikii* are synonyms. The genus *Gerdemannia* published by Walker *et al.* (94) is a synonym of *Pacispora*, and is an illegitimate name based on the publication date.

The spores of the genus *Pacispora* are formed terminally on hyphae, a feature they share only with *Glomus* and *Paraglomus*. An inner, usually three-layered wall is present inside the spores, from which the spores germinate directly through the outer spore wall, which usually is also three-layered. This germination characteristic is shared with *Scutellospora*, *Acaulospora* and *Entrophospora* but not with *Glomus* and *Paraglomus*. The staining characteristics of the root internal fungal structures and the features of the subtending hyphae of the spores are almost similar to species of the genus *Glomus*. All the *Pacispora* species can morphologically be differentiated by the spore surface structure, the characteristics of the spore wall ornamentation, and by the spore colour and spore size (60).

Similarly, another genus erected as a result of split off from *Glomus* is named *Diversispora*, which at present comprises of only one species *viz., Diversispora spurcum* and its erection is mainly based on ribosomal small subunit sequence signatures (95).

Earlier, the 'sporiferous saccule' was thought to be characteristic feature of the family Acaulosporaceae (*Acaulospora* and *Entrophospora*), but now it is known to occur in at least one additional lineage *viz., Archaeospora*.

The characteristic feature of the family Gigasporaceae (*Gigaspora* and *Scutellospora*) is the formation of spores on a "bulbous suspensor" which is well supported by molecular data. However, based on most rDNA sequence analysis, the families Gigasporaceae and Acaulosporaceae form a clade, which is conflicting with the earlier morphology that placed *Glomus* and Acaulosporaceae together in the sub order Glomineae (51). This amounts to not only just complete divergence in the process of spore formation and traits of inner wall development, but also in mycorrhizal structures, timing of their formation, infectivity of those structures, fatty acid profiles, cell wall chemistry, and even timing and abundance of transcript synthesis during spore germination. To resolve this conflict, besides the relationship from rDNA data, would also require congruence with other genes. Although additional genes have begun to be sequenced from some taxa (33, 19), phylogenetic hypothesis based on multilocus DNA sequence data is yet to be incorporated in the classification.

## *Literature Cited*


1. Abbott LK 1982. Comparative anatomy of vesicular-arbuscular mycorrhizas formed on subterranean clover. *Aust. J. Bot.* 30: 485-489.
2. Abbott LK, Robson AD. 1978. Growth of subterranean clover in relation to formation of endomycorrhizae by introduced and indigenous fungi in a field soil. *New Phytol.* 81: 575-585.
3. Abbott LK, Robson AD. 1979. A quantitative study of the spores and mycorrhizas formed by a species of *Glomus* with reference to its taxonomy. *Aust. J. Bot.* 27: 363-375.
4. Abbott LK, Gazey C. 1994. An ecological view of the formation of VA mycorrhizas. *Plant and Soil* 159(1): 69-78.
5. Almeida RT, Schenck NC. 1990. A revision of the genus *Sclerocystis* (Glomaceae, Glomales). *Mycologia* 82(6):703-714.
6. Ames RN, Schneider RW. 1979. *Entrophospora*, a new genus in the Endogonaceae. *Mycotaxon* 8: 347- 352.
7. Benjamin RK. 1979. Zygomycetes and their spores. In: *The whole fungus. Vol. II.* (B.B. Kendrick, ed.) National Museums of Canada, Ottawa, Canada. *pp.* 573-622.
8. Bentivenga SP, Morton JB. 1995. A monograph of the genus *Gigaspora*, incorporating developmental patterns of morphological characters. *Mycologia* 87: 719-731.
9. Berch SM. 1985. *Acaulospora sporocarpa*, a new, sporocarpic

species, and emendation of the genus *Acaulospora* (Endogonaceae, Zygomycotina). *Mycotaxon* 23: 409-418.

10. Berch SM. 1986. Endogonaceae: Taxonomy, specificity, fossil record, phylogeny. *Front. Appl. Microbiol.* 2: 161-188.
11. Berch SM, Koske RE. 1986. *Glomus pansihalos,* a new species in the Endogon-aceae, Zygomycetes. *Mycologia* 78: 832-836.
12. Biermann B, Linderman RG. 1983. Use of vesicular arbuscular mycorrhizal roots, intraradical vesicles and extra-radical vesicles as inoculum. *New Phytol.* 95: 97-105.
13. Berkeley MJ, Broome 1873. Fungi of Ceylon. *J. Linn. Soc.* 14: 137.
14. Bonfante-Fasolo P, Faccio A, Perotto S, Schubert A. 1990. Correlation between chitin distribution and cell wall morphology in the mycorrhizal fungus *Glomus versiforme. Mycol. Res.* 94: 157-165.
15. Brundrett MC. 2002. Coevolution of roots and mycorrhizas of land plants. *New Phytol.* 154: 275-305.
16. Brundrett MC. 2004. Diversity and classification of mycorrhizal associations. *Biol. Rev.* 79: 473-495.
17. Bucholtz F. 1912. Beitrage zur Kenninis der Gattung *Endogone* Link. *Beih. Zum Botan. Centr. Abr.* 2.29: 147-225.
18. Chu-Chou M, Grace LG. 1979. *Endogone flammicorona* as a mycorrhiza symbiont of Douglas fir in New Zealand. *N. Z. J. For. Sci.* 9: 344-347.
19. Corradi N, Kuhn G, Sanders IR, 2004. Monophyly of ß-tubulins and $H^{+}$- ATPase gene variants in *Glomus intraradices*: Consequences for molecular evolutionary studies of AM fungal genes. *Fungal Genet. Biol.* 41: 262-273.
20. Cox GC, Sanders FET. 1974. Ultrastructure of the host - fungus interface in a vesicular-arbuscular mycorrhiza. *New Phytol.* 73: 901-912.
21. Dangeard PA. 1896. Une maladie du peuplicr dans l'ouest de la France. *Botaniste* 58: 38-43.
22. Frank AB. 1885. Ueber die auf Wurzelsymbiose beruhende Ernährung gewisser Bäume durch unterirdische pilze. *Ber. Dtsch. Bot. Ges.* 3: 128-145.
23. Frank AB. 1887. Ueber neue Mycorrhiza-formen. *Ber. Dtsch. Bot. Ges.* 5:395-409.
24. Frank M, Morton JB. 1994. Ontogenetic comparisons of arbuscular mycorrhizal fungi *Scutellospora heterogama* and *Scutellospora pellucida*: Revision of taxonomic concepts, species description and phylogenetic hypothesis. *Can. J. Bot.* 72(1): 122-134.
25. Fries EM. 1849. *Summa Vegetabilium Scandinaveae* 2:261-572.
26. Gallaud I. 1905. Etudes sur les mycorrhizes endophytes. *Revue General de Botanique* 17: 5-48.

27. Gehrig H, Schüßler A, Kluge M. 1996. *Geosiphon pyriforme*, a fungus forming endocytobiosis with *Nostoc* (Cyanobacteria), is an ancestral member of the Glomales: Evidence by SSU rRNA analysis. *J. Mol. Evol.* 43: 71-81.
28. Gerdemann JW, Trappe JM. 1968. Vesicular-arbuscular mycorrhizae and plant growth. *Ann. Rev. Phytopathol.* 6:397-418.
29. Gerdemann JW, Trappe JM. 1974. Endogonaceae in the Pacific Northwest. *Mycol. Mem.* 5, 1-76.
30. Gibson JL, Kimbrough JK, Benny GL. 1986. Ultrastructural observations on Endogonaceae (Zygomycetes). II. Glaziellales ord. nov. and Glaziellales fam. nov.: New taxa based upon light and electron microscopic observat-ions of *Glaziella aurantiaca*. *Mycologia* 78: 941-954.
31. Harley JL, Smith SE. 1983. *Mycorrhizal Symbiosis*. Academic Press, London.
32. Hass H, Taylor TN, Remy W. 1994. Fungi from the lower Devonian Rhynie Chert: Mycoparasitism. *Am. J. Bot.* 81:29-37.
33. Helgason T, Watson IJ, Young JP. 2003. Phylogeny of the Glomerales and Diversisporales (Fungi: Glomeromycota) from actin and elongation factor 1-alpha sequences. *FEMS Microbial. Lett.* 229: 127-132.
34. Jense JM. 1897. Les endophytes radicaux de quelques plantes Javanaises. *Ann. Jardin Bot. Buitenzorg* 14: 53-201.
35. Jones FR. 1924. A mycorr-hizal fungus in the roots of legumes and some other plants. *J. Agric. Res.* 29: 459-470.
36. Kelly AP. 1931. The concept of mycorrhiza. *Mycologia* 23: 147-151.
37. Kelly AP. 1950. Mycotrophy in plants. Chronica Botanica, Waltham, Mass.
38. Link HF. 1809. Observations in ordine plantarum naturales. *Ges. Naturforsch. Freunde Berlin Mag.* 3: 3-42.
39. Lohman ML. 1927. Occurrence of mcyorrhiza in Iowa forest plants. *University of Iowa Studies in Natural History* 11: 33-58.
40. Maia LC. 1991. Morphological and ultrastructural studies on spores and germ tubes of selected arbuscular mycorrhizal fungi (Glomales). Ph.D. dissertation, University of Florida, Gainesville.
41. Mehrotra VS. 1997. Problems associated with morphological taxonomy of AM fungi. *Mycor. News* 9(1): 1-10.
42. Mehrotra VS, Baijal U. 1994. Advances in taxonomy of vesicular-mycorrhizal fungi. In: *Biotechnology in India*. (B.K. Dwivedi and G. Pandey eds.) Allahabad: Bioved Research Society. *pp.* 227-286.
43. Moreau F. 1953. Les Champignons. Tome II. *Systematique.*

*Encyl. Mycol.* 23:941-2212. P. Lechevalier, Paris.

44. Morton JB. 1985. Variation in mycorrhizal and spore morphology of *Glomus occultum* and *Glomus diaphanum* as influenced by plant host and soil environment. *Mycologia* 77: 192-204.
45. Morton JB. 1986. Three new species of *Acaulospora* (Endogonaceae) from high aluminum, low pH soils in West Virginia. *Mycologia* 78: 641-648.
46. Morton JB. 1988. Taxonomy of VA mycorrhizal fungi: Classification, nomenclature, and identification. *Mycotaxon* 32:267-324.
47. Morton JB. 1995. Taxonomy and phylogenetic divergence among five *Scutellospora* spp. based on comparative developmental sequences. *Mycologia* 87: 122-137.
48. Morton JB. 2000. Evolution of endophytism in arbuscular mycorrhizal fungi in Glomales. In: *Microbial Endophytes* (C.W. Bacon and J.H. White, eds.), Marcel Dekker, New York. *pp.* 121-140.
49. Morton JB, Redecker D, 2001. Two new families of Glomales, Archaeosporaceae and Paraglomaceae, with two new genera, *Archaeospora* and *Paraglomus*, based on concor-dant molecular and morpholo-gical characters. *Mycologia* 93(1): 181-195.
50. Morton JB, Bentivenga SP. 1994. Levels of diversity in Endomycorrhiza (Glomales, Zygomycetes) and their role in defining taxonomic and non-taxonomic groups. *Plant and Soil* 159: 47-59.
51. Morton JB, Benny GL. 1990. Revised classification of arbuscular mycorrhizal fungi (Zygomycetes): A new order Glomales, two new suborders Glomineae and Gigasporineae and two new families Acaulosporaceae and Gigasporaceae with an emendation of Glomaceae. *Mycotaxon* 37: 471-491.
52. Morton JB, Bever JD, Pfleger FL. 1997. Taxonomy of *Acaulospora gerdemannii* and *Glomus leptotichum*, synanamorphs of an arbuscular mycorrhizal fungus in Glomales. *Mycol. Res.* 101: 625-631.
53. Morton JB, Bentivenga SP and Bever JD. 1995. Discovery, measurement, and interpretation of diversity in arbuscular endomycorrhizal fungi (Glomales. Zygomycetes). *Can. J. Bot.* 73:25-32.
54. Mosse B. 1956. Fructifications of an *Endogone* species causing endotrophic mycorrhiza in fruit plants. *Ann. Bot.* 20: 349-362.
55. Mosse B. 1970. Honey-coloured, sessile *Endogone* spores. I. Life history. *Archiv. Mikrobiol.* 74: 129-145.
56. Mosse B. 1973. Plant growth response to vesicular-arbuscular mycorrhizae, X. Response of *Stylosanthes* and maize to inoculation in unsterile soils. *New Phytol.* 78: 277-288.
57. Mosse B, Bowen GD. 1968. Key to the recognition of some

*Endogone* spore types. *Trans. Br. Mycol. Soc.* 51: 469-483.

58. Nageli C. 1842. Pilze im Innern von Zellen. *Linnaea* 16: 278-285.
59. Nicolson TH, Gerdemann JW. 1968. Mycorrhizal *Endo-gone* species. *Mycologia* 60: 313-325.
60. Oehl F, Sieverding E. 2004. *Pacispora*, a new vesicular arbuscular mycorrhizal fungal genus in the Glomeromycetes. *J. App. Bot.* 78: 72-82.
61. Petri L. 1903. Ricerche sul signifacto morfologica del prosporoidi (sporangioli di Janse) nelle micorrize endotrofiche. *Nuovo G. Bot. Ital.* 10:541.
62. Peyronel B. 1924. Specie di "Endogone" produttrici di micorize endotrofiche. *Boll. Stn. Patol. Veg. Roma* 5:73-75.
63. Pirozynski KA, Dalpe Y, 1989. Geological history of the Glomaceae with particular reference to mycorrhizal symbiosis. *Symbiosis* 7: 1-36.
64. Powell CL, Bagyaraj DJ. 1984. VA Mycorrhiza. CRC Press, Boca Raton, Fla.
65. Read DJ, Duckett JG, Francis R, Ligrone R, Russell A. 2000. Symbiotic fungal associations in 'lower' land plants. *Phil. Trans. Royal Soc. London, Series B.* 355: 815-831.
66. Redecker D, Kodner R, Graham LE. 2000. Glomalean fungi from the Ordovician. *Science* 289, 1920-1921.
67. Redecker D, Morton JB, Bruns TD. 2000a. Ancestral lineages of arbuscular mycorrh-izal fungi (Glomales). *Mol. Phylogenet. Evol.* 14: 276-284.
68. Rodrigues BF. 1995. The occurrence of VAM fungi in corms of *Amorphophallus commutatus* Engler. *Mycor. News* 7(3): 5.
69. Rodrigues BF.1996. Occurrence of Vesicular-arbuscular mycorrhizal fungi in tubers of *Pueraria tuberosa* (Willd.) DC. *Mycor. News* 8(2): 9.
70. Sawaki H, Sugawara K, Saito, M. 1998. Phylogenetic position of an arbuscular mycorrhizal fungus, *Acaulospora gerdemannii*, and its synanamorph *Glomus leptotichum*, based on 18S rRNA gene sequence. *Mycoscience* 39: 477-480.
71. Schenck NC, Spain JL, Sieverding E. 1986. A new sporocarpic species of *Acaulospora* (Endogonaceae). *Mycotaxon* 25: 111-117.
72. Schenck NC, Spain JL, Sieverding E, Howeler RH. 1984. Several new unreported vesicular arbuscular mycorrhizal fungi (Endogonaceae) from Colombia. *Mycologia* 76: 685-699.
73. Schlicht A. 1889. Beitrag zur Kenntniss der Verbreitung und Bedeutung der Mycorhizen. *Landwirtschaftliche Jahrbüchern* 18: 478-506.
74. Schüßler A, Schwarzott D, Walker C. 2001. A new fungal phylum, the Glomeromycota: phylogeny and evolution. *Mycol. Res.* 105: 1413-1421.

75. Schüßler A, Schwarzott D, Walker C. 2003. Glomeromycota rRNA genes – the diversity of myths? *Mycorrhiza* 13:233-236.

76. Schwarzott D, Walker C, Schüßler A. 2001. *Glomus*, the largest genus of the arbuscular mycorrhizal fungi (Glomales), is non-monophyletic. *Mol. Phylogenet. Evol.* 21: 190-197.

77. Sequeira JO, Sylvia DM, Gibson J, Hubbell DH. 1985. Spores, germination, and germ tubes of vesicular-arbuscular mycorrhizal fungi. *Can. J. Microbiol.* 31:965-972.

78. Simon L, Levesque RC, Lalonde M. 1993. Identification of endomycorrhizal fungi colonizing roots by fluorescent single-strand conformation polymorphism-polymerase chain reaction. *Appl. Environ. Microbiol.* 59: 4211-4211.

79. Smith FA, Smith SE. 1997. Structural diversity in (vesicular)-arbuscular mycorrhizal symbiosis. *New Phyto.* 137:373-388.

80. Smith SE, Read DJ. 1997. *Mycorrhizal Symbiosis. Second ed.* London: Academic Press.

81. Spain SE. 1992. Patency of shields in water mounted spores of four species of *Acaulospora* (Glomales). *Mycotaxon* 43:331-339.

82. Spain JL, Sieverding E, Schenck NC. 1989. *Gigaspora ramisporophora*: a new species with novel sporophores from Brazil. *Mycotaxon* 34: 667-677.

83. Taylor TN, Remy W, Hass H, Kerp H. 1995. Fossil arbuscular mycorrhizae from the early Devonian. *Mycologia* 87: 560-573.

84. Thaxter R. 1922. A revision of the Endogonaceae. *Proc. Am. Acad. Arts Sci.* 57:292-348.

85. Tommerup IC, Sivasithamparam K. 1990. Zygospores and asexual spores of *Gigaspora decipiens*, an arbuscular mycorrhizal fungus. *Mycol. Res.* 94(7): 897-900.

86. Trappe JM, Schenck NC. 1982. Taxonomy of the fungi forming endomycorrhizae. A. Vesicular arbuscular mycorr-hizal fungi (Endogonales). In: *Methods and principles of mycorrhizal research.* (N.C. Schenck ed.) St. Paul Minnesota. *pp.* 1-9.

87. Tulasne LR, Tulasne C. 1845. Fungi nonnulli hypogaei, novi v. minus cogniti act. *Giorn. Bot. Ital.* 2: 35-36.

88. Walker C. 1983. Taxonomic concepts in the Endogonaceae. Spore wall characteristics in species descriptions. *Mycotaxon* 18: 443-455.

89. Walker C. 1986. Taxonomic concepts in the Endogonaceae: II. A fifth morphological wall type in Endogonaceae spores. *Mycotaxon* 25: 95-97.

90. Walker C. 1987. Current concepts in the taxonomy of the Endogonaceae. *In: Mycorrhizae in the next Decade, Practical Applications and Research Priorities.* (D.M. Sylvia, L.L. Hung, and J.H. Graham, eds.)

IFAS, University of Florida. Gainesville, Florida. *pp.* 300-302.

91. Walker C. 1992. Systematics and taxonomy of the arbuscular endomycorrhiza fungi (Glomales) a possible way forward. *Agronomie* 12: 887-897.
92. Walker C, Reed LE, Sanders FE, 1984. *Acaulospora nicolsoni*, a new endogonaceous species from Great Britain. *Trans. Br. Mycol Soc.* 83:360-364.
93. Walker C, Sanders FE. 1986. Taxonomic concepts in the Endogonaceae: III. The separation of *Scutellospora* gen. nov. from *Gigaspora* Gerd. & Trappe. *Mycotaxon* 27: 169-182.
94. Walker C, Blaszkowski J, Schüßler A. 2004. Gerdemannia gen. nov., a genus separated from *Glomus*, and Gerdemanniaceae fam. nov., a new family in the Glomeromycota. *Mycol. Res.* 108: 707-718.
95. Walker C, Schüßler A. 2004. Nomenclatural clarifications and new taxa in the Glomeromycota. *Mycol. Res.* 108:981-982.
96. Weijman AC, Meuzelaar HCL. 1979. Biochemical contributions to the taxonomic status of the Endogonaceae. *Can. J. Bot.* 57:284-291.
97. Wu C. 1993. Glomales of Taiwan. III. A comparative study of spore ontogeny in *Sclerocystis* (Glomaceae, Glomales). *Mycotaxon* 47: 25-39.

*Rev. Plant Pathol. Vol. 4, 2008*
*Indian Society of Mycology and Plant Pathology*
*Scientific Publishers (India), Jodhpur pp. 217 - 260*

# BIOINFORMATICS: MOLECULAR SEQUENCE ANALYSIS APPLICATIONS IN PLANT VIROLOGY – A PRIMER

***D.V.R. Sai Gopal* and A.M. Anthony Johnson***

Department of Virology, Sri Venkateswara University,
Tirupati 517 502, India
*Corresponding author: dvrsaigopal@rediffmail.com

KEY WORDS: Sequence analysis DNA, RNA and protein analysis, phylogenetic analysis, character based and Distance based analysis, confidence limits.

## ABSTRACT

Molecular Characterization of plant viruses is very essential and the present on going research to understand the etiology of plant viral diseases and their management. The present researchers are striving hard to understand the genomics of plant viruses. In order to understand the genomics of the viruses, they have to be characterized at molecular level by sequencing various genes and genomes of the plant viruses and also some of the gene products of importance. Deriving analysis from these sequences is very difficult by using wet lab techniques as it consumes a lot of time as well as economy. In these situations, Bioinformatics plays a major role in deriving outputs of such analysis. Many of Bioinformatics tools area available via www interface, of which some of them are freely available and some of them are commercial. We describe very simple methods by using many of the freely available tools and some of the commercial tools for analysis of DNA sequences of plant viruses. As it is not possible to define all the methods of molecular sequence analysis, we have described only some of the important methods of molecular sequence analysis and their applications in plant virology.

## INTRODUCTION

Bioinformatics is showing a great impact and revolution in the present era of research. It involves a vast application of information technology tools combined with other applications like mathematics, statistics, physics, chemistry *etc.*, to solve various problems of biological research. The present era of research could be stated as post genomic era as the present molecular biology research bags 394 published genomes as per GOLD (Genomes online Database) statistics (Updated on June 21, 2006). 934 Bacterial genome sequencing projects, 56 Archaeal genome project, 46 Meta genomes and 608 Eukaryotic genome sequencing projects (including 12 chromosomes) are ongoing (26) (http:// www. genomesonline.org). Analyzing these sequences and annotating them is practically expensive and time consuming. Bioinformatics tools provide easy solutions to such type of problems. These tools are applied in various fields for different types of biological sequence analysis. Plant molecular virology is an applied science which deals with the molecular characterization of plant viruses. Plant viruses are known as viral pathogens which infect plants and their epidemics lead to threat in agriculture. To manage plant viral diseases, there is a great need to understand their molecular pathogenesis and host interactions at molecular level, which is the core concept of Plant molecular virology. In order to characterize these plant viruses, genome sequencing and characterization of the plant viral genomes is required which applies wide range of bioinformatics tools that can analyze these sequences at molecular level, they also provide an insight to the molecular diversity of these viruses. The main concept of this review is to brief up the different tools that are used in plant molecular virology for the viral sequence analysis. This review is quite different from other reviews which only elucidate the theoretical status of research. In the present review, even some of the protocols for accessing the tools are also defined. In this article we have used Citrus Yellow

Mosaic Badnavirus as a model for illustration for some of the methods. Citrus yellow mosaic virus is a bacilliform virus having dsDNA as genetic material and an RNA mediated replication strategy. It belongs to genus Badnavirus and the family Caulimoviridae. The genome is 7559 bp comprising six open reading frames out of which ORF III is the major one encoding a polyprotein that is post translationally modified into Movement protein, Coat protein, Aspartic protease and Reverse transcriptase/RnaseH (33). The present example 1.1 kb fragment belongs to the ORF III of CYMV overlapping protease and RT/RNase H domains. In case of Phylogenetic analysis, whole genome of CYMV is considered as an example.

### *Sequence Analysis*

The first and foremost step of a researcher after getting the biological sequence is to analyze the sequence of interest. This sequence analysis yields results that are useful in further wet lab experiments. Many of the softwares which are both online and offline are available via www for sequence analysis. Some of the commercial tools and as well as free tools are discussed in this topic.

## (A) Sequence Formats and Format Converter Tools

Databases have their unique form of representation of sequence data available with their databases (6). Most often these formats may not be suitable for an offline sequence analysis package to use because they accept only standard formats. Some of these sequence formats are represented below.

***(a) GenBank DNA sequence Entry:*** It is a unique format in which all the sequences in GenBank are represented. It is otherwise known as GenBank flat file format. Information describing sequence entry is given including literature references, information about function of sequence, location of mRNA's and coding regions. The information is organized into different fields which are again divided into sub fields. The sequence entry is present from the word 'origin' to the '//' mark. The protein database GenPept is also having its sequence entries in the same format as that of a GenBank.

***(b) EMBL Data library format:*** EMBL maintains both DNA and Protein sequences. It is also represented in Flat file format as that of a GenBank Entry but the Fields representing the sequence are different. The sequence is assumed to be lying between the "SEQUENCE" and "//" by the computer softwares.

***(c) Swiss Prot Sequence Format:*** The format of an entry in the Swiss Prot database is most similar to that of EMBL format except that it contains information about physical and biochemical properties of the protein and also it has many links to other databases which furnish information of different aspects about the protein.

***(d) FASTA Sequence Format*:** The FASTA sequence format is a unique and more accessible format by many of the analysis tools. The format includes three parts (1) A comment line preceded by ">" symbol followed by name and origin of the sequence. (2) The sequence standard one-letter symbols and (3) an optional "*" which may or may not be present at the end of the sequence.

***(e) NBRF/PIR Format*:** This sequence format is used by PIR and also many other software tools. Generally the data downloaded from PIR is having additional information about the protein different from this format. PIR format starts with a first line of ">" symbols followed by a two letter code such as "P" for complete sequence of "F" for a fragment, followed by 1or 2 to indicate the type of sequence, then a four to six lettered unique name of the entry. There is an essential second line with full name of the sequence, a hyphen, and then the species of origin.

***(f) Stanford University/Intelligenetics Sequence Format*:** IG format is started by molecular genetics group at Stanford University; it is similar to that of PIR format except that a semicolon is placed before the comment line and also a second line identifier. At the end of the sequence 1 is placed if the sequence is linear or else 2 is placed if the sequence is circular.

***(g) Genetics computer group*:** Older version of GCG sequence analysis package required a specific format of data for analysis known as GCG format. It also has tools to convert other sequence formats into GCG format. Information about the sequence in GenBank is first included, followed by a line of information about the sequence and a checksum value. This checksum value is useful to check if the sequence is changed or intact after conversion. The rest is the sequence and there is presence of line numbers.

***(h) Plain/ASCII.Staden Sequence Format:*** This sequence format includes only the sequence with no other accessory information. This is used by Staden sequence analysis package.

***(i) ASN.1 Format:*** ASN.1 format means Abstract Syntax Notation Sequence Format. Abstract Syntax Notation is a formal data description language that has been developed by the computer industry. This has been adopted by NCBI to encode data such as sequences, Maps, Taxonomic information, molecular structures, and bibliographic information.

***(j) Genetic Data Environment Sequence Format:*** Genetic Data Environment format is used by a sequence analysis system called the Genetic Data Environment, around a multiple sequence alignment editor that runs on UNIX machines. It is similar to ASN.1 format which stores information about colors and sequence in a tagged field format.

## (B) Format Converters

Many Sequence format converters are available among which some are freeware and some are commercial tools. Some of them are READSEQ which is a freeware downloadable from (*http://iubio. bio.indiana.edu/soft/molbio/readseq/java/*), GCG Sequence analysis package programs and Edit Seq module of DNASTAR sequence analysis package, Sequence verter which is a free suite of Gene Studio.

**READSEQ:** READSEQ is a sequence formatting program developed by D.G. Gilbert at Indiana University, Bloomington. It can identify a DNA or a Protein sequence of any format, identify it and then converts it into other formats and writes on to other file.

The following are the different sequence formats that are recognized and inter converted into other sequences in read seq software

1. Abstract Syntax Notation (ASN.1)
2. DNA Strider
3. European Molecular Biology Laboratory (EMBL)
4. Fasta / Pearson

5. Fitch (for phylogenetic analysis)
6. GenBank
7. Genetics Computer Group (GCG)
8. Intelligenetics / Standford
9. Multiple Sequence format (MSF)
10. National Biomedical Research Foundation (NBRF)
11. Olsen (in only)
12. Phylogenetic Analysis Using Parsimony (PAUP) NEXUS format
13. Phylogenetic Inference package (Phylip v3.3, v3.4)
14. Phylogenetic Inference package (Phylip v3.2)
15. Plain text / Staden
16. Pretty format for publication (output only)
17. Protein Information Resource (PIR or CODATA)
18. Zuker for RNA analysis (in only)

*eg:* Copy the input sequence file into the folder containing Read Seq.dos program. Open the READSEQ.dos program by double clicking the read seq.dos icon it will prompt to give an output file name. Then afterwards it will prompt to choose the number corresponding to sequence format and then it will ask you for an input file name. After giving the input file name, the output is written into outfile.

This Read Seq is inbuild in analysis packages such as bioedit.

## (B) Sequence Similarity Searches

The first thing a plant virology researcher does is to compare the present sequences to the available sequences related to that virus. In order to get the sequences related to any virus, a sequence similarity search is to be performed against any sequence database or a collection of sequences. Sequence similarity search is also most useful in case of structural annotation and functional annotations. There are many tools for sequence similarity searches and two main tools accepted by most of the Databases are 1. BLAST and 2. FASTA.

**1. BLAST:** BLAST algorithm was developed by S. Altschul *et.al.,* in early 1990's (1, 2). BLAST algorithm searches for maximal scoring similar subsequences related to the subsequences in the query. Until now, there are many versions of BLAST. The present version of the BLAST running online for database similarity searches is (BLAST V. 2.2.14). NCBI adopted this powerful tool for database searching in different aspects and databases. To define briefly, how a BLAST algorithm works, an example associated with BLASTP tool is illustrated below (25).

eg: Given a Query "AILDTGATGDA" and a database sequence "MQVWGWAILDTVATDAAMLL"

1. The first step involves breaking of the sequence into words of word length as defined with search parameters (Fig. 1).

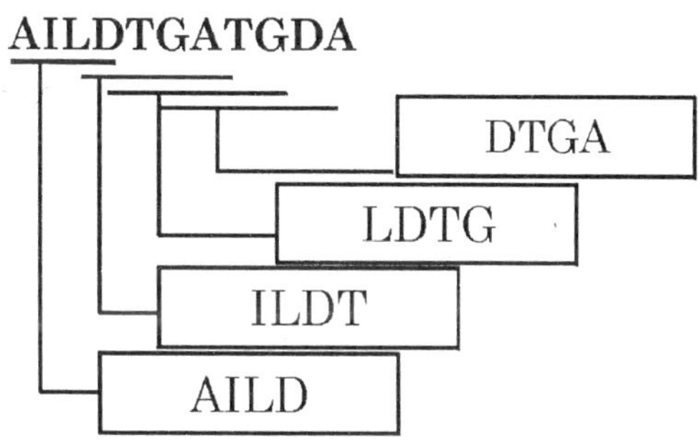


**Fig 1.** Word Size pattern of BLAST search

2. Then the next step allows word matches (also called as High Scoring Pairs or HSPs) in the database sequences.

AILD

MQVWGWAILDTVATDAAMLL...........

3. The matching can be extended with the remaining words until the alignment score falls below a given threshold parameter in the search.

AILDTGAT

MQVWGWAILDTVATDAAMLL...........

The final alignment score and homology percentage is calculated between the given query and the subject is displayed. There are different types of BLAST tools available (37) (Table 1).

**Table 1.** Different types of BLAST tools and their Functions

| Name of the BLAST tool | Function |
|---|---|
| BLASTN | Searches Nucleotide query against a nucleotide database. |
| BLASTP | Searches Protein query against a protein database |
| BLASTX | Searches a translated Nucleotide query against a Protein database |
| TBLASTN | Searches a protein against a translated Nucleotide query from the database |
| TBLASTX | Searches a translated nucleotide query against a translated nucleotide sequence from database |
| PSI-BLAST | Searches iteratively remote homologues from the obtained subjects iteratively. |

Different parameters to be considered about BLAST are Word size 10 for a nucleotide sequence and 4 for a protein sequence, expect value (set by default as 10), threshold score below which the extension of the alignment is terminated in finding a homolog in database. Expect value is a significant score given by the BLAST tools which can be used to state how much apt is the obtained sequence hit. In order to define this parameter that how much apt is the hit, the first thing is observing the alignment score. This alignment score is provided by the tool itself based on the number of matches or mismatches between the query and the subject. But this score (s) differs from tool to tool and also from scoring matrices applied. In order to find the reliability of the results in these sequence similarity searches, we need to ask an appropriate question "Given a set of sequences not related to the query sequence, what is the probability of finding a match with score S simply by chance?" As an answer to this question, database gives a score called Expect Score or E-Value. Lower the value of E, that much more relevant is the sequence found with score S. This BLAST tool is accessible via URL *http://www.ncbi.nlm.nih.gov/BLAST*. The standard input for this search is FASTA format but BLAST also accepts other formats. Generally most of the plant virus related publications show the usage of online

BLAST tool. The procedure involves accessing the home page of BLAST and selecting the appropriate BLAST tool for analysis, then pasting the sequence in FASTA format in the appropriate box provided in the tools. The search is started by clicking on BLAST button. This gets into the next page which gives an option of format by which one can change the parameters of the search used. If everything is found to be ok, format button can be clicked which will lead to the BLAST result page. But some times if the search result has many hits, this will take time and the page is automatically updated into result page. The result includes graphical representation of results based on the score; the graphics are colored, then each and every subject local alignment with the query is displayed and also the resultant values. A version of BLAST known as Standalone BLAST version is available NCBI software downloads. This can be run on any PC or a network server. This is useful for majority of the researchers who want to make their data confidential and also who would like to work offline in absence of Internet. This software provides different tools to make own database of ones interest and also query them with their query of interest. The method of installing, running the software from the corresponding environment is mentioned in detail along with the software backup.

*eg*: Preparing a database of all Badnavirus genomes and searching it with a 1.1kb fragment of CYMV which is a member of Badnavirus using Standalone version of BLAST tool.

- Copy all the sequences of whole genomes of Badnavirus in FASTA format into a text editor and name it as badnagenomes.txt.
- Create a directory named BLAST in any of the drives. Extract all the standalone BLAST files are extracted from its self extracting archive by simply double clicking the file. Then copy the file badnagenomes.txt into the BLAST directory. This will serve as the database for BLAST.
- Copy the query sequence also in FASTA format into a new text editor and save it into the same BLAST directory with the name query.txt

- Access the BLAST directory form command Prompt. The first thing is that the database is to be formatted into an appropriate form which the standalone BLAST tool can access. This is done by a tool called formatdb which is apart of standalone BLAST. It is done by typing a command as below and pressing the enter key to access the command.

  **formatdb –i badnagenomes.txt –p F –o T**

  Now a set of files are formed in the BLAST directory with names as badnagenomes.txt with different extensions. They indicate that the database badnagenomes.txt is ready for use by standalone BLAST. The –i denotes the input file for formatting, –p denotes type of file which provides a values F for nucleotide here, –o denotes parse i.d. options which provides a value T here which denotes that the sequence i.d's may be parsed and the index may be created.

- The next step is to use the query & search this database. For this search, the following command was used.

  **blastall -p blastn -d badnagenomes.txt -i query.txt -o query.out**

  Here blastall is the command to call the standalone blast program and –p is the program name string which is blastn here, –d is the database name which is badnagenomes.txt here, –i is the input query file which is query.txt here and –o is the output file which is query.out here. This will write a result file called query.out that can be accessed by opening it by any text editor like notepad. The result is presented as shown in the figure.

The present result denotes that the query used up here is having 95% identity with CYMV genome, an expect value of 0.0 which shows the relevance of the result.

As it is not possible to provide all the details of the other programs and BLAST, only this example is attempted which is plant virus sequence analysis application (Fig. 2).

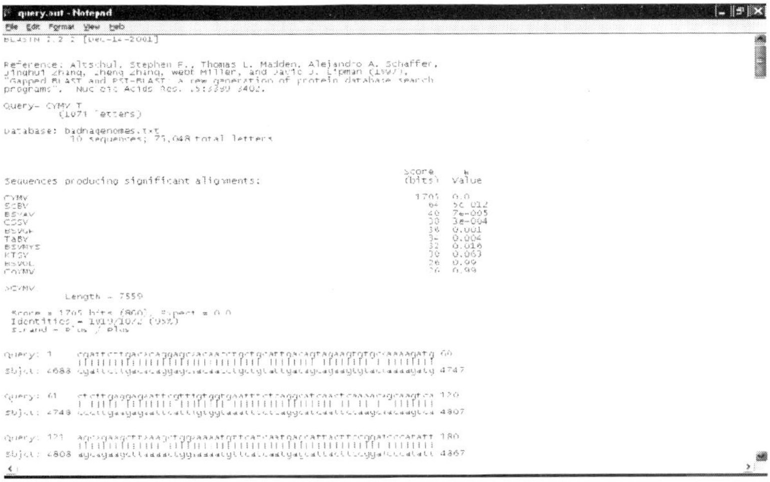


**Fig 2.** Result of Stand alone BLAST search

**2. FASTA:** FASTA algorithm is designed by Lipman, Pearson *et. al.,* for fast and heuristic database searches (27, 31). This tool can be accessed form EBI or form FASTA server at University of Virginia online. As that of BLAST, a standalone version of FASTA tool is also available which can be downloaded from URL (*ftp:/ftp.virginia.edu/pub/fasta*). It is having different types of FASTA tools that can be used for different types of database searche. Some of them are listed below.

| Name of the FASTA tool | Application |
|---|---|
| *fasta3* | Compare a protein sequence to a protein sequence database or a DNA sequence against a DNA database by using FASTA algorithm |
| *Ssearch3* | Compare a protein sequence to a protein sequence database or a DNA sequence against a DNA database by using Smith-Waterman algorithm More sensitive than *fasta3* but slower than *fasta3* |
| *fastx3/fasty3* | Compare a DNA sequence to a protein sequence database by comparing the translated DNA in all 3 frames, allowing gaps. |

| | |
|---|---|
| *tfastx3 / tfasty3 / tfasta3* | Compare a protein sequence to the DNA database, calculating similarities with frame shifts to forward and reverse orientations |
| *fastf3* | Compares a mixed peptide sequence to a protein database. (a mixed set of sequences from Edman degradation sequencing) |
| *tfastf3* | Compare a mixed peptide sequence against a DNA database |

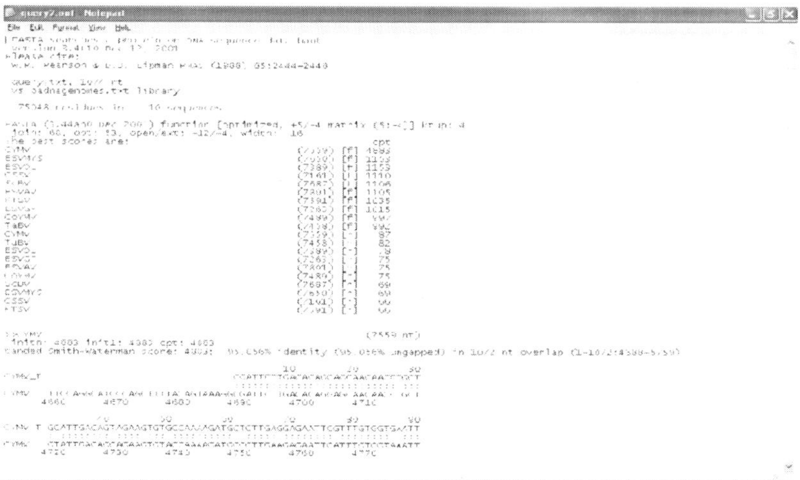

**Fig 3.** Result of Stand alone FASTA search.

Searching using a standalone FASTA tool is comparatively easy that BLAST tools as there is no need of formatting the sequences that are used as database for searching. A similar value to that of the E-value of BLAST is given by FASTA tools called P value. Lower the P value, best is the reliability of the resultant hit. The word size in case of FASTA is defined as *ktup* which as a default is 4 for DNA and 2 for Protein sequences. To perform a Standalone FASTA search, the programs are copied into a directory and the database (file) containing sequences and the query is also copied into the same directory. Then the appropriate program is run by double clicking the icon. It will ask for the input query or test sequence and the database library of sequences to be searched and also name for a results file into which it writes the results, it also asks the *ktup* you want to use, then the number of alignments you want to display. If you input the appropriate file names along with their extensions, then the search is completed an the results are written into the results file which can be read by using any of the text editors (Fig. 3).

**(C) Sequence Analysis**

Sequence analysis is another part of bioinformatics analysis in plant viruses. This analysis involves different analysis of both protein and nucleic acid sequences like complementation, reverse complementation, reverse, primer design, RNA secondary structure prediction, ORF analysis, Translation, Hydrophobicity, Isoelectric pH, Protein Secondary structure etc., Many of the Sequence analysis packages and as well as software suites, some free wares and some commercials are available via www. Some of them include DNASTAR, DNASIS MAX, DNASIS (old version), DSGENE, OMIGA (old version), GCG, EMBOSS, BIOEDIT **(18)**, GENEDOC, Win gene, Win Pept, Vector NTI Suite etc.,

- BioEdit is a freeware downloaded from

  *www.mbio. ncsu.edu / BioEdit / bioedit.html*
- DNASIS MAX is a commercial tool but however, a one month trial version is available from *www. miraibio.com*
- Genedoc is also a freeware available via *www.psc. edu / biomed / genedoc*
- DSGENE is a commercial tool available from Acclerys, a trial version is available from the site upon request *www.accelrys.com / products / dstudio*
- EMBOSS (European Molecular biology open software suite) is also a free software suite available from *http: / / emboss.sourceforge.net*
- Vector NTI Suite is also a commercial package but a DEMO version is downloadable from *www.invitrogen.com*
- GCG is a fully commercial suite, no trial is available
- DNASTAR is also a commercial suite, trial is provided by the company upon request.

**Sequence Complement, Reverse and Reverse Complementation:** Sequence complement, reverse and reverse complementation is essential when a sequence assembly of the sequences derived from sequencer is done. It is also useful in deriving the anti sense strand of DNA. This can be done using tools like EDIT SEQ of DNASTAR, Bioedit, Win Gene, DNASIS max etc.

*eg*: using Bioedit to complement the given sequence, reverse complement them.

The query which is to be used for analysis is saved in the FASTA format in any of the text editors like notepad. This is opened using file open menu of Bioedit. Other formats can also be used and they are acceptable by Bioedit as it has in built read seq program which can convert the sequence into different formats. The opened sequence can be selected in the left side panel and then the following steps are followed for each analysis.

**COMPLEMENT:**

Sequence menu > Nucleic acid > Complement

**REVERSE COMPLEMENT:**

Sequence menu > Nucleic acid > Reverse Complement.

**REVERSE:** This can be done in DNASIS package (older version). Here the sequence is to be submitted in raw text format. The query is opened in to the tools and then the reverse option of the Sequence menu of DNASIS. If reverse is to be done with Bioedit, first complement and then reverse complement is to be done which will yield reverse of the sequence.

The sequence below shows is a 1.1kb sequence of CYMV (Citrus yellow mosaic virus) which is analyzed in the above methods; only a part of sequence about 60bp's is used for analysis as an example here.

**Query**

CGATTCTTGACACAGGAGCAACAACCTGCTGCATTGA
CAGTAGAAGTGTGCCAAAAGATGC

**Complement derived from Bioedit**

GCTAAGAACTGTGTCCTCGTTGTTGGACGACGTAACT
GTCATCTTCACACGGTTTTCTACG

**Reverse complement derived from Bioedit**

GCATCTTTTGGCACACTTCTACTGTCAATGCAGCAGGT
TGTTGCTCCTGTGTCAAGAATCG

**Reverse derived from Bioedit**

CGTAGAAAACCGTGTGAAGATGACAGTTACGTCGTCC
AACAACGAGGACACAGTTCTTAGC

**DNA to RNA and RNA to DNA:** This is a simplest analysis that can be done with Bioedit by selecting DNA → RNA option and RNA → DNA option of the sequence menu of the Bioedit software. This can be of use in deriving RNA sequence of RNA viruses and then moving for RNA secondary structure prediction in order to find secondary structure of RNA of sub viral agents like Viroids.

*eg*: Derivation of RNA sequence from the DNA sequence using Bioedit software:

**RNA Sequence derived from Bioedit from the query mentioned in the previous analysis:**

CGAUUCUUGACACAGGAGCAACAACCUGCUGCAUUG
ACAGUAGAAGUGUGCCAAAAGAUGC

**RNA Secondary structure prediction:** RNA Secondary structure prediction is a theoretical method of prediction of RNA secondary structure using mathematical and bonding patterns of the structure based on the possible base pairings (29).

*eg*: This can be done by using RNA secondary structure of the DNASIS software and also an online server in bioedit software. Load the RNA sequence in the DNASIS software in raw sequence format. Then select the sequence which is to be analyzed and follow these steps:

Open Function menu > Prediction > RNA Secondary structure

The prediction in this case is done for a short stretch of RNA sequence from the previous section which yields a structure (Fig. 4).

**ORF Prediction:** Open reading frame is a region flanked by a start codon and a stop codon. In case of many of the viruses, overlapping open reading frames are more common as the virus has a very small genome which it has to use efficiently to synthesize all of its proteins. Detection of these ORF's is very essential in case of plant viruses in order to characterize the genome.

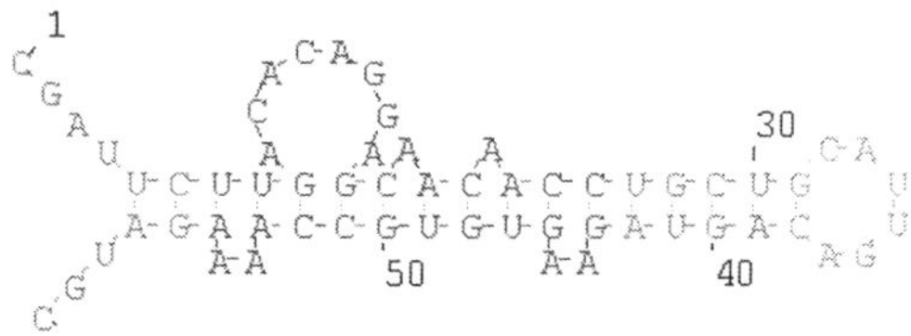

**Fig 4**. Predicted RNA secondary structure for a small stretch of fragment of 1.1 kb of CYMV

Eg: This ORF prediction is done by using software like DNASIS and Bioedit. In order to predict the ORF in Bioedit software, the query is to be opened in the tool and then the following steps are to be followed

Sequence > Nucleic acid > find next ORF.

**Translation of DNA:** The translation of DNA using computation tools is an important method of obtaining Amino acid sequence from the DNA with out any protein sequencing experiments. The amino acid sequence can be represented either in one letter code or a three letter code. This amino acid sequence is of use in prediction of different parameters of protein corresponding to the DNA. This translation theoretically done in six frames, among which three frames are in +1, +2, +3 orientations *i.e.*, they start from 1st base, 2nd base and 3rd base of the given sequence. The other three frames are in -1, -2, -3 orientations *i.e.*, they start form 1st base, 2nd base and 3rd base of the reverse sequence. This can be done in all six frames. Among the six frame translations, longest frame that can encode maximum length of protein can be chosen as the translated amino acid sequence. The translation can be done using different sequence analysis softwares. The present analysis shows translation of a sequence using Bioedit software. The translation is done in all the six frames but only the frame with the stretch of amino acids above the mentioned threshold are displayed in the results.

*eg*: The following translation denotes the translation of a1.1kb fragment of CYMV in all six frames using Bioedit and getting an amino acid sequence from one of the translated frames. In order to translate the following sequence, follow these steps. Open the file in Bioedit, select sequence menu > nucleic acid > sorted six frame translation.

**Sequence of 1.1kb fragment of CYMV**

CGATTCTTGACACAGGAGCAACAACCTGCTGCATTGACAGTA
GAAGTGTGCCAAAAGATGCTCTTGAGGAGAATTCGTTTGTGG
TGAATTTCTCAGGCATCAACTCAAAACAGCAAGTCAAGCAGA
AGCTTAAAGCTGGAAAAATGTTCATCAATGAGCATTACTTCCG
GATCCCATATTGTTACAGCTTTGAGATGCAAATTGGTGATGGC
ATCCAACTTATCCTCGGGTGCAACTTTATACGAAGTATGTACG
GTGGTGTACGATTAGAAGGTAATACAGTAACCTTCTACAAGC
AGATCACAAGTATCAACACCAGGCTTGCTGCACCTCTCCTTAA
GCAAGAAAAAGAGGAGAAAGAAGAAGAGTTTAACCTAGATGA
GCACAGGCTAATTCAGGAAATGGTTGCATATTCTACTGAGCG
GCCATTTGTTCATTTCCAACAGAAATTTGCAGGGCTTATTCAA
GAGTTAAAAGCCCAGGGATACATTGGGGAAGAGCCTATGAAG
TATTGGGCCAAAAACAAAGTTGTTTGCCATCTGGATATTAAAA
ATTCAGATATGGTAATTGAAGATCGCCCACTGAAGCATGTGA
CACCCCAGATGGAAGAAAGCTTTCGCAAGCATGTGGAAGCCC
TGTTAAAAATAGGAGCAATCCGGCCCAGTAAAAGTCGACACA
GAACCACAGCTATAATGGTCAGCTCTGGAACCAGCATAGACC
CTATCACAGGGAAAGAGGTTAAGGGAAAGGAGCGAATGGTCT
TCAACTATAAAAGGTTAAATGACCTAACTAATAAAGATCAGTA
CAGCCTTCCTGGAATCCAGACGATCCTACAAAGATTAAAGGG
GAGTACTATATTCTCCAAATTTGACCTTAAAAGTGGTTTTCAT
CAGGTAGCAATGCATCCAGATTCAATAGAATGGACAGCTTTTT
GGGTGCCCAGCGGTCTTTATGAATGGTTAGTTATGCCACTCG
GATTAAAGAATGCTCCAGCAGTTTTTCAGAGGAAAATGGATC
ACTGTTTCAAAGGCACGGAGGCCTTTATTGCCGTCTACATCGA
CGACATCCTAGTTC

**Translated Six Frames**

>CYMV_T: 3 to 1073: Frame 3 357 aa

ILDTGATTCCIDSRSVPKDALEENSFVVNFSGINSKQQVKQ
KLKAGKMFINEHYFRIPYCYSFEMQIGDGIQLILGCNFIRS
MYGGVRLEGNTVTFYKQITSINTRLAAPLLKQEKEEKEEE
FNLDEHRLIQEMVAYSTERPFVHFQQKFAGLIQELKAQGY
IGEEPMKYWAKNKVVCHLDIKNSDMVIEDRPLKHVTPQM

EESFRKHVEALLKIGAIRPSKSRHRTTAIMVSSGTSIDPITG
KEVKGKERMVFNYKRLNDLTNKDQYSLPGIQTILQRLKG
STIFSKFDLKSGFHQVAMHPDSIEWTAFWVPSGLYEWLV
MPLGLKNAPAVFQRKMDHCFKGTEAFIAVYIDDILV

>CYMV_T: 88 to 267: Frame 1 60 aa

ISQASTQNSKSSRSLKLEKCSSMSITSGSHIVTALRCKLVM
ASNLSSGATLYEVCTVVYD

>CYMV_T: 266 to 108: Frame -2 53 aa

SYTTVHTSYKVAPEDKLDAITNLHLKAVTIWDPEVMLIDE
HFSSFKLLLDLLF

>CYMV_T: 766 to 467: Frame -3 100 aa

LKTIRSFPLTSFPVIGSMLVPELTIIAVVLCRLLLGRIAPIFN
RASTCLRKLSSIWGVTCFSGRSSITISEFLISRWQTTLFLAQ
YFIGSSPMYPWAFNS

As the minimum ORF size is 50 as a screening point, only some of the ORF's which have more than 50 amino acids are displayed in the results.

### Nucleic acid content and Codon Usage

Nucleic acid content and codon usage in DNA sequences is useful in prediction of the nature of template *i.e.*, whether it is GC rich or AT rich and also the stability of template. This can be done by using any common sequence analysis tools like Bioedit, DNASIS *etc.*,

The Nucleic acid content for a 1.1kb viral sequence mentioned in the above sequence determined by using Bioedit yields following results (Fig. 5).

**Restriction Enzyme analysis:** Restriction Enzymes have specific recognition sites on DNA they recognize and cleave at specific positions. This can be found in any DNA sequence with the present day sequence analysis tools like Bioedit, DNASIS, DNASTAR *etc.*, There are different Restriction enzymes, some are sticky end cutters, some are blunt end cutters. This restriction enzyme analysis helps in the identification of RFLP patterns between isolates, in designing primers with restriction sites for cloning of expression clones of different proteins.

```
Protein: CYMV_T
Length = 357 amino acids
Molecular Weight = 40949.31 Daltons
Amino Acid  Number   Mol%
  Ala  A      18     5.04
  Cys  C       6     1.68
  Asp  D      16     4.48
  Glu  E      26     7.28
  Phe  F      20     5.60
  Gly  G      22     6.16
  His  H      10     2.80
  Ile  I      28     7.84
  Lys  K      31     8.68
  Leu  L      28     7.84
  Met  M      12     3.36
  Asn  N      14     3.92
  Pro  P      14     3.92
  Gln  Q      18     5.04
  Arg  R      16     4.48
  Ser  S      22     6.16
  Thr  T      18     5.04
  Val  V      22     6.16
  Trp  W       4     1.12
  Tyr  Y      12     3.36
```

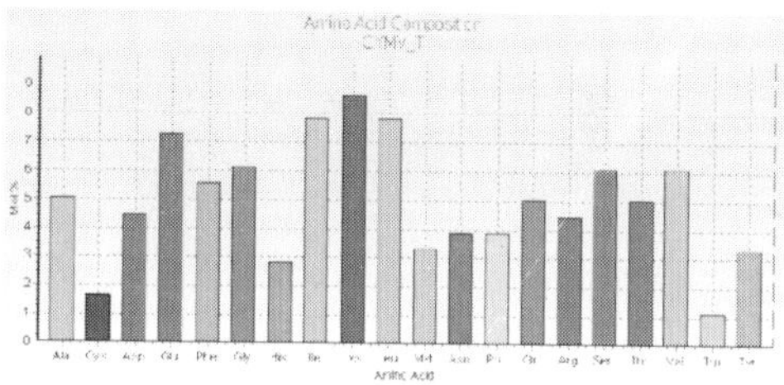

**Fig. 5.** Graphical representation of nucleotide base count

Eg: The following example represents Restriction pattern of a 1.1 kb clone of CYMV with defined restriction enzymes in Bioedit. In order to achieve this, open the sequence in the Bioedit tool, then sequence menu, nucleic acid, Restriction map. Only some of the enzymes are tried for mapping.

```
1    CGATTCTTGACACAGGAGCAACAACCTGCTGCATTGACAGTAGAAGTGTG       80
     CCAAAAGATGCTCTTGAGGAGAATTCGTTT
1    GCTAAGAACTGTGTCCTCGTTGTTGGACGACGTAACTGTCATCTTCACACG      80
     GTTTTCTACGAGAACTCCTCTTAAGCAAA                       EcoRI
81   GTGGTGAATTTCTCAGGCATCAACTCAAAACAGCAAGTCAAGCAGAAGCT       160
     TAAAGCTGGAAAAATGTTCATCAATGAGCA
81   CACCACTTAAAGAGTCCGTAGTTGAGTTTTGTCGTTCAGTTCGTCTTCGAA      160
     TTTCGACCTTTTTACAAGTAGTTACTCGT                  HindIII
161  TTACTTCCGGATCCCATATTGTTACAGCTTTGAGATGCAAATTGGTGATGG      240
     CATCCAACTTATCCTCGGGTGCAACTTTA
161  AATGAAGGCCTAGGGTATAACAATGTCGAAACTCTACGTTTAACCACTACCG     240
     TAGGTTGAATAGGAGCCCACGTTGAAAT            BspEI
                                                BamHI
241  TACGAAGTATGTACGGTGGTGTACGATTAGAAGGTAATACAGTAACCTTCTA     320
     CAAGCAGATCACAAGTATCAACACCAGG
241  ATGCTTCATACATGCCACCACATGCTAATCTTCCATTATGTCATTGGAAGATG    320
     TTCGTCTAGTGTTCATAGTTGTGGTCC                        BsbI
321  CTTGCTGCACCTCTCCTTAAGCAAGAAAAAGAGGAGAAAGAAGAAGAGTTTA     400
     ACCTAGATGAGCACAGGCTAATTCAGGA
321  GAACGACGTGGAGAGGAATTCGTTCTTTTTCTCCTCTTTCTTCTTCTCAAATTG   400
     GATCTACTCGTGTCCGATTAAGTCCT                         AflII
```

401 AATGGTTGCATATTCTACTGAGCGGCCATTTGTTCATTTCCAACAGAAATTTG 480
CAGGGCTTATTCAAGAGTTAAAAGCCC
401 TTACCAACGTATAAGATGACTCGCCGGTAAACAAGTAAAGGTTGTCTTT 480
AAACGTCCCGAATAAGTTCTCAATTTTCGGG BsrBI

481 AGGGATACATTGGGGAAGAGCCTATGAAGTATTGGGCCAAAAACAAAGTT 560
GTTTGCCATCTGGATATTAAAAATTCAGAT
481 TCCCTATGTAACCCCTTCTCGGATACTTCATAACCCGGTTTTTGTTTCAACA 560
AACGGTAGACCTATAATTTTTAAGTCTA

561 ATGGTAATTGAAGATCGCCCACTGAAGCATGTGACACCCCAGATGGAAGA 640
AAGCTTTCGCAAGCATGTGGAAGCCCTGTT
561 TACCATTAACTTCTAGCGGGTGACTTCGTACACTGTGGGGTCTACCTTCTTT 640
CGAAAGCGTTCGTACACCTTCGGGACAA HindIII

641 AAAAATAGGAGCAATCCGGCCCAGTAAAAGTCGACACAGAACCACAGCTATA 720
ATGGTCAGCTCTGGAACCAGCATAGACC
641 TTTTTATCCTCGTTAGGCCGGGTCATTTTCAGCTGTGTCTTGGTGTCGATATTA 720
CCAGTCGAGACCTTGGTCGTATCTGG
BmrI SalI DrdII DrdII

721 CTATCACAGGGAAAGAGGTTAAGGGAAAGGAGCGAATGGTCTTCAACTA 800
TAAAAGGTTAAATGACCTAACTAATAAAGAT
721 GATAGTGTCCCTTTCTCCAATTCCCTTTCCTCGCTTACCAGAAGTTGATATT 800
TTCCAATTTACTGGATTGATTATTTCTA

801 CAGTACAGCCTTCCTGGAATCCAGACGATCCTACAAAGATTAAAGGGGAGT 880
ACTATATTCTCCAAATTTGACCTTAAAAG
801 GTCATGTCGGAAGGACCTTAGGTCTGCTAGGATGTTTCTAATTTCCCCTCAT 880
GATATAAGAGGTTTAAACTGGAATTTTC ScaI

881 TGGTTTTCATCAGGTAGCAATGCATCCAGATTCAATAGAATGGACAGCTTTT 960
TGGGTGCCCAGCGGTCTTTATGAATGGT
881 ACCAAAAGTAGTCCATCGTTACGTAGGTCTAAGTTATCTTACCTGTCGAAAAA 960
CCCACGGGTCGCCAGAAATACTTACCA NsiI

961 TAGTTATGCCACTCGGATTAAAGAATGCTCCAGCAGTTTTTCAGAGGAAAATG 1040
GATCACTGTTTCAAAGGCACGGAGGCC
961 ATCAATACGGTGAGCCTAATTTCTTACGAGGTCGTCAAAAAGTCTCCTTTTAC 1040
CTAGTGACAAAGTTTCCGTGCCTCCGG StuI

1041 TTTATTGCCGTCTACATCGACGACATCCTAGTTC 1074
1041 AAATAACGGCAGATGTAGCTGCTGTAGGATCAAG 1074

**Homology plot Analysis:** This is one of the important analyses performed in plant viruses. It is rather a tabular representation of the percentage of homology between the viruses at molecular level. This can be achieved by application of Bioedit software. The sequences whose homologies are to be calculated are to be aligned and loaded into Bioedit, then open the alignment menu and then select sequence identity matrix which will derive the homology between the given set of sequences into an output file in decimals when enhanced to hundred times will give the homology percentage. An example of the analysis of a set of Badnavirus genomes is given below. The homology determination can vary based on method of alignment and software. So it is to be carefully calculated. The same type of homology plot can be analyzed for Amino acid / Protein sequences as well.

**Table 2.** Homology Plot of Badnavirus genomes

| | ScBV | KTSV | BSVMY | CoYMV | CSSV | CYMV | TaBV |
|---|---|---|---|---|---|---|---|
| **ScBV** | 100 | 48.1 | 47.5 | 42.2 | 35.5 | 34.2 | 36.7 |
| **KTSV** | | 100 | 48.9 | 40.7 | 35.5 | 35.6 | 36.1 |
| **BSVMY** | | | 100 | 42.2 | 36.5 | 35.8 | 37 |
| **CoYMV** | | | | 100 | 36.6 | 35.6 | 34.7 |
| **CSSV** | | | | | 100 | 50.4 | 43.5 |
| **CYMV** | | | | | | 100 | 43.2 |
| **TaBV** | | | | | | | 100 |

**Primer Designing:** Polymerase chain reaction (PCR) as known has a very great impact on Molecular biology research. It has revolutionized the present research trend. In PCR the major issue is designing of specific primer that will amplify a desired gene of interest. Primer design is a very critical step of PCR (39). Many conditions are essential to be considered while designing a primer.

## Basic principles to be followed in PCR primer design

**1. Primer length**: Primer length is very important in PCR primer design. Generally the primer lengths vary from 18-22 mers or base pairs.

**2. Melting Temperature:** Melting temperature of the primers is very important as it decides the annealing ability of the primers with that of the template. Tm defines the stability of a DNA-DNA hybrid formation or annealing of the primer to the template. This annealing temperature should be always less than the melting temperature of the primer because above the melting temperature, the primer cannot anneal to the template and lower annealing temperature leads to mis amplifications because at this temp, the primer can bind to the template at non specific points owing to stability even in less number of matches. So while designing primers one should consider that the annealing temperatures of primer should be with in a range of 55-65°C and also the difference in melting temperatures of both the primers should not exceed 5°C. If the difference is far more in range, at the given annealing temperature only one primer can anneal to the template and this may not lead to amplification.

**3. GC content**: The GC content of the primers decides the melting temperature of primers. So optimal GC should be at least 40-60% of the primer so that the desired Tm (melting temperature) can be achieved. The GC content of the primers should be approximately equal as the vast difference may lead to different melting temperatures and in order to avoid such differences, the length of the primers should be varied in order to obtain the optimality in primer melting temperatures.

**4. GC Clamp**: Presence of G/C at 3' end last 5 bases increases efficiency of primer binding (Not more than 5 bases at 3' end). If the clamp exceeds more than 5 bases that may lead to self loops and other problems.

**5. Repeats**: Di-nucleotide repeats can mis-prime. ATATATATATAT (limit for di-nucleotide repeats is 4 di-nucleotides). If the limit of repeats exceeds 4 di nucleotide repeats, it can again lead to mis annealing of the primer in the template to palindromic repeats.

**6. Runs:** Long runs of the same base are to be avoided (limit is 4 bases repeated). If there are more than 4 runs of the same base this may also lead to mis annealing of the primer to low complexity regions of the template.

**7. 3' End stability**: Maximum, $\Delta G$ = - ve to avoid false priming. $\Delta G$ indicates the thermodynamic binding energy of the primer with template, defined for a specific region. This should be more negative at 3' ends of a primer in order to ensure more efficient binding of the primer to the template at 3' end because 3' end OH is very essential in strand extension.

**8. Secondary structure formations in primers**: These secondary structures in the primers are formed because of intra and inter molecular interaction of the primers.

Hair pins or self loops on the same primer: This results a bonding of nucleotides at the ends of the primer with that of them within the primer.

*Homodimer:* This is a secondary structure pattern formed due to inter molecular bonding of the two molecules of the same primer. This may lead to hinderance in amplification as only one primer will be available for the reaction in this case.

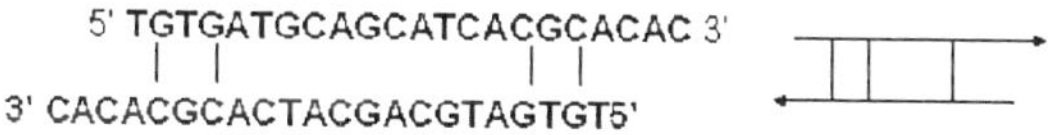


*Heterodimer*: This secondary structure is formed due to inter molecular bonding between the two primers i.e the reverse and forward primers. This also can arrest amplification as none of the primers will be available for the template in the reaction. This is also called primer dimer formation.

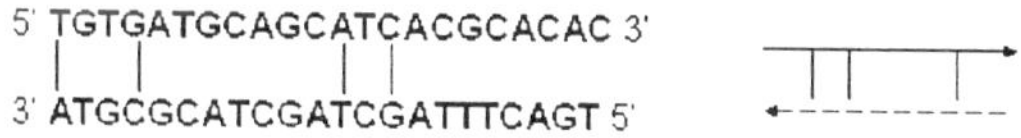


Tolerance for a homodimer is upto $\Delta G$ = -5 Kcal/Mole

Tolerance for a heterodimer is upto $\Delta G$ = -6Kcal/Mole

**9. Annealing to the target**: After designing of the primer following the above rules then the primer is to be cross checked for binding with the target. Care should be taken that there is no mis annealing with any of the other host DNA while using a direct tissue lysate for PCR.

**10.** While designing primers with restriction site for cloning, care must be taken that the restriction sites are not present in the insert region and also two or more extra bases are to be added to the 5' of the primers so that the restriction enzyme should hold the primer to cut at that over hang site.

Analysis of "CYMVF" a 26-mer DNA Oligonucleotide (Sense)

**5' CCA TCG ATC GAT TCT TGA CAC AGG AG 3'**

| Oligonucleotide Analysis | | Analysis Parameters | |
|---|---|---|---|
| Molecular weight | 8035.3 | Delta G Temperature | 25.0 degrees C |
| Tm thermodynamic | 75.4 degrees C | Probe concentration | 0.6 pMol |
| Filter Tm | 67.8 degrees C | Salt concentration | 1000.0 mMol |
| % GC Tm | 76.0 degrees C | Formamide concentration | 0.0 % |
| AT+GC Tm | 78.0 degrees C | 3' End length | 7 bases |
| Absorbance | 4.0 nMol/A260 | Run length | 4 bases |
| Absorbance | 31.9 ug/A260 | Palindrome length | 8 bases |
| Percent GC | 50.0 % | Hairpin loop stem length | 3 bases |
| Delta G | -42.9 kCal/Mol | | |
| Delta H | -188.6 kCal/Mol | | |
| Delta S | -482.4 eu | | |
| 3' End Delta G | -6.1 kCal/Mol | | |

Structural Analysis Summary

| | | |
|---|---|---|
| Number of base runs | / palindromes | 0 / 1 |
| Number of hairpin loops | | 1 |
| Number of dimers | / 2-oligo dimers | 2 / 11 |
| Number of bulge loops | / 2-oligo bulges | 0 / 14 |
| Number of internal loops | / 2-oligo internals | 0 / 0 |

*Largest Base Runs and Palindromes*

Pal: 5' cc**ATCGATCGAT**tcttgacacaggag

*Most Stable Hairpin Loops*

```
5'              CCATCGA
                  |||  |
3' GAGGACACAGTTCTTAGCT
Stem at 3 is 3 bp long. Loop = 4. Delta G = 2.8 Tm = 0.0
```

*Most Stable Dimers*

```
           5' CCATCGATCGATTCTTGACACAGGAG 3' Sense
                  ||||||||||
3' GAGGACACAGTTCTTAGCTAGCTACC 5' Sense
Stack at 3 is 10 bp long. Delta G = -13.1 Tm = 14.0

       5' CCATCGATCGATTCTTGACACAGGAG 3' Sense
             ++   ||||||    ++
3' GAGGACACAGTTCTTAGCTAGCTACC 5' Sense
Stack at 3 is 6 bp long. Delta G = -4.8 Tm = -38.9
```

*Most Stable 2-Oligo Dimers*

```
5' CCATCGATCGATTCTTGACACAGGAG 3' Sense
    ||||||       ++ +     +  +++
 3' GTAGCTGCTGTAGGATCAAGATCTCG 5' Antisense
Stack at 2 is 6 bp long. Delta G = -5.2 Tm = -40.3

       5' CCATCGATCGATTCTTGACACAGGAG 3' Sense
             ||||          +   +
3' GTAGCTGCTGTAGGATCAAGATCTCG 5' Antisense
Stack at 2 is 4 bp long. Delta G = 0.0 Tm = -106.2
```

*Most Stable 2-Oligo Bulge Loops*

```
5' CCATCGA-------TCGATTCTTGACACAGGAG 3' Sense
    ||||||             |||| ++
 3' GTAGCTGCTGTAGGATCAAGATCTCG 5' Antisense
Stack at 2 is 14 bp long. Delta G = -1.9 Tm = -12.5

5' CCATCGATCGATTCTTGACACAGGAG 3' Sense
    |||||| |||        +
 3' GTAGCT-GCTGTAGGATCAAGATCTCG 5' Antisense
Stack at 2 is 9 bp long. Delta G = -5.8 Tm = -15.1
```

**Fig 6a.** Results of Primer design analysis

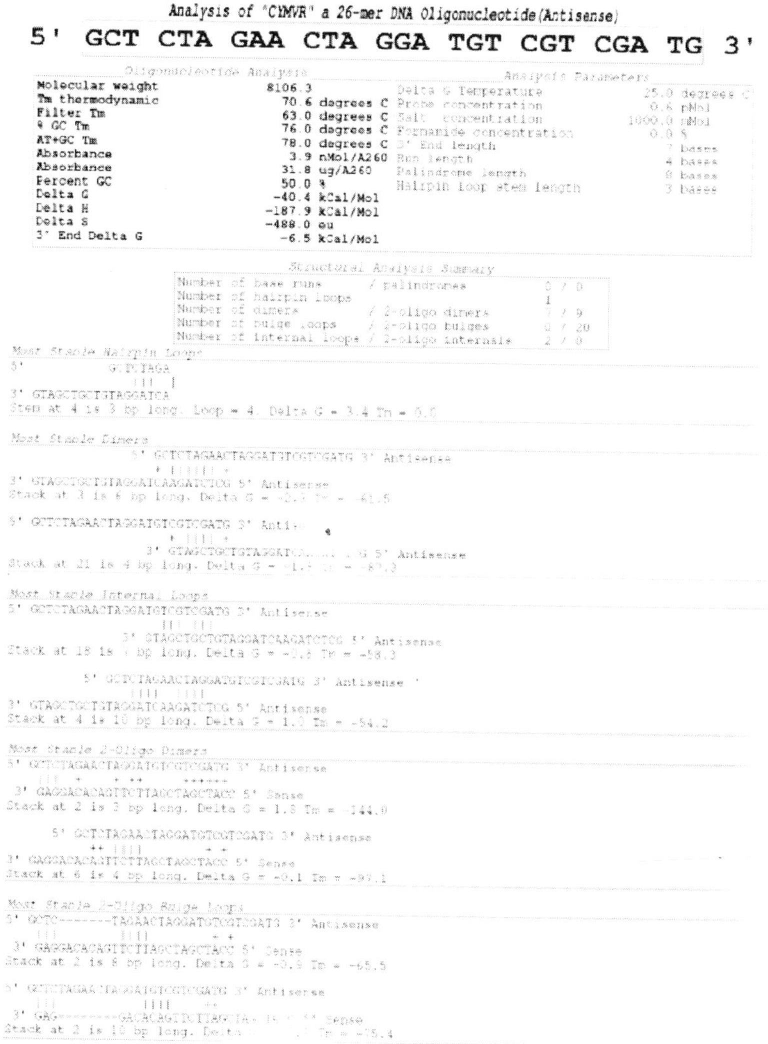


**Fig 6b.** Results of Primer design analysis

PCR primer design is achieved by many of the free as well as commercial tools. Free tools include Gene Runner (V.3.05) (Downloadable at *www.generunner.com*), Integrated DNA technologies Primer quest *etc.*, Commercial tools include

DNASTAR's PRIMER SELECT, OLIGO, Vector NTI suite (free for academics upon request), DNASIS *etc.*

Eg: Design of a set of primers for amplification of 1.1 kb fragment of CYMV

Generally the first few bases of the sequence itself can act as forward primer with slight modifications as per primer rules and last few bases of the sequence in reverse complementation will act as the reverse primer. The modifications should be done according to the primer design rules. Then they should be cross checked using the software. The present set of primers are designed with the Gene Runner software. The advantage with the gene runner is efficacy easily usable environment. For a dimer or loop formation, it will define the melting temperature of its formation along with the ΔG (energy of formation) which is useful in deciding whether to ignore that secondary structure formation or not. In order to access the gene runner, open the gene runner software, then the oligo analysis window, paste the forward primer sequence in the oligo column provided in the window. And also paste the reverse primer sequence in the column provided after clicking the edit anti sense button. Now click on ok or press enter. This will show you the Tm's of the primers and also the different secondary structure formations like self loops, dimmers etc., and the result below is the sequence of the primer obtained for the 1.1 kb fragment of CYMV described in the above sections. The dimmers formed here are theoretical and their Tm is far low *i.e.* 0°C and even too low which is an extreme condition not used in PCR and these dimer formations will not interfere the common PCR. So these set of primers can be used in PCR as they will full fill all the PCR primer design principles. And these primers have been used in reality to amplify the viral DNA and we got the amplification. The above sequence was an experimental sequence derived from this experiment (Fig. 6).

**Amino acid composition:** Amino acid composition is necessary in order to report the nature of the protein. This is to be predicted for any protein sequence in order to say about its characteristics whether it is a basic protein or acidic protein. This also helps to calculate its isoelectric pH of the given protein. This can be done using sequence analysis packages like Bioedit, DNASIS *etc.*

Eg: The present example states the analysis of a protein of CYMV with Bioedit. In order to perform this analysis, open the protein sequence in Bioedit, open sequence menu > protein > amino acid composition (Fig. 7).

Protein: CYMV_T
Length = 357 amino acids
Molecular Weight = 40949.31 Daltons

| Amino Acid | | Number | Mol% |
|---|---|---|---|
| Ala | A | 18 | 5.04 |
| Cys | C | 6 | 1.68 |
| Asp | D | 16 | 4.48 |
| Glu | E | 26 | 7.28 |
| Phe | F | 20 | 5.60 |
| Gly | G | 22 | 6.16 |
| His | H | 10 | 2.80 |
| Ile | I | 28 | 7.84 |
| Lys | K | 31 | 8.68 |
| Leu | L | 28 | 7.84 |
| Met | M | 12 | 3.36 |
| Asn | N | 14 | 3.92 |
| Pro | P | 14 | 3.92 |
| Gln | Q | 18 | 5.04 |
| Arg | R | 16 | 4.48 |
| Ser | S | 22 | 6.16 |
| Thr | T | 18 | 5.04 |
| Val | V | 22 | 6.16 |
| Trp | W | 4 | 1.12 |
| Tyr | Y | 12 | 3.36 |

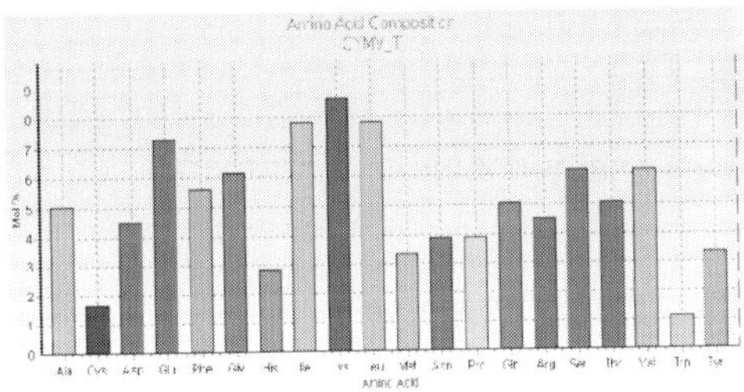

**Fig 7.** Amino acid count

**Hydrophobicity index and helical wheel:** Hydrophobicity of a protein is very essential and is to be calculated for a protein to know its property in *Invitro*. As some of the more hydrophobic proteins are not expressed by common expression vectors. So in order to express them, specialized strategies are to be standardized. This index helps to know about the hydrophobicity of the protein. There are many methods for calculation of hydrophobicity profile, each method is chosen as per ones analysis is required.

Eg: The present example deals with the calculation of mean hydrophobicity of a protein by Kite and Doolittle method from CYMV using Bioedit (Fig. 8). In order to perform this task, open the sequence in Bioedit > open sequence menu > protein > Kyte and Doolittle mean hydrophobicity profile.

Eg: In order to get helical wheel diagram of a protein based on mean hydrophobic movement using Bioedit, the following steps are to be performed. Open the sequence in Bioedit >sequence menu > protein > Helical Wheel Diagram (Fig. 9).

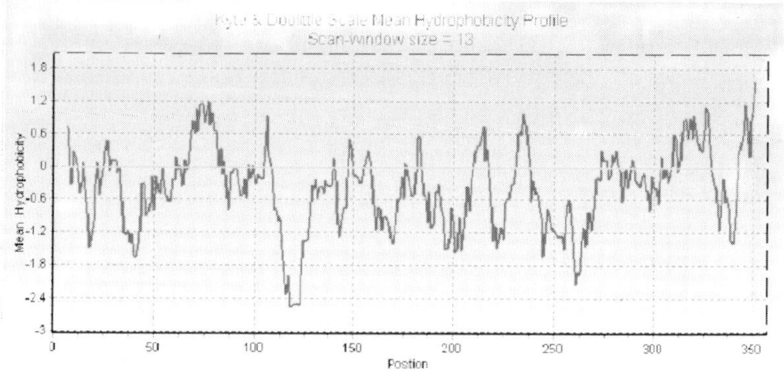


**Fig 8.** Hydrophobicity index profile of the query

**Isoelectric Point:** Isoelectric point of the protein is very important to characterize the protein using 2D page. In 2D page, the protein is run in isoelectric focusing in the first front and in SDS PAGE in the second front to obtain proper separation of even smaller proteins. So calculation of isoelectric point of protein is essential in this method of characterization of proteins.

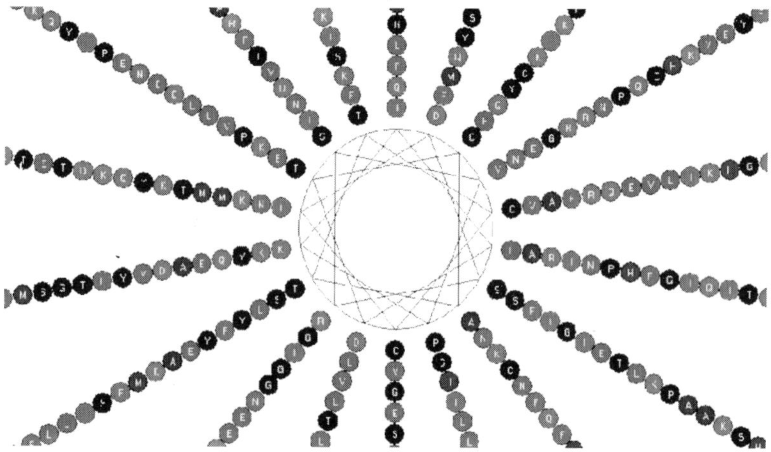

**Fig 9.** Helical wheel of the query

Eg: The following example is analysis of isoelectric point of a protein of CYMV. To obtain it, DNASIS is used and the analysis follows these steps. Open the sequence in DNASIS > open function menu > Prediciton > Isoelectric Point (Fig. 10).

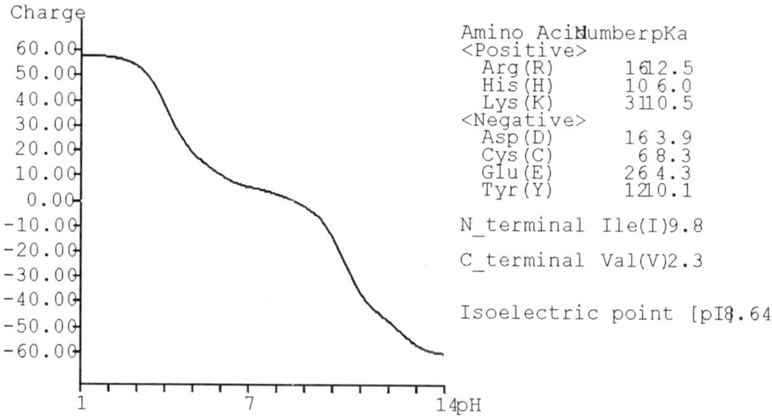


**Fig 10.** Isoelectric pH profile of the query

**Proteolytic cleavage sites:** These proteolytic enzymes are the enzymes which specifically cleave peptides at specific recognition sites as restriction enzymes do with the nucleic acid. The analysis of these cleavage sites help to distinguish the boundaries of some of the proteins which are translated as polyproteins and post translationally modified into individual functional proteins. This can be done using any of the protein sequence analysis tools like DNASIS, Win Pept, etc.,

Eg: This example gives an analysis of CYMV protein with some of the proteolytic enzymes like proteinase-K, CNBr, Chymosin. In order to get the protease cleavage sites in DNASIS, open the sequence in DNASIS > open function menu > search > proteolytic enzymes (Fig. 11).

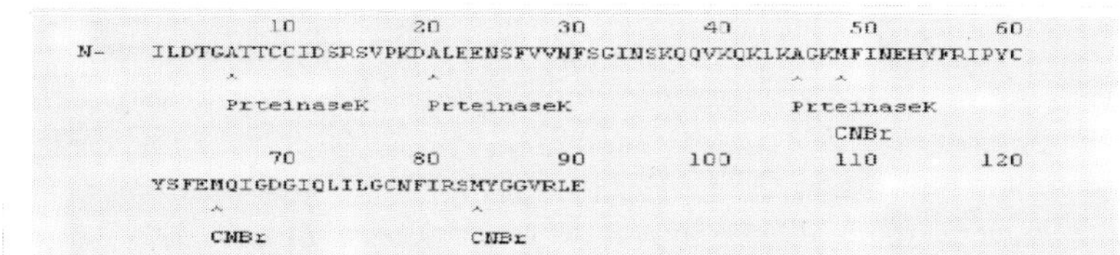


**Fig 11.** Protease cleavage site profile of the query

**Protein Secondary structure:** Protein secondary structure prediction is one of the important aspects of protein sequence analysis. The secondary structure prediction of the protein secondary structure involves different statistical methods and also some of the pre available structural data from structural studies. This is very essential in moving a step towards modeling the 3D structure of a protein. There are different methods of theoretical protein secondary structure prediction such as chou fasman, chou fasman rose, Robson etc. **(32)**.

Eg: The example shows a secondary structure pattern of a CYMV protein predicted from the DNASIS software. This is done by the following steps:

open the sequence in DNASIS > open function menu > select prediction > protein secondary structure.

Many of the online tools are also available for the prediction of secondary structure of a protein from Expasy server (*http://www.expasy.org*) such as PSIPRED, JPRED, SOPMA, GOR 4, etc.

It is very difficult to represent each and every method of sequence analysis, only some of the methods that are useful to analyze the plant virus are represented here.

**Phylogenetic Analysis:** Phylogenetic analysis is most important in order to know the evolutionary aspects of Plant viruses. This is a very primitive study which starts as a tool to distinguish between every genus, every species and also gives taxonomical status of an organism. Phylogenetics as a primitive classical study uses many parameters to study the evolutionary aspects which included morphological, physiological and other characters which are very vague. In usage of these characters, there are a lot of errors in the analysis. For example presence of an eye, a morphological character if taken into consideration in Phylogenetics, Drosophila, Fish, Dog and Homo sapiens can be placed in the same group which phylogenetically unrelated. This Phylogenetics evolved as the biological research grew over years into advanced Molecular Phylogenetics involves analysis of evolutionary aspects of the organisms at the level of both DNA and Protein sequences. This level of Molecular Phylogenetics is highly advantageous in analyzing the evolutionary aspects of organisms. The Concept of Phylogenetic tree is another important representation of Phylogenetic relationships. A Phylogenetic tree is a two dimensional graph which represents Phylogenetic relationships of organisms where the root of the tree represents the ancestral or primitive organisms and the top of the branches represent the higher or more evolved species. The branches and nodes are the connecting points and they show the path of evolution, how the organisms evolved from the primitive ones. These Phylogenetic trees are of different types which are broadly classified into gene tree and species tree.

Gene tree is represented when the organisms are related taking into consideration of single genes which are either conserved or diverged, for example RNA polymerase gene in different species. Species tree is represented when whole genomes of different species or organisms as a whole are considered in deriving Phylogenetic tree. The Phylogenetic tree can also be classified based upon presence of ancestral species into rooted tree and unrooted tree. Rooted tree is said to have a ancestor or root from which the organisms have been said to evolve. Unrooted tree represents only phylogenetic evolutionary relationships between the given species but does not show any level of organization of the species with respect to its ancestor. Dendrogram is a phylogenetic tree that represents phylogenetic relationships between the organisms and where as phylogram is a phylogenetic tree that represents not only the phylogenetic relationships but also the times required for the evolution to take place. This is very important in analyzing the evolutionary pattern of Plant viruses either at gene level or at genome level. The other aspect called Molecular clock hypothesis (14) is important in the phylogenetic analysis which assumes that the time taken for the number of changes between the organisms in the course period of evolution is constant and is not dependant on the number of change in a family of conserved sequences. A normal phylogenetic tree is obtained by using the following strategy.

1. To obtain a proper sequence alignment.
2. Tree Building.
3. Tree Evaluation.

**1. Sequence alignment:** Sequence alignment is a method of aligning two sequences by mapping residues of one sequence over other in order to obtain maximum number of matches. In order to obtain best alignment in case of highly diverged sequences, to obtain optimal or best alignment, a character called gap is introduced which represents Insertions or Deletions (Indels) as shown below in a sample alignment. In this representation, the linear bars show the matched nucleotides between two DNA sequences and the other regions represent nucleotide mismatches. The "_" character represents the gap introduced to increase the number of matches in the alignment. Depending upon the introduction of Gaps, the alignments are classified into two types 1. **Gapped Alignment Alignment** (Gaps are introduced) and 2. **Ungapped Alignment** (No gap is introduced). But the character gap which is introduced is considered as disadvantage if it is introduced for many times above a particular threshold as it denotes unwanted character such as Insertions or Deletions (as insertions or deletions are unfavorable events in evolution). So penalty is paid for introduction of gap in an alignment while scoring the alignment. Thus if the number of gaps increase, the penalty increases and the score of the alignment decreases. Score of an alignment is a numerical value given to the alignment to say how much reliable is the alignment. For example, If you give a 10 score for a match, -1 for a mismatch and -5 for a gap (Gap penalty) for the alignment as shown in the Fig. 12, the score of the alignment can be stated as addition of all the scores f residues obtained in the alignment. Score in this case is 8, as the number of matches are 14, mismatches are 2, gaps introduced is 1.

```
ATGCCAACAGTCTGGTA
|||| ||||| |||  |
ATGC_AACAGCCTGGAA
```

**Fig. 12** Example Alignment

Depending upon the number of sequences used in the alignment, the alignment is classified into 2 types. 1. **Pairwise alignment** and 2. **Multiple alignment**. If two sequences are aligned, it is known as pairwise alignment and if more than two sequences are aligned, it is termed as multiple sequence alignment. If the whole sequences are aligned, then the alignment is termed as **Global alignment** and if only a part of sequences is aligned, it is known as **Local alignment**. Thus the alignment is done by many methods like Dot matrix analysis, Dynamic Programming methods and Heuristic searches. Dot matrix analysis is a simplified analysis to obtain local alignment between a pair of sequences can be obtained by many of the sequence analysis tools like DNASIS, EMBOSS etc., and also by using online tools like DOTTER, DOTLET etc., Dynamic Programming methods are accurate alignment methods but they consume a lot of time also need a lot of computational memory. There are two chief Dynamic Programming methods 1. Needleman and Wunsch algorithm and 2. Smith and Waterman algorithm. Needleman and Wunsch algorithm **(30)** is used to obtain the multiple sequence alignment and Smith and Waterman algorithm **(36)** is used to obtain local alignment between a pair of sequences. Heuristic methods are very fast alignment methods to give local alignments between a pair of sequences. Generally, heuristic methods are used for database searches and they need fast and local alignments. Some of the heuristic methods include BLAST (Basic Local Alignment Search Tool) algorithm **(1, 2)** and FASTA algorithm **(27, 31)**. Generally the sequence alignment used for phylogenetic analysis is the Multiple Sequence Alignment (MSA). This is obtained by using software like Clustal X **(20)**, Clustal W **(23)**, MEGALIGN of DNASTAR *etc.*, Clustal X is one of the freely available and friendly user interface tool that can be used to obtain multiple sequence alignment *ftp://ftp-igbmc.u-strasbg.fr/pub/ClustalX/*. This uses a method known as Progressive multiple sequence alignment which is slightly different and easy from the traditional approaches used for multiple sequence alignment. This method involves the first step of scoring between the sequences and then the next step is first to align the sequences more closely related and then adding the other sequences one by one into the alignment until the alignment score is with in a given threshold (a predefined value), aligning them into groups and finally displaying the alignment. This is an easier and faster method to obtain the multiple sequence alignment in a short time. This method also produces a guide tree based upon the alignment groups. This tree may also be used to infer the phylogenetic relationship between the given sequences. As it is very difficult to define all the sequence alignment methods within this paper, a brief note of the methods is given in order to give a brief idea of these methods. In order to obtain a MSA of a set of sequences, an input file should be produced with all FASTA formats of the sequences in any text editors like notepad. Open the Clustal X window and then follow the following steps.

Open File menu > Load sequences > open the input file (this will load the sequences)

Open Alignment Menu > Do complete Alignment > (this will prompt the location to save the output files) > Align.

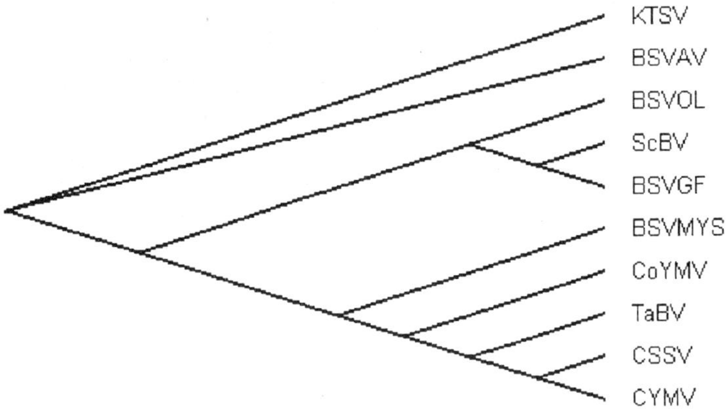


**Fig 13.** NJ guide tree produced from Clustal alignment

This will produce two files with extensions .aln (alignment file) and .dnd (dendrogram or tree file). The alignment file can be viewed by any of the text editors and the tree file is visualized using tree viewing tools like tree view, Tree explorer etc., a guide tree obtained from alignment of Badnavirus whole genomes in Clustal X is represented in Fig. 13.

**Tree Building Methods:** In phylogenetic analysis, this is a very important step as it decides the phylogenetic tree. There are chiefly two types of tree building methods.

1. Character based methods and 2. Distance based methods.

Many commercially and free available software are there in the WWW which can be used to build trees which one requires. Among them Phylip (Phylogenetic inference package by Dr. Felenstein group) is a freely available package **evolution. genetics.washington.edu/phylip.html** which can be used for this purposes. The other packages include GCG interfaces, PAUP **(35)** *etc.* **(4, 7, 10, 28)**

**1. Character based methods:** These methods treat nucleotides or amino acids as individual characters and based on the changes or patterns of these characters, these methods predict the phylogenetic relationships between the sequences. The two important character based methods are Maximum Parsimony method and Maximum likelihood method.

**Maximum Parsimony method:** This method states that the evolutionary model is perfect when it invokes very less number of changes or mutations that are rare events in evolution **(16, 19, 21)**. This considers two types of sites in a sequence alignment Informative and uninformative sites. Mutations that are only informative are of use than the other ones which commonly occur. For a change to be informative, it should at least occur at the same point in at least half of the species represented. The possible unrooted trees are drawn from these informative sites and the tree which represents the least number of changes is considered as the best tree. The limitation with this method is that it can be only performed between the sequences with high level of homology. As the number of differences increase, it will be difficult to draw all the possible trees for all the informative sites and also to choose the tree with number of changes. The program of the Phylip package which builds tree with this type of method is DNAPARS (Fig. 14) (for nucleotide) and PROT PARS (for protein). Many other programs of the same package with modification of method are available for DNA sequences which include DNAPENNY, DNAMOVE. In order to run Phylip programs, an input file of the sequence alignment in Phylip format is required that can be obtained by using READSEQ or also from Clustal X by choosing the Phylip output format. The infile or input file is to be copied into the exe directory of Phylip package where the executables of Phylip package are available. Then the appropriate executable named DNAPARS is run by double clicking the icon, then the program is executed in a dos window where it asks input file name and executes the input if it is named as infile. After the execution, the program requests the acceptance to complete the program, If it is provided by typing the letter Y, the program writes the output into an outfile. If already an outfile exists, the program asks option whether to append the existing file or to replace or else the user wishes to write the output into another file, the same with the outtree file also. If the user wants to change any of the parameters, the corresponding letter to the parameter is type and the parameter is changed.

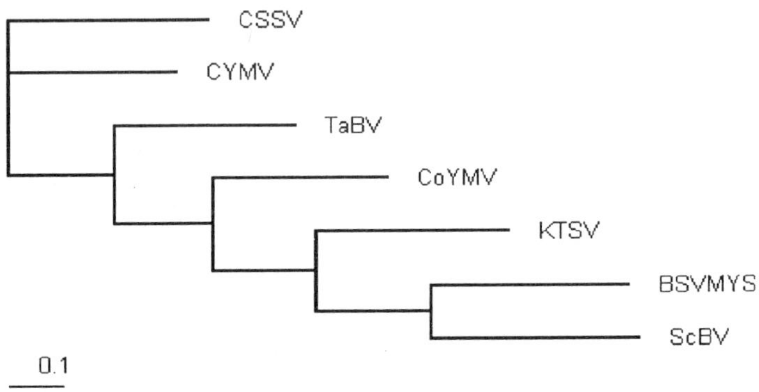


**Fig. 14** Phylogram representing DNAPARS tree for Badnaviruses

**Maximum Likelihood method:** Maximum likelihood method is a character based method of phylogenetic prediction which is exclusively based on statistical basis of prediction **(9, 11, 12)**. In this method, probabilities are considered for every individual nucleotide substitution for prediction. Generally, base transitions are most common than the base transversions in case of the biological sequences. This is regarded as a feature in deriving the evolutionary relationships between the organisms. Basing on the number of substitutions and their pattern, a statistical approach is derived to predict the possible trees and the best tree with highest probability is chosen to predict the phylogenetic evolutionary relationships. The Phylip programs that use the Maximum likelihood method for tree building is DNAML (Fig. 15) (without Molecular clock hypothesis) and DNAMLK (Fig. 16) (with molecular hypothesis) for DNA sequences and PROTML (without Molecular clock hypothesis) and PROTMLK (with molecular hypothesis) for amino acid sequences. The method of building a tree by maximum likelihood tools is same as that of defined in the Maximum parsimony methods in the PHYLIP software except that the parameters are different. This method is applied in case of highly diverged sequences as this method needs to calculate all the probable changes.

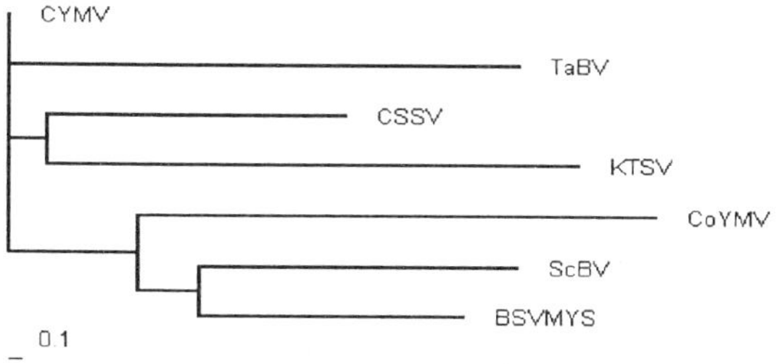


**Fig. 15** Phylogram representing DNAML tree for Badnaviruses

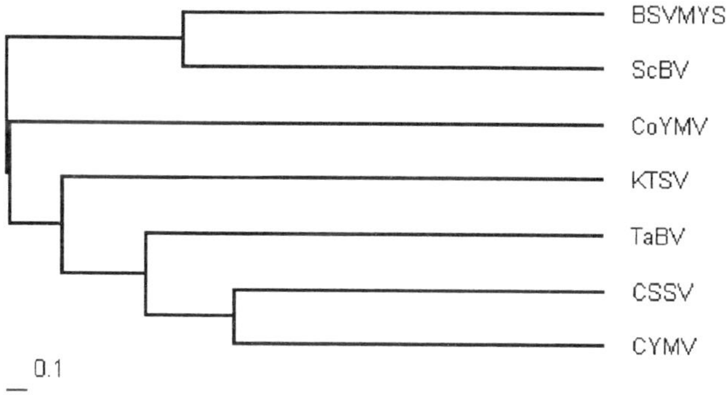


**Fig. 16**. Phylogram representing DNAMLK tree for Badnaviruses

**2. Distance based methods:** Distance based methods estimate phylogenetic relationships by estimating a distance parameter based on nucleotide differences between the sequences **(8, 13, 22, 24)**. Here the changes in nucleotides are considered as distances. The only change in this distance based method phylogenetic prediction is first the distance measures between the sequences in the sequence alignment are to be calculated and then using these distances, phylogenetic tree is constructed. The programs that calculate the distances in Phylip software are DNADIST for DNA sequences and PROTDIST for Amino acid sequences. The input for these predictions is the sequence alignment file in Phylip format. The output distance in outfile is used as input for the distance based phylogenetic predictions. There are three chief methods of distance based tree building. They are

**1. Fitch and Margoliash method (FM):** This method uses a distance table and the sequences are defined in threes to define the branches of predicted tree and to calculate the branch lengths of tree. There are programs in Phylip which build the phylogenetic tree based on Fitch and Margoliash method. The program that constructs phylogenetic trees by using Fitch and Margoliash method **(17)** is Fitch (Fig. 17) (does not use molecular clock hypothesis) and Kitsch (Fig. 18) (uses molecular clock hypothesis).

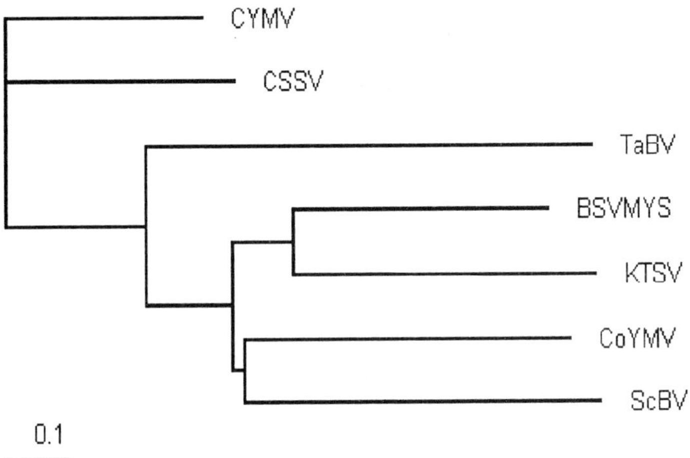


**Fig. 17** Phylogram representing FM tree for Badnaviruses

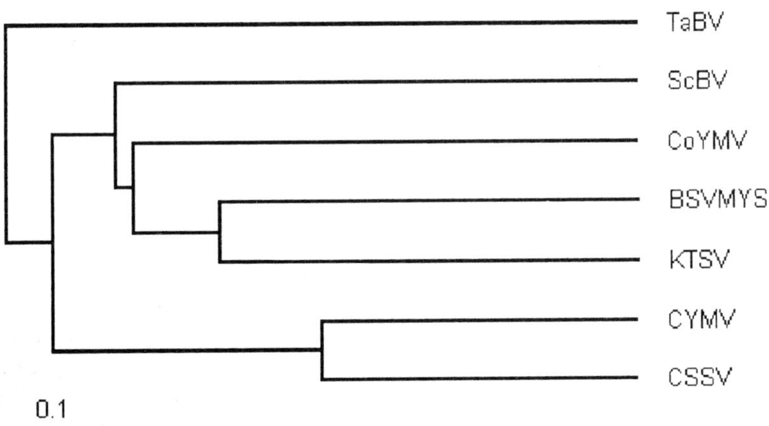


**Fig. 18.** Phylogram representing FM tree with molecular hypothesis (Kitsch) for Badnaviruses

**2. UPGMA (Unweighed pair group method with arithmetic mean):** UPGMA method is a simple method for tree construction that assumes the rate of change along the branches of the tree is a constant and the distances are approximately ultrametric **(34, 38)**. UPGMA method starts by calculating branch lengths between the most closely related sequences, then averages the distance between this pair or sequence cluster and the next sequence cluster and continues until all the sequences are included in the tree. The program that builds phylogenetic tree in Phylip package applying UPGMA method principle is Neighbor (Fig. 19).

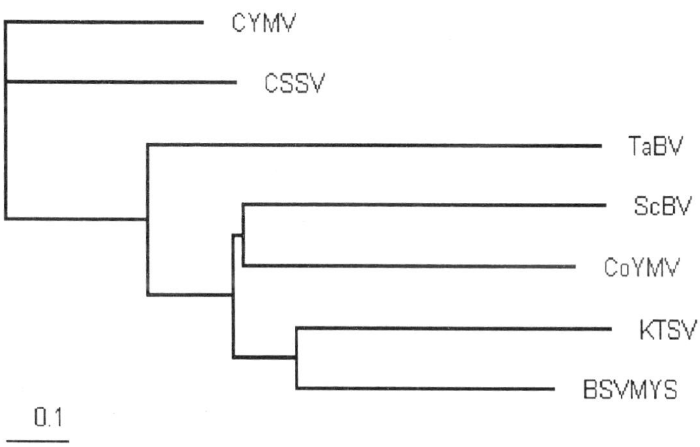


**Fig. 19.** Phylogram representing UPGMA tree for Badnaviruses

**3. Neighbor Joining method (NJ method):** Neighbor Joining method (Saitou and Nei 1987) **(34, 38)** is very much like the Fitch-Margoliash method except that the choice as to which sequences to pair is defined by a different algorithm. The tree starts as a star topology and grows up into an unrooted tree based on the distances between the species. The program that builds a phylogenetic tree based on Neighbor Joining principle is Neighbor (Fig. 20).

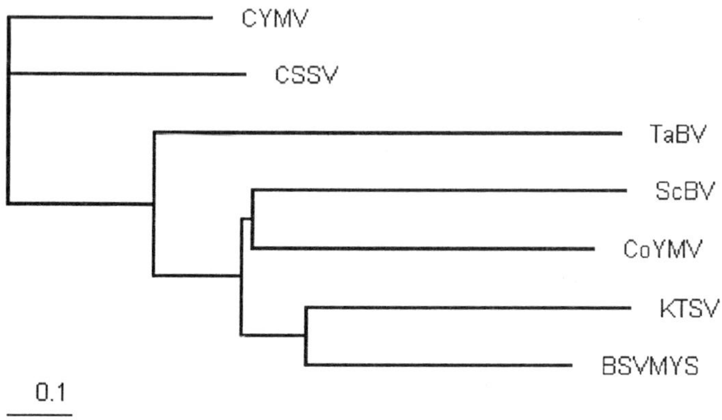


**Fig. 20.** Phylogram representing NJtree for Badnaviruses

The above three Distance based tree building methods are common for both the DNA and Protein sequences except the distance estimating programs like DNADIST and PROTDIST. And the trees illustrated are only based on DNA sequences in phylogram view.

**Tree Evaluation Methods:** The tree evaluation methods are generally used to evaluate the reliability of the phylogenetic prediction. This is generally done by using some of the statistical methods which are used to evaluate different types of datasets. One of such methods is Bootstrap analysis **(40, 15)**. Bootstrap is a statistical based random method of sampling of all the data sets or species of a phylogenetic tree in order to state the reliability of phylogenetic prediction. In this method, first the data set or the phylogenetic tree is copied into multiple datasets and the original data set is preserved, the random sampling method is used on the data sets to obtain the reliability of phylogenetic prediction from the total number of data sets represented as Bootstrap values. They differ from the taxa to taxa. The method which does the boot strap analysis in Phylip is Seqboot. The other method of doing boot strapping is by deriving a bootstrap tree by using Clustal X software tool. Where the option boot strap NJ tree of the tree menu is used to draw a boot strap NJ tree that can be visualized by common tree viewing softwares like NJPLOT, UNROOTED TREE, TREEVIEW, TREEVIEW, TREE EXPLORER.

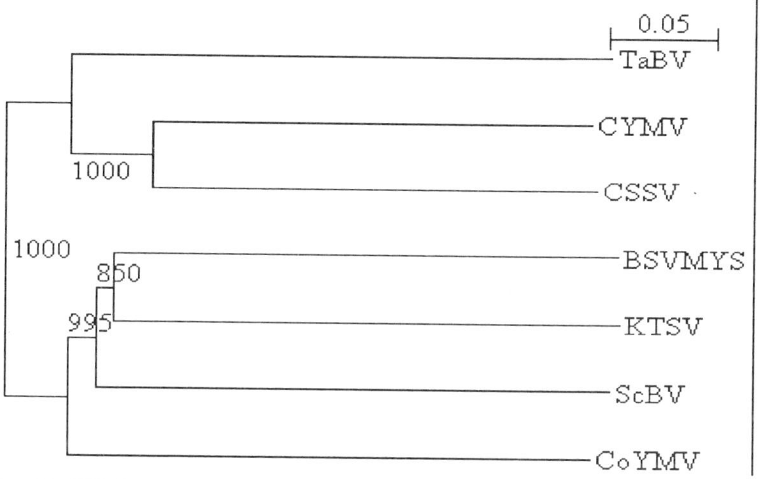


**Fig. 21.** NJ Tree representing Boot strap values derived using Clustal X

## *Literature Cited*


1. Altschul, Gish W, Miller W, Myers EW, Lipman DJ, 1990, A basic local alignment search tool, *J. Mol. Biol.* 215: 403-410.
2. Altschul, Stephen F, *et al.* 1997. "Gapped BLAST and PSI-BLAST: a new generation of protein database search programs", *Nucleic Acids Res.* 25:3389-3402.
3. Baum BR. 1989. PHYLIP: Phylogeny Inference Package. Version 3.2. (Software review). *Quarterly Review of Biology* 64: 539-541.
4. Cavalli-Sforza LL, Edwards AWF. 1967. Phylo-genetic analysis: models and estimation procedures. *Evolution* 32: 550-570 (also Amer. J. Human Genetics 19: 233-257).
5. Dayhoff MO, 1979. *Atlas of Protein Sequence and Structure*, Vol. 5, Supplement 3, 1978. National Biomedical Research Foundation, Washington, D.C.
6. David W. Mount, Bioinformatics: Sequence and Genomes analysis. Cold Spring Harbor Laboratory Press. New York. PP. 27-44.
7. Ewens J Warren, Grant R Gregory, 2001. Statistical Methods in Bioinformatics: An Introduction. Springer Verlag. New york Inc. PP. 385-422.
8. Farris JS. 1981. Distance data in phylogenetic analysis. pp. 3-23 in *Advances in Cladistics*: Proceedings of the first meeting of the Willi Hennig Society, ed.

V.A. Funk and D.R. Brooks. New York Botanical Garden, Bronx, New York.

9. Felsenstein J. 1973a. Maximum likelihood and minimum-steps methods for estimating evolutionary trees from data on discrete characters. *Systematic Zoology* 22: 240-249.
10. Felsenstein J. 1979. Alternative methods of phylogenetic inference and their interrelationship. *Systematic Zoology* 28: 49-62.
11. Felsenstein J. 1981a. Evolutionary trees from DNA sequences: a maximum likelihood approach. *J. Molecular Evolution* 17: 368-376.
12. Felsenstein J. 1981c. Evolutionary trees from gene frequencies and quantitative characters: finding maximum likelihood estimates. *Evolution* 35: 1229-1242.
13. Felsenstein J. 1984a. Distance methods for inferring phylogenies: a justification. *Evolution* 38: 16-24.
14. Felsenstein J. 1985a. Confidence limits on phylogenies with a molecular clock. *Systematic Zoology* 34: 152-161.
15. Felsenstein J. 1985b. Confidence limits on phylogenies: an approach using the bootstrap. *Evolution* 39: 783-791.
16. Felsenstein J, Sober E, 1986. Parsimony and likelihood: an exchange. *Systematic Zoology* 35: 617-626.
17. Fitch WM, Margoliash E. 1967. Construction of phylogenetic trees. *Science* 155: 279-284.
18. Hall TA. BioEdit: a user-friendly biological sequence alignment editor and analysis program for Windows 95/98/ NT. *NUCLEIC ACIDS SYMPOSIUM SERIES* 1999, VOL 41, pages 95-98.
19. Hendy MD, Penny D. 1982. Branch and bound algorithms to determine minimal evolutionary trees. *Mathematical Biosciences* 59: 277-290.
20. Thompson JD, Gibson TJ, Plewniak F, Jeanmougin F, Higgins DG. The CLUSTAL_X windows interface: flexible strategies for multiple sequence alignment aided by quality analysis tools *Nucleic Acids Research*, Vol 25, Issue 24 4876-4882.
21. Jin L, Nei M. 1990. Limitations of the evolutionary parsimony method of phylogenetic analysis. *Molecular Biology and Evolution* 7: 82-102.
22. Jukes TH, Cantor CR. 1969. Evolution of protein molecules. pp. 21-132 in Mam-malian Protein Metabolism, ed. H.N. Munro. Academic Press, New York.

23. Julie D Thompson, Desmond G Higgins, Toby J Gibson CLUSTAL W: improving the sensitivity of progressive multiple sequence alignment through sequence weighting, position-specific gap penalties and weight matrix choice. *Nucleic Acids Research*, 1994, Vol. 22, No. 22 4673-4680

24. Kimura M. 1980. A simple model for estimating evolutionary rates of base substitutions through comparative studies of nucleotide sequences. *Journal of Molecular Evolution* 16: 111-120.

25. Krane E Dan, Raymer L Michael. 2003. Fundamental concepts of Bioinformatics. Pearson Education publications. India. PP. 48-51.

26. Liolios K, Tavernarakis N, Hugenholtz P, Kyrpides NC. The Genomes On Line Database (GOLD) v.2: a monitor of genome projects worldwide. *Nucleic Acids Res.* 2006 Jan 1; 34 (Database issue):D332-4

27. Lipman DJ, Pearson WR. 1985, Rapid and sensitive protein similarity search, *Science* **227:** 1435-1441.

28. Cavalli-Sforza LL, Edwards AWF. 1967. Phylogenetic Analysis: Models and Estimation Procedures. *Evolution*, 21(3): 550-570

29. Zuker M, Mathews DH, Turner Algorithms and Thermodynamics for RNA Secondary Structure Prediction: A Practical Guide In *RNA Biochemistry and Biotechnology*, J. Barciszewski & B.F.C. Clark, eds., NATO ASI Series, Kluwer Academic Publishers, (1999).

30. Needleman SB, Wunsch CD. 1970, A general method applicable to the search for similarities in the amino acid sequences of two proteins, *J. Mol. Biol.* **48:** 443-453.

31. Pearson WR, Wood T, Zhang Z, Miller W. Comparison of DNA sequences with protein sequences. *Genomics* (1997) 46:24-36

32. Prevelige, P Jr, Fasman GD. 1989, Prediction of protein secondary structure and Principles of protein confirmation, pp. 1-91, Plenum Press, New York.

33. Qi Huang, John S Hartung, 2001. Cloning and sequence analysis of an infectious clone of *Citrus yellow mosaic virus* that can infect sweet orange via *Agrobacterium*-mediated inoculation. *Journal of General Virology,* 82, 2549-2558.

34. Saitou N, Nei M. 1987. The neighbor-joining method: a new method for reconstructing phylogenetic trees. *Molecular Biology and Evolution* 4: 406-425.

35. Sanderson MJ. 1990. Flexible phylogeny reconstruction: a

review of phylogenetic inference packages using parsimony. *Systematic Zoology* 39: 414-420.

36. Smith FF, Waterman, MS. 1981, Identification of common molecular subsequences, *J. Mol. Biol.* 147**:** 195-197.

37. Stephen Misener, Stephen A, Krawetz 2000. Bioinform-atics : Methods and Protocols, Humana Press, Totowa, New Jersey. PP. 191.

38. Studier JA, Keppler KJ. 1988. A note on the neighbor-joining algorithm of Saitou and Nei. *Molecular Biology and Evolution* 5: 729-731.

39. Thomas Kämpke, Markus Kieninger, and Michael Mecklenburg 2001. Efficient primer design algorithms. *Bioinformatics* 17(3): 214-225.

40. Wu CFJ. 1986. Jackknife, bootstrap and other resampling plans in regression analysis. *Annals of Statistics* 14: 1261-1295.

*Rev. Plant Pathol. Vol. 4, 2008*
*Indian Society of Mycology and Plant Pathology*
*Scientific Publishers (India), Jodhpur pp. 261-299*

# MYCOTOXINS : TOXICITY, DIAGNOSIS, REGULATIONS AND CONTROL THROUGH BIOTECHNOLOGY

***Kaushal K. Sinha*[1] *and Ajoy K. Choudhary*[2]**

[1] University Department of Botany, T.M. Bhagalpur University, Bhagalpur – 812 007, India.
[2] Botany Department and Dept. of Biotech., T.N. B. College, Bhagalpur – 812 007, India.

KEY WORDS : Mycotoxins, diagnosis, regulations, control, biotechnology

## ABSTRACT

Mycotoxins are produced as contaminants on various food and feed commodities under suitable environmental conditions. These chemical contaminants are known to induce a variety of acute and/or chronic disease syndromes i.e. mycotoxicoses in affected animals and human beings. Aflatoxins, ochratoxins, citrinin, fumonisins, zearalenone, trichothecenes and ergot toxins are the important mycotoxins which have been analysed as natural contaminants in various agricultural commodities.

Present review deals with the symptoms and diagnosis of the above mycotoxicoses along with the nature of their treatment in each case. Because of the variations in health risk caused by mycotoxin contaminated food, many developed countries have laid down specific regulations for the import/export of those items in terms of economic implications. Various measures have also been discussed to minimize the risk of mycotoxin contamination during pre-harvest, harvest and post-harvest conditions utilizing HACCP Concept.

Biotechnology also plays a great role in the control of mycotoxins. Enhancement of host resistance through breeding programmes as well as identification of Resistance Associated Proteins (RAP) in a particular seed variety will enable to prevent mycotoxin production at the outset. Control of mycotoxins through micro-organisms and other suitable methods has also been discussed in detail.

Mycotoxins are secondary metabolites produced by certain toxigenic fungi under suitable environmental conditions. Their production is unavoidable in most of the agricultural commodities because of prevalence of those moulds in the surroundings as well as conducive climatic conditions prevailing in most parts of the world. These groups of chemical compounds have always been found hazardous to men and domestic animals. However, the technologies evolved during the last forty years have allowed researchers to isolate mycotoxins and study their effects on food crops, livestock as well as on humans. Mycotoxins have been classified on the basis of the main organ system where they affect (Table I). Mycotoxin contamination is also unpredictable, which makes it unique challenge to food safety (43).

It is now established that mycotoxicoses (the diseases caused by mycotoxins) have been responsible for major epidemic in man and animals at least during recent historic times. The most important has been the ergotism, which killed thousand of people in Europe during feudal days; Alimentary Toxic Aleukia (ATA) which was responsible for death of several thousands of people in USSR in 1940's; stachybotryotoxicosis, which killed thousands of horses and cattle in USSR in 1930's and aflatoxicosis which killed 1,00,000 young turkeys in England in 1960. In late 1974 and early 1975, over 200 villages in Western India experienced an outbreak of a disease affecting men and dogs which characterized by jaundice, rapidly developing ascites, portal hypertension and a high mortality rate, with death resulting from massive GI tract bleeding. The disease was confined to very poor, who were forced to consume badly moulded corn containing aflatoxins between 6.25 to 15.6 ppm, an average daily intake per victim of 2 to 6 $\mu$g aflatoxins.

**Table 1.** Important groups of mycotoxins based on target organs.

| | |
|---|---|
| (A) Hepatotoxins | Aflatoxins |
| | Sporidesmin |
| | Luteoskyrin |
| | Cyclochlorotine |
| | Rubratoxins |
| | Sterigmatocystin |
| (B) Nephrotoxins | Ochratoxins |
| | Citrinin |
| (C) Neurotoxins | Penitrem |
| | Patulin |
| | Citreoviridin |
| | Miscellaneous Neurotoxins |
| | Ergots |
| (D) Cytotoxins | Trichothecenes |
| (Alimentary Tract Toxins) | T-2 toxins |
| | Diacetoxyscripenol |
| | Neosolaniol |
| | Nivalenol |
| | Diacetynivalenol |
| | Deoxynivalenol (DON, vomioxin) |
| | HT – 2 toxin |
| (E) Estrogenic mycotoxins | F – 2 toxin (Zearalenone) |

Aflatoxin may be a factor in Reyes syndrome, a common cause of death in South East Asian children. Shank *et al.* (70) found significant levels of aflatoxins (1.4 μg/Kg) in liver of 23 Thai children who died of Reyes syndrome. Children who died from Reyes syndrome in Czechoslovakia and in New Zealand have also been found to contain aflatoxins in their livers at autopsy. Kwashiorkor, a disease of children in Northern Africa, and elsewhere in undernourished populations, which is usually attributed to nutritional deficiency, may also be related to aflatoxin intake (46). Aflatoxin induced liver damage may make these children less able to cope with high protein diets usually recommended as cure for Kwashiorkor (57). Each of these diseases is now known to have been caused by the growth of specific moulds, which produced one or more potent toxins, usually in one specific kind of commodity or feed.

Mycotoxic moulds generally attack kernels of grain robbing the nutrients and lowering the fat, protein and vitamin of the grain. The mould also often changes the colour of the kernels, consistency (texture) and often impart odour that causes feed refusal in animals. These effects lead to economic losses due to impaired health in animals and humans, reduced productivity (reduced production of eggs, milk, weight gain) and in severe cases death of animals and humans. When the grain is processed into final products (flour or feed), the visible moulds may be removed, but the majority of toxins persist and can still cause poisoning. Since mycotoxins are chemically diverse, it would be impossible to develop one single control method that would ensure reduction of every mycotoxin present in every agricultural commodity. In addition, mycotoxin contamination is heterogeneous in nature, so sampling and analysis are complicated by the presence of 'hot spots'. Considering all these factors, it can be concluded that development of food safety programmes for mycotoxins control is not a simple issue (60,61).

Mycotoxins can be acutely or chronically toxic or both, depending on the kind of toxin and the dose. In animals, acute diseases include liver and kidney damage, attack on central nervous system, skin disorder and hormonal effects. Neurotoxin may cause trembling or even death. Skin disorders may be necrotic lesions or photosensitive, while hormonal effects include abortion in cattle, swollen genitals and a variety of poorly defined disorders including vomiting, feed refusal and failure to thrive. Toxins which act on liver and kidney are especially difficult to detect and levels much lower than those producing acute affects are often carcinogenic. When eaten in minute quantities in the daily diet, they can cause cancers in experimental animals long after the time of feeding. It is probable that humans can be affected in the same way.

Diagnosis of mycotoxicoses is usually not very easy. Exposure cannot be established by detection of mycotoxins in tissues from animals suspected of being poisoned by mycotoxins because analytical methods for detection of mycotoxins in animal tissues are not commonly available. We rely on detection of mycotoxins in grain or feed to establish the diagnosis. Investigation of suspected mycotoxicosis shuuld begin by obtaining through history. Clinical signs are very important because they can be used to select appropriate diagnostic tests to confirm or refute mycotoxicosis. Specific histopathological lesions may be evidence of mycotoxicosis. If there are no lesions or no evidence of pathology in organs known to be affected by mycotoxins, then likelihood of mycotoxicosis is reduced.

Confirmation of suspected mycotoxicosis is assisted by either reproducing the clinical symptoms with feeding trial using suspected feed, or by detection of a known mycotoxin in the ration or tissues of the animals consuming the ration. Feeding trials can not be routinely performed because they are difficult to conduct, expensive and slow to provide results. One of the challenging aspects of diagnosing cases of suspected mycotoxicosis is collecting a feed sample that adequately represents the feed suspected of being contaminated as mycotoxins are not evenly distributed in feeds. Further the contaminated part of the feed may be consumed before the disease is evident.

Treatment of animals suffering from mycotoxicosis usually is supportive and often not very effective. Antidotes for mycotoxins are generally not available. It is not uncommon for animals to improve after they quit eating feed suspected or known to be contaminated with mycotoxins. Such an occurrence supports the suspicion of the feed as causal factors, but does not prove that the animals suffered from mycotoxicosis.

## (I) SYMPTOMS AND DIAGNOSIS OF MYCOTOXICOSES

Health effects, clinical symptoms, lesions and treatment of selected mycotoxicoses are discussed below.

### I.1 Aflatoxicoses

Aflatoxicosis is caused by aflatoxins, produced by *Aspergillus flavus*, *A. parasiticus* and *A. nomius* (18,35,53). There are five important aflatoxins *viz.*, aflatoxin $B_1$, $B_2$, $G_1$, $G_2$ and $M_1$. Aflatoxin $M_1$ is a metabolite of aflatoxin $B_1$ found in milk and urine. Aflatoxin is vague sentence hepatotoxin in all animal species that consume.

### I.1.1 *Carcinogenicity*

Aflatoxin $B_1$ is confirmed as potential carcinogen. Using LD parameters, aflatoxin $B_1$ is 1000 times more potent carcinogen when compared to benzopyrene (37). IARC has classified aflatoxin as a group one carcinogen (50).

Metabolism plays a major role in deciding the degree of toxicity (38). After ingestion, aflatoxin is metabolised by cytochrome P 450 group of enzymes in the liver, where it is converted to many metabolic products like aflatoxicol, aflatoxin $Q_1$, aflatoxin $P_1$ and aflatoxin $M_1$ depending on genetic predisposition of the species. In addition, another metabolite (aflatoxin 8, 9 epoxide) is also formed. The amount of metabolite decides species susceptibility as this can induce mutations by intercalating into DNA, by forming an adduct with guanine moiety in the DNA. This intercalation of epoxide causes G to T transversion at codon 249 in p 53 gene in liver, which may lead to hepatic carcinoma. It is now established that this is the major cause for aflatoxin carcinogenicity. Moreover, species susceptibility to aflatoxin mainly depends on its liver detoxification system, genetic make up, age and other nutritional factors (48). Further, analysis of $M_1$ in the urine can be used as an index for human exposure to aflatoxin $B_1$ in an epidemiological study. A significant correlation was found between the mortality rates of primary liver cancer and intake of aflatoxin $B_1$.

Recently, antibodies to aflatoxin $B_1$ have been reported in humans. At present, these antibodies are considered indicative of exposure and may not be related to disease (66,78). Workers handling animal feed exposed to mycotoxins, including aflatoxins were examined for alterations in plasma alpha tumour increasing factor (TNF alpha). Exposed workers had increased levels of TNF, alpha and concomitant serum lactate dehydrogenase isoenzyme activity (59).

There are particular geographical areas, where the annual liver cancer rate is reported to be well above the normal rate (2 cases per 10,0000 people). Certain populations of Africa, Southern India, Japan and South East Asia have usually high incidence of liver cancer. The evidence for the association of aflatoxins in the liver cancer has been considered strong enough to justify intervention in the food contamination cycle. However, other factors such as the role played by hepatitis, must be assessed. Several field studies revealed the association of consumption of aflatoxins with human liver cancer (1,50,51,63). Studies relating to aflatoxin exposure in humans in India are not being persued actively. However, this problem may break loose at anytime in near future as incidences of aflatoxin contamination in foods and feeds is increasing.

### I.1.2 *Clinical Signs*

**Acute exposure :** Causes depression, anorexia, reduced gain or milk production, sub-normal body temperature. These symptoms are common in all species.

*Chronic exposure :*

**Poults :** Causes decreased growth rate, reduced feed efficiency, steatorrhea (fat feces), bruising.

**Cattle:** Shows rumen mortality for 24 to 48 hours.

**Swine:** Anorexia, unthriftness, slow growth, icterus, mild anemia, ascites, increased susceptibility to infection.

### I.1.3 *Lesions*

**Acute exposure**: Hemorrhage, ascites

**Chronic exposure:** Pale, soft, clay-coloured liver, mild anemia, icterus, ascites

**Histopathology:** Hepatocyte degeneration and necrosis, centrilobular hemorrhagic hepatic necrosis, fatty changes and regeneration of hepatocytes, bile duct epithelial proliferation progressing to interlobular fibroplasia and extensive proliferation, karyomegaly, vacuolization, bile retention.

### I.1.4 *Diagnostic Aids*

**A. Blood work up :** Check up for anemia, elevated liver enzymes, serum bile acids, albumin-globulin ratio, prothrombin activity.

**B. Tissue or fluid analysis :** Aflatoxin $M_1$ may be present in milk or urine while other aflatoxins may be present in kidney or liver.

**C. Grain or feed analysis:** Dietary aflatoxin concentration at which performance or clinical effects become noticeable depends upon species and effects. Decreased performance may occur at concentration as low as 200 ppb in young sensitive species. Immunity may become impaired at concentration of about 200 ppb. Hepatic lesions may become noticeable at 200-400 ppb. Clinical illness may become obvious at about 400 ppb. Violative residues in milk can occur at concentrations at or near 50 ppb aflatoxin $M_1$.

### I.1.5 *Treatment*

**Stop exposure:** Feeding of contaminated ration should immediately be stopped and it should be replaced by non-contaminated ration.

**Supportive treatment :** Supportive treatment may also be provided as per the clinical situation.

### I.2 Fumonisins Toxicoses

Fumonisins are produced by *Fusarium moniliforme* and *F. proliferatum* and is found primarily in corn. It is of three kinds – Fumonisins $B_1$, $B_2$ and $B_3$. Fumonisin $B_1$ is most prevalent in naturally contaminated corn, sorghum and related grains and it is most toxic. Horses and pigs are most sensitive animals. Equine leukoencephalomalacia (ELE) is a fatal disease of horses caused by fumonisins. Porcine pulmonary syndrome is the form of disease in swine. The mechanism of action is believed to be the inhibition of enzymes (ceramide synthase) involved in the production of sphingosine from sphinganine. Sphingosine is an important component of cell membrane, especially neurons. Fumonisin is nephrotoxic and hepatotoxic in many animals of recent studies linked fumonisins to high incidences of human esophageal cancer in the Transkei region of Southern Africa (58,67).

### I.2.1 Clinical Signs

**A. Swine :** Dyspnea, cyanosis and weakness which develop 4-7 days after fumonisin contaminated feed begins. Death may occur within a few hours of onset of the syndrome.

**B. Horses :** Depression, blindness, ataxia, aimless wande-ring, facial paralysis, which may rapidly progress to coma and death. Death may occur 1-7 days after onset of sign.

**C. Poults :** Appears to be more resistant than other species. Inappetence and skeletal abnormalities may develop at concentrations of 200-400 ppm.

### I.2.2 *Lesions*

**A. Horses :** Leukoencephalomalacia – massive softening and liquefaction of cerebral white matter, ranging from discrete focal areas to large cavitations and inward collapse of the cortical gray matter. Haemorrhage is prominent. Microscopically, liquefa-ction and proliferation of macrophages in response to necrosis.

**B. Swine :** Acute pulmonary edema, which is characterized by marked to massive interlobular pulmonary edema and marked hydrothorax. Lungs are distended and turgid. Thoracic cavity is filled with straw coloured proteinaceous fluid. Microscopically, interstitial and interlobular edema may include multiple areas of focal pancreatic necrosis and hepatic lesions characterized by disorganized hepatocytes, increased mitotic figures, necrosis of single hepatocytes and mild bile retention Icterus.

### I.2.3 *Diagnostic Aids*

**A. Blood work up:** Evidence of hepatic dysfunction.

**B. Tissue analysis:** Sphinganine : Sphingosine ratio is increased. Analysis of tissues for sphinganine and sphingosine is not readily available.

**C. Analysis of grain for fumonisin :** Concentration in excess of 3 ppm fumonisin may be significant for horses, while concentrations above 5ppm may be significant for pigs.

### I.2.4 *Treatment*

**A. Stop exposure :** Contaminated feed from the ration may immediately be removed. Because of the long time between consumption of feed and onset of signs, oral detoxification is not recommended.

**B. Supportive treatment :** Prognosis for horses and pigs suffering from advance stages of the disease is poor.

## I.3 Zearalenone Toxicoses

Zearalenone (F-2) is produced by *Fusarium roseum* (*F. graminearum*), *F. moniliforme*, *F. tricinctum* and *F. oxysporum*. This anabolic compound is generally not lethal to the consuming animal but it induces a hyperestrogenism. Excessive exposure does not cause death or abortions, but it can disrupt the estrus cycle in females, cause infertility and feminization in males and precocious puberty in sexually immature females. The visible symptoms of this mycotoxicosis include swelling and reddening of the vulva, increase in size of uterus and growth and lactation in mammary glands. Unusual environmental conditions during growing season or insufficiently dried grain put up for storage may increase zearalenone production.

Since *F. roseum* also produces vomitoxin (Deoxynivalenol, DON), cocontamination of vomitoxin may also be found along with zearalenone.

### I.3.1 *Clinical Signs*

Clinical signs vary with species, sex and age of the animal.

**A. Swine, sexually immature gilts :** Behavioral estrus, swollen and edematous vulva, enlarged mammary glands, tenesmus (spasmodic contraction of and or bladder sphincter), sometimes vaginal or rectal prolapse, clinical signs appear 2-7 days after exposure begins and subsides 4-10 days after exposure ends.

**B. Swine, mature :** Exposure early in estrus cycle, suppression of ovulation and signs of estrus that are severe and prolonged.

**C. Exposure mid-cycle:** Pseudo-pregnancy, anestrus which may persist for 40-60 days after exposure stops.

**D. Swine, castrated boars:** Enlarged prepuce and nipple

**E. Swine, immature boars :** Reduced libido, retarded testicular development.

**F. Swine mature boars :** Not affected unless dietary concentration reaches 200 ppm or higher.

### I.3.2 *Lesions*

Lesions are present only in reproductive system.

**A. Prepubertal gilts :** Swollen and edematous vulva, enlarged mammary glands; enlarged hypertrophic and edematous uterus. Histopothologically, uterine and vaginal metaplasia and follicular atresia.

**B. Mature sows:** Retained and functional corporlutea with anestrus. Mammary alveolar development and ductular squamous metaplasia.

### I.3.3 *Diagnostic Aids*

**A. Blood work up:** Serum analysis for estrogen can help to rule out an organic hormonal problem.

**B. Tissue or fluid analysis:** Not available.

**C. Feed or grain analysis :** Since clinical affects are delayed several days after ingestion, feed analysis may be of limited value. Other estrogenic chemicals from plants (phytoestrogens) or other sources could also produce some of the signs of zearalenone toxicoses. Estrogenic assays of feed is of course expensive, but may be valuable to assess estrogenic activity of feed.

### I.3.4 *Treatment*

**A. Stop exposure:** Stop contaminated feed and replace with uncontaminated feed.

**B. Decrease zearalenone absorption :** Activated charcoal may help to limit absorption of ingested zearalenone by individual animals. Several commercially available feed additives are available to bind zearalenone and to prevent excessive exposure.

## I.4 Vomitoxin (Deoxynivalenol, DON) Toxicoses

Vomitoxin is produced by *Fusarium roseum* (*F. graminearum*) and *F. moniliforme*. It is found in corn, wheat, barely, milo and oats. Vomitoxin is a chemical that belongs to trichothecenes group.

Vomitoxin is associated with feed refusal and decreased feed consumption, which can affect animal performance, concentrations ranging from 5 to 10 ppm are associated with vomiting in pigs, hence its name. Threshold for feed intake are about 10 to 20 ppm in ruminants. The mechanism by which vomitoxin acts has not been elucidated. In addition, trichothecenes also inhibit protein and nucleic acid synthesis. Cattles are very resistant to the effects of vomitoxins. However, pigs are more sensitive. There have been reports of health effects in dogs from food contaminated with vomitoxin.

### I.4.1 *Clinical Signs*

**Swine, dogs, cats :** Feed refusal, vomiting.

### I.4.2 *Lesions*

No gross or histopathological lesions have been reported in animals consuming vomitoxin contaminated feed.

### I.4.3 *Diagnostic Aids*

**A. Analysis of grains or feed :** For the contamination of vomitoxin.

**B. Analysis of tissues from affected animals:** Not available.

### I.4.4 *Treatment*

To remove vomitoxin contaminated grain/ feed from ration.

## I.5 Ergot Toxicoses

Ergot toxicosis is a disease caused by ingestion of ergot alkaloids contained in the sclerotia of *Claviceps* species. They are commonly found in cereal grains, especially rye. Tall fescue (*Festuca arundinaceaes*) may contain endophytic fungus (*Acremonium coenophialum*), which can make ergot alkaloids.

Sclerotia are dark brown, brownish-purple or black coloured bodies that stick out from the seed heads of infected plants. They are visible to the naked eyes and look similar to rodent droppings when removed from seed heads. They may become quite large upto an inch in length.

Several alkaloids are present in sclerotia. They cause two syndromes, a central nervous system disorder and a peripheral vascular disorder. They can cause agalactia in lactating females. Ergot alkaloids are potent smooth muscle stimulants. Any organ with smooth muscle in it may be affected, especially arterioles. Uterine contraction may be stimulated.

### I.5.1 *Clinical Signs*

**A. Peripheral vascular syndrome**: Lameness, swelling of feet and fetlocks, sharply demarcated necrosis of feet, ears or tail. In severe case, hooves or feet or tail may slough. Dry gangrene is also common. These effects are due to ischemia resulting from constriction of arterioles in peripheral vascular beds.

**B. Lactating females**: Cassation of milk production; agalactia in post partum females.

**C. Central Nervous System syndrome:** Hyperexcitability, hypermetria tremors, heat tolerance in cattle.

### I.5.2 *Lesions*

Peripheral vascular syndrome. Small and flaccid mammary glands of females in late pregnancy with no evidence of secretions, are indicators of probable agalactia.

**A. Histopathological :** Endothelial damage, thrombosis, coagulative necrosis.

**B. Analysis of stomach or rumen content for ergot alkaloids :** Detection of ergot alkaloids in stomach or rumen content is evidence of exposure.

**C. Analysis of feed or grains for ergot alkaloids ingested by affected animals :** Sclerotia may be visible if the feed has not been ground.

### I.5.3 Treatment

**A. Stop exposure**: Remove contaminated grain or feed from ration. Lactation may commence 5-7 days afterwards.

**B. Control secondary infection:** In limbs suffering from dry gangrene it is advisable to place the animals in a warm, clean and stress free environment.

## I.6 Ochratoxin Toxicoses

The ochratoxins are a group of iso-coumarin derivatives produced by a number of *Aspergillus* species and *P. viridicatum* (76). Van der Merwe *et al.* (75) crystallized and identified the most potent derivative, ochratoxin A, from a toxic *A. ochraceus* strain. The primary target organ is kidney. Of the seven ochratoxins known, only ochratoxin A is thought to be significant in outbreaks of ochratoxicosis in domestic animals. In the early 1970, high incidence of nephritis (kidney inflammation) in pig at slaughter was observed in Denmark. Analysis of pig feeds showed that 50% of samples contained ochratoxin A at levels upto 27 μg/Kg. The mould responsible was reported to be *Penicillium verrucosum* (65). The risk of humans is difficult to assess but as pig meats are important part of food in some parts of the world, risk certainly exists. Death rates from kidney failure are high in some Danish rural areas and it is reasonable to suppose the cause of ochratoxicosis. The significance of ochratoxin A has yet to be assessed in tropical countries.

### I.6.1 *Clinical Signs*

**Pigs** : Renal tubular function decreased, kidney enlarged, grey-yellow, markedly increased in firmness. Emesis, depression, reduced feed intake, weight loss, diarrhoea, polyuria polydipsia, dehydration and subsequently death.

Destructive action of renal tubular epithelium is evident by elevated blood urea, nitrogen, glucosuria, proteinuria, and changes in specific gravity and urinary sedimentation and mainly by concentration of urinary enzymes. Guts, liver, and lymphoreticular tissues get damaged.

### I.6.2 *Histopathological findings*

Renal changes were most prominent histopathological findings of pigs. Nephropathy characterized by tubular atrophy and interstitial fibrosis. Ochratoxin A is found in increasing concentration through the tissues, muscular, adipose, liver and kidney.

**A. Rats and mice :** They are sensitive to renal toxicity. Oral $LD_{50}$ for Wistar rats is 20 to 30 (males) and 20 to 21 (female). μg/Kg. Develop proximal renal tubular necrosis and hepatic periportal necrosis. Besides, fatty degeneration of liver, hyaline degeneration of liver cells with focal necrosis, enteritis, pale enlarged liver.

**B. Chicken and poults :** Most sensitive visible indicator is kidney enlargement resulting from ecdema. During acute ochratoxicosis they become restless, huddled and suffered from diarrhoea, tremors, renal disturbances, retarded growth rate. Nephropathy symptoms include pale swollen kidney and necrosis of proximal tubule of kidney.

### I.6.3 *Lesions*

Ranged from dilatation and focal degeneration of proximal tubules to severe tubular atrophy with increased thickness of basement membrane, extensive interstitial fibrosis and glomerular atrophy and sclerosis.

### I.6.4 *Diaganostic Aids*

**A. Blood work up :** Elevated blood urea, nitrogen, glucosuria, proteinuria, change in specific gravity can be evidence for destructive renal function. Urinary sedimentation mainly by concentration of urinary enzymes.

**B. Tissue analysis :** Detection of ochratoxin A in kidney may be evidence of exposure.

**C. Grain or feed analysis:** For ochratoxin A contamination. In areas where pig meat is important part, these should be analysed for ochratoxin A.

### *I.6.5 Treatment*

**A. Stop exposure:** Stop feeding contaminated ration.

**B. Supportive treatment :** Can be provided as per clinical situation dictates.

## II. (A) REGULATIONS

In terms of economic implications, mycotoxins are one of the most important contaminants of food-chain. Grain produces grain handlers, grain at processors are not getting proper orice due to moulds and consumers are paying higher price due to consumption of contaminated products.

According to FAO estimates, world losses of foodtuffs due to mycotoxins are in the range of 1000 million tonnes a year (64). Mycotoxins of most concerns are aflatoxins, ochratoxin A, deoxynivalenol, zearalenone, T-2 toxin and fumonisins. Stringent mandatory standard, voluntary regulations or guidelines, tolerance levels have been laid down for these mycotoxins by many countries. At the International level, the CAC (Codex Alimentarius Commission) through its Committee on Food Additives and Contaminants (CCFAC) and other relevant commodity committee is considering the establishment of international guidelines for levels of mycotoxins in food. Till date about 77 countries are known to have specific regulations for some important mycotoxins. Besides, after the formation of European Community (EC) with the association of 15 nations in 1995, they have issued joint European Commission Regulation (39) and have common standards.

Recently, Vishwanath (77) reviewed an update status of existing regulations and has also discussed the role of CAC in providing a platform to reach a consensus on the maximum permitted levels in food. Since mycotoxins can occur both in raw products and finished byproducts, this has necessitated the testing of mycotoxins at various stages in the food-chain. Important commodities under regulations include – food and food products : cereal grains and cereal based products, peanuts and their products, nuts and other oil-seeds, milk and milk products, infant and children food, fruits and fruit juices, miscellaneous food products.

Risk of health from mycotoxin contaminated food varies with average daily intake of a particular category of food and also with particular target population. Therefore, many developed countries have laid down maximum limits for particular categories of food in addition to general standards. Some countries have laid down mandatory standard for aflatoxins and some other mycotoxins in all foods.

## (i) AFLATOXINS

Many countries have regulations for total sum of aflatoxins $B_1$, $B_2$, $G_1$ while others have $G_2$ and others for aflatoxin $B_1$ ($AFB_1$) alone. These standards vary from 0 to 35 $\mu$g/Kg of $AFB_1$. Romania, Poland, Singapore have set zero tolerance level in food *i.e.* total absence of aflatoxin. Most countries have regulations laid down only for aflatoxins in cereals. Range of regulatory limits varying from 0 $\mu$g / Kg of Netherlands to 50 $\mu$g/Kg of Taiwan.

Story of aflatoxin began in 1960, with the discovery of aflatoxin in peanut meal and Turkey- X disease in U.K. Consequently, most countries have laid down regulatory limits for aflatoxins in these foods (peanuts, nuts, oilseeds) which range from 0 ($B_1$ $G_1$) of Dominican Republic to 30 $\mu$g/Kg total aflatoxins of Brazil, Uruguay.

Aflatoxin $M_1$, a compound closely related to $AFB_1$ (cattle are able to metabolize $AFB_1$ in their feed to $AFM_1$) is secreted into the milk and ends up in various dairy products. Except for China and Russia, most countries have regulatory limits only for $AFM_1$ and not $AFB_1$, in these products. Most of the Western European countries have laid down a tolerance level of 0.05 $\mu$g/Kg while USA, Czechoslovakia and others have a tolerance limit of 0.5 $\mu$g/Kg.

Regulatory limits for aflatoxins in fruits and dried fruit products such as raisins, dates and figs have also been laid down by some countries. In figs regulatory limits range between 2 and 10 $\mu$g/Kg.

Germany has set limits of 0.005 $\mu$g/Kg for enzymes intended for food production and China for fermented feed at 5 $\mu$g/Kg for $AFB_1$ (43).

The Codex Commission met in 1998 and accepted the CCFAC recommendation of maximum limit of 15 $\mu$g/Kg total aflatoxin in peanuts and a sampling plan that uses a single 20 Kg sample (2).

## (ii) OCHRATOXIN A

Ochratoxin A has exhibited nephrotoxicity in all mammalian species tested. It has been associated with Balken endemic nephropathy. The International Agency for Research on Cancer (IARC) has classified it in group 2B (a possible human carcinogen). It is teratogenic and also been shown to have immunosuppressive effects in a number of species. A joint committee of Food Additives and Contaminants has agreed that it would be generally acceptable and practicably achievable to establish a maximum level of 5 $\mu$g/Kg of ochratoxin A in wheat, barley and rye and derived products (17).

Israel has a limit of 50 $\mu$g/Kg of ochratoxin A in peanut and peanut products. Czech Republic has limits for ochratoxin – A at 1 μg/Kg for infant foods and 5 μg/Kg for children food. Greece has standards for ochratoxin A in apple juice and apple products as well, at the level of 20 μg/Kg. The EC (39) has recently introduced regulatory limit of 10 μg/Kg ochratoxin A for dried vine fruits (currants, raisins, sultanas). Italy has laid a limit of 0.2 μg/Kg of ochratoxin A in beer.

### (iii) DEOXYNIVALENOL (DON)

DON a trichothecenes is also called vomitoxin. IARC (51) has placed DON in Group – 3 (Not classified as to its carcinogenicity to humans). DON and Nivalenol have been associated together with acetyl DON and T-2 toxin in a disease outbreak in Kashmir, India due to consumption of mould damaged wheat (4). DON is frequently found in cereal grains, such as wheat, maize, oat, barley and rice. Currently there is no legislation regarding maximum levels of DON and zearalenone in cereals in EC. Canada has laid down a level of 1000 μg/Kg DON in uncleaned wheat for infant foods.

### (iv) ZEARALENONE

Zearalenone has been implicated as a causative factor of porcine hyperestrogenism. In Russia, the regulatory limit for zearalenone in nuts is 100 mg/Kg. A zearalenone test was carried out in Canada in 1987. However, considering tentative tolerable daily intake and estimated intake from corn in young children, it was concluded that no adverse health effects were anticipated from exposure to zearalenone in corn in Canada and thereby, no regulatory action was recommended.

### (v) PATULIN

It is a toxic substance with suspected carcinogenic properties. However, it does not accumulate in body. PMDTI (Provisional maximum tolerance daily intake) was determined at 0.4 μg/Kg. Apple juice and its products are most susceptible commodities. In the 24$^{th}$ session of the CAC (16), maximum level of 50 μg/Kg was proposed for patulin in apple juice and apple juice ingredients in other beverages. Although it was favoured by US and Mexico, EC blocked the adoption with the argument that this level was inadequate to protect children and, therefore, the resolution was not adopted. Czech Republic has framed a limit of 20 μg/Kg for infant food.

### (vi) T - 2 TOXIN

T – 2 toxin is a potent inhibitor of protein synthesis, both *in vivo* and *in vitro*. There is substantial evidence of the immunotoxicity and haematotoxicity of T-2 toxin in several species. IARC (51) has placed it under Group 3 with regard to its carcinogenicty. Rice, barley, oat and wheat are the cereals most contaminated with T-2 toxin. Russia is the only country in the world to have a legal tolerance level for T-2 toxin (100 μg/g) in grain (56).

### (vii) FUMONISINS

During last two decades fumonisins, have been most researched. Pure fumonisins cause nephrotoxicity. A correlation between the intake of fumonisins and oesophageal cancer has been established. A joint FAO/WHO committee established PMTDI of 0.2 μg/Kg fumonisin $B_1$, $B_2$ and $B_3$ alone or in combination of body weight per day on the basis of NOEL (No-observed effect level).

Since mycotoxins can occur both in raw products and in finished byproducts, this has necessitated the testing of mycotoxins at many points in the food chain. Total diet (Market basket) surveys should be used to determine the overall trend of dietary exposure within the population as a whole or in specific groups of concerns. National Authorities are expected to give greater attention to the development of consistent and standardised approaches to regulations and their enforcement, including sampling and methods of analysis. As technology has progressed in recent years, testing has become simplified, faster and cheaper allowing more testing to be accompanied in all phases of food handling.

## (B) MINIMIZING RISKS OF MYCOTOXINS UTILIZING HACCP (HAZARD ANALYSIS AND CRITICAL CONTROL POINT) CONCEPT

"An ounce of prevention is worth a pound of cure". HACCP is a system that identifies, evaluates, and controls hazards which are significant for food safety. It is structured , systematic approach for the system from plough to plate. It requires a good understanding of the relationship between cause and effect in order to be more pro-active and it is key element in Total Quality Management (TQM). HACCP builds on the foundations of well established Quality Management Systems such as Good Manufacturing Practices (GMP), Good Hygienic Practices (GHP), Good Agricultural Practices (GAP) and Good Storage Practices (GSP).

HACCP was originally developed as a microbiological safety system in the early days of US manned space programme in order to guarantee a safety of astronauts food. Until that time most of food safety systems were based on end product testing and could not fully assure safe products as 100% testing focused system was needed and HACCP concept was borne. Presently, Regulatory bodies have recognized the usefulness of this tool and its principles have been incorporated into legislative requirements by both the EU (in General Hygiene Regulations) for managing food safety (93/43/EEC) and the United States Department of Agriculture (CPR-123). The National Advisory Committee on Microbiological Criteria for Foods (NACMCF) provided guidelines on HACCP including generic plans and decision tree in 1992, and Codex Alimentarius Commission adopted the HACCP system at its session in 1993.

It is much easier to apply a HACCP system in a factory, where there is a single management or 'owner' and where it is possible to completely prevent a food safety hazard or climate or reduce it to an acceptable level. In the commodity system, there are often many disparate 'owners' of the commodity, as it passes from the farm to the consumer and complete control may be unavoidable.

According to FAO, at least 25 per cent of world's food crops are contaminated with mycotoxins, at a time when the production of agricultural commodities is barely sustaining the increasing population (9). The destruction of contaminated products or their diversion to non- human uses is not always practical and could seriously compromise the world food supply. In order to design and develop effective HACCP based integrated mycotoxin management programmes, a given country has to consider such factors as the climate, farming system, pre- and post-harvest technologies, public health significance of the contaminants, producer and processor compliance, availability of analytical resources and also economy.

Prevention through pre-harvest management is the best method for controlling mycotoxin contamination. However, should contamination occur, the hazard associated with the toxins must be managed through post-harvest procedures, if the product is to be used for food and feed purposes. In an ideal integrated management system, mycotoxin hazards would be minimized in every phase of production, harvesting, processing and distribution (41,54).

**HACPP Programme**

In the terminology of HACCP systems, 'hazards' refer to conditions or contaminants in foods that can cause illness or injury. It does not refer to certain other undesirable conditions such as insects, hair, filth and spoilage, which would be considered in the context of a broader *Quality assurance system.* HACCP concept is built on seven principles and actions –

1. Conduct hazard analysis & identify preventive measures.
2. Identify critical control (CCPs).
3. Establish critical limits.
4. Monitor each CCP.
5. Establish corrective action in the event of a deviation from the critical limit.
6. Establish record keeping.
7. Establish verification procedures.

The development and application of HACCP programmes are complex matters and not all countries have required technical expertise to establish effective integrated mycotoxin management systems that are based on HACCP approach. Owing to the importance of HACCP in food safety programmes, FAO has given high priority to the provision of training to professionals in developing countries on the HACCP approach and its application (42).

**A. Pre-Harvest Procedures**

Significant levels of mycotoxins can occur in food crops in the field. Some of the strategies for prevention of mycotoxins in the field are:

**a. Pre-harvest procedures**

- Reduction in plant stress through irrigation, mineral nutrition, protection from insect damage.
- Avoidance of environmental conditions that favour infection in the field e.g. Drought, insect infestation, primary inoculum, delayed harvesting etc.

**b. Potentially effective**

- Good cultural practices viz. crop rotation, cropping pattern, irrigation, timely planting and harvesting and use of pesticides are protective actions that reduce mycotoxin contamination of field crops.
- Breeding of maize cultivars resistant to fungal infection.
- Use of crop protection chemicals that are antifungal agents.

**c. Developmental**

- Development of transgenic plants resistant to fungal infection.
- Development of transgenic cultivars capable of catabolism/interference with toxin production.
- Development of crops genetically engineered to resist insect damage.
- Development of crop seeds containing endophytic bacteria that exclude toxigenic fungi.
- Exclusion of toxigenic fungi by pre-infection of plants with bio-competitive non-toxigenic fungal strains.

**B. Harvesting Procedures**

Harvesting may inflict mechanical damage on commodities. When damage is kept to a minimum during this phase, subsequent contamination is significantly reduced. Field crops should be harvested in a timely manner to reduce moisture or water activity (Aw) level to a point where mycotoxin formation will not occur.

## C. Post-Harvest Procedures

Even if the contamination occurs or persists after this phase, the hazards associated with toxin must be managed through post-harvest procedures, if the product is to be used for food and feed purpose. Storage and processing are the major areas where contamination can be prevented.

### a. Strategy

1. Removal of damaged grain and drying of the grain to the minimal moisture content.
2. Control of insect and rodent activity and maintenance of appropriate moisture levels and temperature.
3. Appropriate packaging is often successful way of excluding insects and moulds.
4. Frequent cleaning of feed delivery systems and short term storage.
    - Use of antifungal agents such as propionic and acetic acid.
    - Thermal inactivation is also one of the alternatives for products that are usually heat processed. Fumonisins and ochratoxin levels have been shown to be lower in thermally processed maize and wheat products.

Effective integrated mycotoxin management programmes not only cover prevention of mycotoxin formation in agricultural products or their detoxification / decontamination but also involve routine surveillance, regulatory measures to control, the flow of mycotoxin contaminated materials in national and international trade and information, education and communication activities. Implementation of HACCP cannot be expected to succeed in the absence of regulatory activities. Regulatory limits are law; violation of which has legal consequences. Regulatory limits or standards provide a benchmark against which the effectiveness of food safety programmes can be tested.

## b. Surveillance and Monitoring Programmes

Surveillance and monitoring activities fall within HACCP principles 4 to 7. In the absence of regulation and surveillance, voluntary compliance with any system may not fully be achieved. Our safety inspection should focus on preventing problems rather than chasing horses after they are out of the barn. Safety inspections are the responsibility of Govt. Regulatory Agencies that ensure the adequacy of industry food safety programmes. In order to set up a monitoring and management programme the following data and policy decisions must be acquired and made –

1. Identify the mycotoxins and products or commodities that are to be included in the programme.
2. Set up a system of inspection and sample collection.
3. Set up sampling plan.
4. Establish a policy guide for end use of products.
5. Divert to decontamination procedures of lower risk use.

Once the regulatory limits has been set, monitoring programmer play an important role in determining compliance.

### *III. CONTROL OF MYCOTOXINS*

Pre-harvest prevention, especially through host resistance, is probably the best and widely explored strategy for the control of mycotoxins. The preeminence of the host resistance strategy is also due to several research advances, such as the identification of resistant genotypes in corn through plant breeding. Since last decade high impact researches involving biotechnological approaches have been accomplished that have enhanced a host resistance strategy for aflatoxin contamination.

For the sake of convenience we shall discuss biotechnological approaches for the control of mycotoxins under three broad headings :-

a. Enhancement of host resistance
b. Biological control through microorganisms.
c. Other methods.

## (A) ENHANCEMENT OF HOST RESISTANCE

### (a) Breeding

Resistance against pre-harvest aflatoxin contamination in corn is aimed at incorporation of antifungal resistance against mycotoxigenic fungi. Strategy for enhancing host resistance through breeding involves successful identification of germplasm resistant to aflatoxin contamination and also identification of natural resistance mechanisms and traits (13,24). It is now evident that aflatoxin contamination involves multiple chromosome regions and several genes (32). Therefore, attempt to select resistant traits, while maintaining desirable agronomic characteristics, have slowed down due to failure to identify expressed genes and proteins involved in resistance. So further research is needed to elucidate the biochemical mechanisms that confer resistance in corn kernels that are vulnerable to aflatoxin contamination. These resistance mechanisms could be used to enhance germplasm through marker assisted breeding and/or genetic engineering (12).

Different techniques have been employed to facilitate corn breeding for developing germplasm resistant to fungal growth and/or aflatoxin contamination. Inoculation methods include the pinbar inoculation technique (for inoculating kernels through husks with *A. flavus* conidia), the silk inoculation technique, and infesting corn ears with insect larvae infested with *A. flavus* conidia. Two resistant inbreeds ($M_P$ $420_E$ and $M_P$ 313 $_E$) were discovered and tested in field trials (69,82).

For preliminary investigation kernel screening assay (KSA) under the laboratory condition was developed and used to study resistance to aflatoxin production in mature kernels (12,72). These findings suggested the presence of two levels of resistance i.e. at pericarp level and at the sub pericarp level. The sub pericarp level of resistance was shown to require a viable embryo (12). KSA has advantages owing to the rapidity of assay. However, field trials are necessary for confirmation of any resistance.

An improved technique of KSA was developed recently in order to quantify fungal biomass by using $\alpha$–glucuronidase (GUS) or Green fluorescent protein (GFP) (36,83) reporter gene containing *A. flavus* tester strains. *A. flavus* tester strains were genetically engineered with a gene construct consisting of the GUS reporter gene linked to an *A . flavus* α– tubulin gene promoter for monitoring fungal growth (12,14) or with the reporter gene linked to an aflatoxin biosynthetic pathway gene, which could also provide a quick and economical way to indirectly measure afflatoxin levels (12,13,62). Therefore, now it is possible to accurately assess fungal infection levels and to predict the corresponding aflatoxin levles in the same kernels, as a result of fungal infection. With this, it is now possible to determine whether kernel resistance mechanisms are affecting fungal growth or aflatoxin biosynthesis. Chromosomal regions with the resistance to *A. flavus* and inhibition of aflatoxin production in corn have been identified through Restriction Fragment Length Polymorphism (RFLP) analysis in three Ilinois resistant lines (Rool, LB 31 and Tex 6). After mapping populations were developed using B73 and/or $M_0$ 17 elite inbreds as susceptible parents (79,80). RFLP analysis or RNA should provide an efficient means for development of resistant germplasm. However, limiting the growth of aflatoxigenic fungi might at times not be enough to maintain aflatoxins at 'acceptable' levels in corn crops. Therefore, the identification of compounds that block aflatoxin biosynthesis may represent 'magic bullet' needed to ensure resistance.

### (b) Resistance Associated Proteins (RAP)

Studies demonstrating sub pericarp (wounded kernel) resist-ance in corn kernels have led to identification of sub pericarp resistance mechanisms. Examination of kernel proteins of several genotypes revealed differences between genotypes resistant and susceptible to aflatoxin contamination (45). Imbibed susceptible kernels, for example, showed decreased aflatoxin levels and contained germination – induced ribosome inactivation protein (RIP) and zeamatin – both inhibiting *A. flavus* growth *in vitro* (44).

In another investigation, two kernel proteins, one 28 KDa and inhibitory to *A. flavus* growth, the other over 100 KDa in size and primarily inhibitory to toxin formation were identified from resistant corn inbred (49). When a commercial corn hybrid was inoculated with aflatoxin and non-aflatoxin producing strains of *A. flavus* at milk stage, one induced chitinase and one α-1-3 glucanase isoform was detected in maturing infected kernels, while another isoform was detected in maturing uninfected kernel (52).

Chen *et al.* (19) observed kernel protein profiles of 13 corn genotypes which revealed a constitutively expressed 14 KDa Trypsin inhibitor protein (TI), present at relatively high concentrations in seven resistant corn lines, but at low concentration or absent in six susceptible lines. The mode of action of TI against fungal growth may be partially due to inhibition of fungal amylase, limiting *A. flavus* growth but also for toxin production (84). TI also demonstrated antifungal activity against other mycotoxigenic species (20). The identific-ation of these proteins and their corresponding genes may provide markers for plant breeders and facilitate the introduction of antifungal genes through genetic engineering into other aflatoxin susceptible crops such as cotton (68).

Kernel proteins from several resistant and susceptible genotypes were compared using large format 2-D gel electrophoresis, and over a dozen such protein spots, either unique or 5 fold up regulated in resistant lines, were identified, isolated from preparative 2-D gels and analyzed using ESI – MS/MS after in gel digestion with trypsin. (21, 22). These proteins can be grouped into three categories based on their peptide sequence homology viz.

a. Storage proteins (such as globulins – $GLB_1$, $GLB_2$) and late embryogenesis abundant proteins (LEA 3, LEA 14);

b. Stress responsible proteins such as Aldose reductase (ALD), Glycoxylase 1 (GLX–1) and heat shock protein and,

c. Antifungal protiens, including TI.

Increased temperature and drought stress, which occur together are major factors associated with aflatoxin contamination of corn kernels (6,62). However, no investigation has been conducted to determine the possible direct involvement of stress related proteins in host-fungal resistance. Possession of unique or of high levels of hydrophilic storage or stress related proteins, may put resistant lines in an advantageous position over susceptible genotypes in the ability to synthesize proteins and defend against pathogens under stress conditions. Future studies are necessitated in relation to physiological and biochemical characterization, genetic mapping, plant transformation using resistance associated protein gene (RAP gene), RNA; gene silencing experiments and marker assisted breeding which will elucidate the role of stress related RAP in kernel resistance.

Aflatoxin biosynthetic pathway and gene cluster comprising genes that govern this pathway, including the key regulatory gene (aflR) may provide a sound basis for studies of the efficacy and mode of action of putative pathway blocking compounds (e.g.. Tex 6 100 KDa protein) (73). On the other hand, regulation of these genes during invasion of the plant host is investigated using a genomic approach. This approach is based on the fact that certain plant derived natural products apparently have regulatory effects on aflatoxin biosynthesis (5).

## (B) BIOLOGICAL CONTROL THROUGH MICRO-ORGANISMS

Biological control of mycotoxins can be achieved by the enzymatic degradation or biotransformation of the mycotoxins either by the whole cell or an enzyme system. Other approach that can be regarded as biological include the use of biocompetitive agents and genetically engineered plants for reducing mycotoxin contamination.

### (a) Aflatoxins

Microorganisms have often been suggested as agents of control of mycotoxin contamination. The interference of other fungi with *A. flavus* in its elaboration of aflatoxin in laboratory culture on maize kernels was reported by Wicklow *et al.* (81). Horn and Wicklow (47) later demonstrated that *Aspergillus niger*, when cultured with *A. flavus* on autoclaved maize, suppressed aflatoxin production by lowering the pH of the substrate. Inhibition of aflatoxin production was also attributed to components of water extracts of *A. niger* fermented kernels. At water activity of $a_w$ 0.95, Devi and Polasa (33) observed interference in toxin production when three toxigenic Aspergilli, *A. flavus*, *A. ochraceus*, and *A. versicolor* were grown in different combination. Similarly Choudhary (23) also reported that *Fusarium*, *Trichoderma* and *Rhizopus* sp., co-inhabiting maize kernels were responsible for inhibiting aflatoxin production by *A. flavus*.

## Non-toxigenic *A. flavus* and *A. parasiticus*

There is a great diversity of phenotypes of *A. flavus* in agricultural fields and the common occurrence of atoxigenic strains (28, 34). Further, toxigenicity is apparently unrelated to a strains ability to colonize and/or infect living or dead plant tissues. These observations led to the conclusion that atoxigenic strains can be used to displace toxigenic strains (26). Cotty and Bayman (30) tested competitive ability of an atoxigenic *A. flavus* strain to inhibit the aflatoxin contamination of developing cotton bolls. Competitive exclusion was found to contribute to the effect of the atoxigenic strain on contamination, and the results also suggested that a second mechanism might also have been in effect. Now, Biocontrol of aflatoxin producing strains with atoxigenic strains of *A. flavus* is being developed for corn, cotton seeds, peanuts, rice kernels and wheat seed (8,11).

Reduction in aflatoxin contamination of cotton seed has been achieved by application of a non-aflatoxin strain of *A. flavus* to soil around developing cotton plants (29) and a patent was sanctioned (US patent number 5, 171, 686) for the use of non-aflatoxigenic *A. flavus*. Application of competitive, non-aflatoxigenic strain of *A. parasiticus* reduced aflatoxin contamination of peanuts over a three year period and a patent was granted for use of non-aflatoxigenic strains of *A. parasiticus* for controlling aflatoxin contamination (27; US Patent Number 5, 292, 661). Presently, population biologists seek to understand the genetic structure and diversity of *A. flavus* population in crop fields and distribution of relevant phenotypic characters (e.g. Aflatoxin and cyclopiazonic acid production, enzyme production, sclerotium formation, and so on) among genetically isolated clonal populaion.

## *Trichoderma* spp.

*Trichoderma* spp. a mycoparasite of plant pathogens, has been accepted as the most potent biological control agent for certain plant diseases. Its mycopararitism involves a complimentary action of antibiosis, nutrient competition and cell-walls degrading enzymes such as $\alpha$–1,3-glucanase, proteases, and chitinases. Since chitin is the major component of most fungal cell-walls, has been attributed to its chitinase biocontrol activity of *Trichoderma*

(40). Several hydrolases of *Trichoderma* were recently identified and purified. Some of the genes coding them were cloned and sequenced, and transformants were obtained, which confirmed that over production of a single protease or chitinase resulted in better biocontrol agent. Culture filtrates of *T. harzianum* and *T. viride* were inhibitory to *A. flavus* growth. *T. viride* was also found to inhibit the production of aflatoxin $B_1$ (73.5%) and aflatoxin $G_1$ (100%) when cultured with *A. flavus* (23).

### *Bacillus*

*Bacillus pumilus* is reported to inhibit the growth and aflatoxin production by *A. flavus* to the extent of 99.2%. Later on, an active compound being produced by *B. pumilus* was identified. *Aspergillus* spore germination and abort elongating hyphae, presumably by inducing a cell-wall lesion (10). Another lactic acid bacterium, *Bacillus subtilis* isolated from groundnut was also found to inhibit the growth of *A. flavus* in ground nuts.

The inocluation of *A. flavus* spores into a culture of *Streptococcus lactis* in tryptone broth medium resulted in little or no aflatoxin production even though the growth of the fungus was not effected (25). Aflatoxin levels were also markedly reduced when *S. lactis* was inoculated into a growing *A. flavus* culture. In addition to inhibiting synthesis of aflatoxin, *S. lactis* also degraded preformed toxin.

### (b) Fumonisins

Two species of black yeast fungi (*Exophiala spinifera, Rhinocladiella atrouirens*) and a Gram-negative bacterium (*Caulobacter* spp.) isolated from mouldy maize kernels have been found to be extensively metabolize fumonisins to $CO_2$. These microorganisms produce fumonisin catabolizing enzymes, such as esterase, which lead to the formation of hydrolyzed fumonisin $B_1$ ($AP_1$) plus tricarboxlylic acid and deaminase, which in turn form N-acetyl $AP_1$ plus 2-$Op_1$ hemiketal. $AP_1$ has reduced toxicity in comparison to the intact $FB_1$.

Fumonisin esterase enzymes were expressed in transgenic maize plants. Lower levels of fumonisin $B_1$ accompanied by an accumulation of $AP_1$ were observed in kernels of transgenic maize plants as compared to conventional plants.

Maize kernel injuries caused by insects, such as the European corn borer and pink borer, promote *Fusarium* infection with consequent accumulation of fumonisins. Levels of fumonisins in transgenic maize hybrids with kernel experession of insecticidal *Bacillus thuringiensis* protein, Cryl A(b) were lower than in conventional maize hybrids.

### (c) Trichothecenes

12,13-Epoxide ring is essential for the toxicity of these mycotoxins, and removal of this ring results in a significant loss of toxicity. Ruminal or intestinal microflora are capable to detoxifying deoxynivalenol (DON) by enzymatic reduction of the expoxide ring resulting in the metabolite DOM – 1 that is known to be non-toxic. Pure strain of *Eubacterium*, capable of biotransformation of DON to DOM – 1, was isolated from an enriched mixed culture obtained from bovine rumen content. This bacterium transformed DON and other trichothecenes within 24-28 hr *in vitro* experiments using pieces of pig culture.

*Agrobacterium – Rhizobium* group of bacteria, isolated from soil was also found to transform 70% of DON to 3-Keto-DON after a 1 day incubation.

### (d) Zearalenone

Different microorganisms (bacteria, fungi and yeasts) are reported to convert zearalenone (ZEA) to $\alpha$ and $\beta$-zearalenol. However, oestrogenicity of these metabolites are similar to ZEA and thereby, this transformation cannot be regarded as detoxification.

### (e) Ochratoxin - A

Microorganisms like *Acinetobacter calcoaceticus, Phenylobacterium immobile* and a non-toxigenic strian of *Aspergillus niger* are reported to convert OTA to the less toxic $\alpha$–OTA.

## (C) OTHER METHODS

Control of mycotoxins can also be achieved by some other biological methods. Use of Green Fluorescent Protein (GFP) is reported to monitor gene expression and food colonization by *Aspergillus flavus* (36). These studies demonstrated that the modified GFP encoded by pNucEm2 and pGAP33 was highly expressed in *A. flavus*. The intensity of fluorescence is sufficient to allow the visualization of a GFP – containing strain under a standard laboratory UV-light. The use of these constructs in strains of *A. flavus* could facilitate the detection of the fungus in substrates such as soils or foods and can be useful in screening corn genotype for resistance to aflatoxin accumulation and making screening faster and more commercial.

Essential oils of plants for one in use since hundreds of years controlling fungus in food (15). Cinnamon, pippermint, basil, origanum, epizote, clove, and thyme caused total inhibition of fungal development on maize kernels and the optimal dosage varied from 3 to 8 per cent. O-methoxycynamaldehyde from cinnamon has been reported to be highly effective against *A. flavus* and *A. parasiticus*. Bilgrami *et al.* (7) have observed inhibition of mycelial growth and aflatoxin production on maize grains when treated with onion, garlic and eugenol. Caffeine appears to inhibit aflatoxin synthesis by restricting the uptake of carbohydretes, which are ultimately used by the mould to synthesize aflatoxins. Capsanthin of *capsicum annum* (red chilli) was also reported to check both growth and toxin production at all tested concentrations.

Sinha (71) screened 25 chlorophyllous plants against aflatoxin production in cereals and oil seeds. Qualitative analysis of these plant extracts revealed the presence of ten identified and ten unidentified phenolics. Phenolics like tannic acid, caffeic acid and phloroglucinol at 0.01 M concentration prevented aflatoxin production by more than 55 percent. Recently, Bankole and Joda (3) observed efficacy of lemongras (*Cymbopogon citrates*) powder and essential oil on *A. flavus* growth and aflatoxin contamination of melon seeds.

## FUTURE OUTLOOK

In areas where conditions are frequently favourable to fungal invasion and mycotoxin production in the field, the grains / seeds should be tested for mycotoxin contamination before use. Moreover, care showed also be taken to reduce the source of inoculum and to minimize plant stress and insect damage in the fields. Further research is needed to develop the ability to predict when and where environmental conditions may make mycotoxin contamination probable and to develop the means to disseminate warnings to farmers and processors. Regular monitoring and surveillance programmes play an important role in determining compliance of regulatory limits.

## *Literature Cited*


1. Alpert ME, Hutt MSR, Wogan GN, Davidson CS. 1971. Association between aflatoxin content of food and hepatoma frequency in Uganda. *Cancer* 28 : 253-260.
2. Anonymous 2002. General reference reports. *J. Assoc. Off. Anal. Chem. Inst.* 83: 443.
3. Bankole SA, Joda AO. 2004. Effect of lemon grass (*Cymbopogon citratus*) powder and essential oil on mould deterioration and aflatoxin contamination of melon seeds (*Colocynthis citrullus* L.). *African J. Biotech.* 3(1): Available at. http :www.academicjournal.org/AJB
4. Bhat RV, Beedu SR, Ramakrishna Y, Munshi KL. 1989. Outbreak of trichothecene mycotoxicosis associated with consumption of mould damaged wheat production in Kashmir valley, India. *Lancet* I:35-37.
5. Bhatnagar D, Cotty PJ, Cleveland TE. 2001. Genetic and biological control of aflatoxigenic fungi. In : Wilson C.L. and Droby, S. (Eds.). Microbial Food Contamination. CRC Press, New York, pp. 207-240.
6. Bilgrami KS, Choudhary AK. 1998. Mycotoxins in preharvest contamination of agricultural crops. In : Sinha, K.K. and Bhatnagar. D. (Eds.). Mycotoxins in Agriculture and Food Safety. Marcel Dekker, New York., pp. 1-43.
7. Bilgrami KS, Sinha KK, Sinha AK. 1992. Inhibition of aflatoxin production and growth of *Aspergillus flavus* by engenol, onion and garlic extracts. *Indian J. Med. Research* 96 : 171-175.
8. Bock CH, Cotty PJ, 1999. The effect of harvest date on aflatoxin contamination of cottonseed in Arizona. *Plant Disease* 83 : 279-283.
9. Boutrif, E. and Carnet, C. 1998. Mycotoxin prevention and control : FAO programmes. *Revue Med. Vet.* 149 : 681-694.
10. Botone EJ, Peluso RW. 2003. Production by *Bacillus pumilus*, (MSH) of an antifungal compound that is active against Mucoraceae and *Aspergillus* species : Preliminary Report. *J. Med. Microbiol* 52 : 69-74.

11. Brown RL, Cotty PJ, Cleveland TE. 1991. Reduction in aflatoxin content of maize by atoxigenic strains of *Aspergillus flavus. J. Food Prot.*, 54 : 623-626.
12. Brown RL, Jenco CS, Obrian GR, Solan S, Payne GA. 1998. Identification of the DNA binding site for *Aspergillus flavus* AFLR in the NOR-p promotor. In : Robens, J. and Wiclow, D. (Eds.). Proceedings of the USDA-ARS Aflatoxin Elimination Workshop, held at St. Louis, Mo, P.102.
13. Brown RL, Brown Jenco CS, Bhatnagar D, Payne G. 2003. Construction and preliminary evaluation of an *Aspergillus flavus* reporter gene construct as a potential tool for screening aflatoxin resistance. *J. Food. Prot*. 66: 1927-1931.
14. Brown RL, Chen ZY, Menkir A, Cleveland TE. 2003. Using biotechnology to enhance host resistance to aflatoxin contamination of corn. *African J. Biotech*. 2(12) : 557-562.
15. Bullerman LB, Lieu Y Seier SA. 1977. Inhibition of growth and aflatoxin produ-ction by cinnamon and eugenol. *J. Food Sci*. 46 : 1107-1109.
16. CAC 1998. Position paper on patulin. 30th session of Codex Committee of Food Additives and Contaminants. Joint FAO / WHO Food Standards Programme, (Rome : Food and Agricultural Organization), item 12 (d) Alinorm 01 / 12, Codex Alimentarius Commission. p. 1-4.
17. CAC 2001. 24th session of the Joint FAO/WHO Food Standards Programme, at Beijing, People's Republic of China. Report of the 32nd meeting and contaminants (Rome : Food and Agricultural Organization), Alinorm 01/12A, Appendix XII, Codex Alimentarius Commission, pp. 145, 122-126.
18. CAST 1989. Mycotoxins : Economic and health risks. Council for Agricultural Sciences and Technology, Ames, IA, 50010.
19. Chen ZY, Brown RL, Lax AR, Guo BZ, Cleveland TE, Russian JS. 1998. Resistance to *A. flavus* in corn kernels is associated with a 14 KDa protein. *Phytopathology*, 88: 276 –281.
20. Chen ZY, Brown RL, Lax AR, Cleveland TE and Russian JS. 1999. Inhibition of plant pathogenic fungi by a corn trypsin inhibitor over expressed in *Escherichia coli. Appl. Environ Microbiol*. 65 : 1320-1324.
21. Chen ZY, Brown RL, Damann KE, Cleveland TE. 2000. Proteomics analysis of kernel embryo and endospesm proteins of corn genotypes resistant or susceptible to *Aspergillus flavus* infection. In : Roben, J., Cary, J.W. and Campbell, B.C. (Eds.). Proceedings of the USDA-ARS Aflatoxin Workshop held at Yosemite, C.A., P.88.

22. Chen ZY, Brown RL, Damann KE, Cleveland TE. 2002. Identification of unique or elevated levels of kernel proteins in aflatoxin resistant maize genotypes through proteome analysis. *Phytopathology* 92:1084-1094.

23. Choudhary AK. 1992. Influence of microbial co-inhabitants on aflatoxin synthesis of *Aspergillus flavus* on maize kernels. *Letters Appl. Microbiol.* 14 : 143-147.

24. Cleveland TE, Dowd PF, Desjardins AE, Bhatnagar D, Cotty PJ. 2003. United States Dept. of Agriculture-Agricultural Research Service, Research on preharvest prevention of mycotoxins and mycotoxigenic fungi in U.S. crops. *Pest Manag. Sci.* 59 : 629-642.

25. Coallier-Ascab J, Idziak ES. 1985. Interaction between *Streptococcus lactis* and *Aspergillus flavus* on production of aflatoxin. *Appl. Environ. Microbiol.* 49 : 163-167.

26. Cole RJ, Cotty PJ, 1990. Biocontrol of aflatoxin production by using biocompetitive agents. In : J.R. Robens (Ed.). A perspective on aflatoxin in field crops and animal food products in United States. US Deptt. of Agric.Res. Sen:, ARS, 83 :62-66.

27. Cole RJ, Dorner JW, Blankenship PD. 1994. Non-aflatoxigenic *Aspergillus parasiticus* strains and their use in controlling aflatoxin contamination. US patent 5, 292,661.

28. Cotty PJ. 1989. Virulence and cultural characteristics of two *Aspergillus flavus* strains pathogenic on cotton. *Phytopathology* 79 : 808-814.

29. Cotty PJ. 1992. Use of native *Aspergillus flavus* strains to prevent aflatoxin contamination. US patent 5, 171, 686.

30. Cotty PJ, Bayman P. 1993. Competitive exclusion of a toxigenic strain of *Aspergillus flavus* by an atoxigenic strain. *Phytopathology* 83 : 1283-1287.

31. Daigle DJ, Cotty PJ. 1997. The effect of sterilization, pH, filler and spore inoculation concentration on the preparation of alginate pellets. *Bicontrol Sci. Technol.* 5 : 175-184.

32. Davis GL, William WP. 1999. QTL for aflatoxin reduction in maize. Maize Genetic Conference 41 : 22.

33. Devi GR, Polasa H. 1987. Interference in toxin production among toxigenic *Aspergillus* spp. *J. Stored Products Research* 23 :149-150.

34. Diener UL, Davis ND. 1966. Aflatoxin production by isolates of *Aspergillus flavus. Phytopathology* 56 : 1390-1393.

35. Diener UL, *et al.* 1987. Epidemiology of aflatoxin formation by *Aspergillus flavus. Ann. Rev. Phytopathol.* 25 : 249:270.

36. Du W, Huang Z, Flasherty JE, Walls K, Payne GA. 1999.

Green fluorescent protein as a reporter to monitor gene expression and food colonization by *Aspergillus flavus*. *Appl. Environ. Microbiol.* 65 : 834-836.

37. Eaton LD. 1997. Aflatoxin toxicology. Chapter 9.26. Comprehensive Toxicology.Pergamon Publications.

38. Eaton LD, Gallhager EP. 1994. Mechanism of aflatoxin carcinogenesis. *Ann. Rev. Pharmacol. Toxicol.* 34 :135-172.

39. European Commission 2002. Commission Regulations. No-472/2002 of 12 March 2002. Setting maximum levels for certain contaminants in foodstuffs. Official European Communities L – 75/20.

40. El – Katatny MH, Gudelj M, Robra KH. 2001. Characterization of a chitinase and an endo.-β/1,3 glucanases from *Trichoderma harzianum* Rifai T24 involved in the control of the phytopathogen *Sclerotium rolfsii*. *Appl. Microbiol Biotechnol.* 56 : 137-143.

41. FAO 1979. Recommended practices for the prevention of mycotoxins in food, feed and their products. FAO Food and Nutrition paper, No. 10. Rome.

42. FAO 1995. The use of hazard analysis critical control point (HACCP) principles in food control. FAO Food and Nutrition Paper No. 58. Rome.

43. FAO 1997. Worldwide regulations for mycotoxins, 1995, FAO Nutrition Paper-64. Rome, pp.7-28.

44. Guo BZ, *et al.* 1997. Germination induces accumulation of specific proteins and antifungal activities in corn kernels. *Phytopathology* 87 : 1174-1178.

45. Guo BZ, *et al.* 1998. Protein profiles and antifungal activities of kernel extracts form corn genotypes resistant and susceptible to *Aspergillus flavus*. *J. Food Prot.* 61: 98-102.

46. Hendrickse RG, *et al.* 1982. Aflatoxins and Kwashiorkor; a study in Sudanese children. *Brit. Med. J.* 285 :843-746.

47. Horn BW, Wicklow DT. 1983. Factors influencing the inhibition of aflatoxin production in corn by *Aspergillus niger* *Canad. J. Microbiol.* 29 :1087-1091.

48. Howard S, David L. 1990. Species susceptibility to aflatoxin $B_1$ carcinogenesis. *Cancer Research* 50 : 615-620.

49. Huang Z, White DG, Payne GA. 1997. Corn seed proteins inhibitory to *Aspergillus flavus* and aflatoxin biosynthesis *Phytopathology* 87 : 622-627.

50. IARC 1987. Monographs on the evaluation of the carcinogenic risk of chemicals to man, Geneva: World Health Organization, International Agency for Research on cancer, 1972- Present (Multivolume work, P. 5783)

51. IARC 1993. Aflatoxins. In: IARC Monographs on the

evaluation of carcinogenic risk to humans, 56. IARC, Lyon, France, pp.245-395.

52. Ji C, Norton RA, Wiklow DT, Dowd F. 2000. Isoform patterns of chitinase and α-1,3 glucanase in maturing corn kernels (*Zea mays* L.) associated with *Aspergillus flavus* milk stage infection *J. Agric. Food Chem.* 48: 507-511.

53. Kurtzman CP, Horn BW, Hesseltine CW. 1987. *Aspergillus nomius*, a new aflatoxin producing species related to *Aspergillus flavus* and *Aspergillus tamarii*. *Antoni Van Leeuwenhoek* 53 : 147-158.

54. Lopez-Garcia R, Park DL. 1998. Management of mycotoxin hazards through post harvest procedures. In : Sinha K.K and Bhatnagar, D. (Eds.). Mycotoxins in Agriculture and Food Safety. Marcel Dekker, New York, pp. 407-433.

55. Marasas WFO, Van Rosenburg SJ, Mirocha CJ. 1979. Incidence of *Fusarium* species and mycotoxins deoxynivalenol and zearalenone in corn produced in esophageal cancer areas in Trapskei, Southern Africa. *J. Agric. Food Chem.* 27.

56. Marasas WFO. 1994. *Fusarium*. In : Hui, Y.H., Gorhan, J.R., Musrel, K.K. and Cliver, D.O. (Eds.). Food borne disease handbook. Vol.2 pp. 534-541.

57. Newell J. 1983. Treatment for starvation may kill. *New Scientist* 99: 471.

58. National Toxicology Program 1999. Technical Report on the Toxicology and Carcinogenesis studies of fumonisin $B_1$ (CAS No-11655-83.0) in F 344/4 rats and $B6C3F_1$ mice. NTP TR 496. NIH Publication No. 99-3955.

59. Nuntharathanpong N. 2001. Increase in tumor necrosing factor-alpha and a change in lactate dehydrogenase isoenzyme pattern in plasma of workers exposed to aflatoxin-contaminated feeds. *Arh. Hig. Rada. Toksikol.* 52 : 292.

60. Park DL. 1993. Controlling aflatoxins in food and feed. *Food Technol.* 47 (10) : 92.

61. Park DL, Liang B. 1993. Perspectives on aflatoxin control for human food and animal feed. *Trends Food Sci. Technol.* 4 : 334.

62. Payne GA. 1997. Characterization of inhibitors from corn seeds and the use of a new reporter construct to select corn genotype resistant to aflatoxin accumulation. In : Robens. J. and Dorher, J. (Eds.). Proceedings of the USDA – ARS Aflatoxin elimination workshop held at Memphis, TN, pp. 66-67.

63. Peers FG, Linsell CA. 1973. Dietary aflatoxin and liver cancer-a population based study in Kenya. *Brit. J. Cancer* 27 : 473.

64. Pineiro M. 2000. Mercosur and Latin American Standards for Mycotoxins. In: Global approaches to mycotoxins – Risk assessments and standards. Paper presented at the 2000 Institute of Food Technologists (IFT) Annual Meeting. pp. 32-34.

65. Pitt JI. 1987. *Penicillium viridicatum*, *P. verrucosum* and production of ochratoxin-A. *Appl. Environ. Microbiol.* 53: 266-269.

66. Ryan, N.J. 1979. Aflatoxin $B_1$-its role in the etiology of Reye's syndrome. *Pediatrics* 64:71.

67. Reeder JP, *et al.* 1992. *Fusarium momiliforme* and *fumonisins* in corn in relation to human esophageal cancer in Transkei. *Phytopathology* 82: 333-357.

68. Rajasekaran K, *et al.* 2000. High frequency stable transformation of cotton *(Gossypium hirsutum* L.) by particle bombardment of embroygenic cell-suspension culture. *Plant Cell Rep.* 19:539-545.

69. Scott GE, Zummo N. 1988. Sources of resistance in maize to kernel infection by *Aspergillus flavus* in the field. *Crop Sci.*, 28 : 505-507.

70. Shank RC, Bourgeois CH, Keschamras N, Chandavinol P. 1971. Aflatoxin and autopsy specimens from Thai children with an acute disease of unknown acetiology. *Food Cosmet. Toxicol.* 9 : 501-507.

71. Sinha KK. 1990. Prevention of aflatoxin production in some cereals and oil-seeds by some natural plant constituents. *Irish J. Food Sci. Technol.* 14(2): 109- 120.

72. Sinha KK. 2005. Investigations on chemo-morphological factors responsible for preventing aflaotoxin production in maize varieties. Final Technical Report, ICAR Project, T.M. Bhagalpur University.

73. Sinha KK, Bhatnagar D. 1998. Mycotoxins in Agriculture and Food safety, Marcel Dekker, New York.

74. Ueno Y 1983. Trichothe-cenes : Chemical, Biological and Toxicological Aspects. Elsevier Sci. Pub. Co., New York, P. 313.

75. Vander Merwe, Steyn KJ, Fourie PS, Scott L, Theron JJ. 1965. Ochratoxin-A, a toxic metabolite produced by *Aspergillus Ochraceus* with. *Nature* (London), 205 : 205:1112.

76. Van Walbeek W, Scott PM Tatcher FS. 1968. Mycotoxins from food borne fungi. *Can. J. Microbiol.* 14 : 131.

77. Vishwanath Prema 2004. Mycotoxin regulations for foods : A global update. *J. Food Sci. Technol.*, 41 (2) : 115-123.

78. Wang JS 2001. Development of aflatoxin $B_1$ lysin adduct monoclonal antibody for human exposure studies. *Appl. Env. Microbiol.* 67 : 2712.

79. White DG, Rochaford TR, Kanfman B, Namblin AM.

1995. Chromosome region associated with resistance to *Aspergillus flavus* and inhibition of aflatoxin production in maize. In : Robens J. (Ed.). Proceedings of the USDA –ARS Aflatoxin elimination workshop held at Atlanta, GA, P.8.

80. White DG, *et al.* 1998. Inheritance of molecular markers associated with and breeding for resistance to *Aspergillus* ear rot and aflatoxin production in corn using Tex 6 In: Robens, J. and Wicklow, D. (Eds.). Proceedings of the USDA. ARS Aflatoxin elimination workshop held at St. Louis, Mo, pp. 4-6.

81. Wicklow DT, Hesseltine CW, Shotwell OL, Adams GL. 1980. Interference, competition and aflatoxin levels in corn. *Phytopathology.* 70 : 761-764.

82. Windham GL, William WP. 1998. *Aspergillus flavus* infection and aflatoxin accumulation in resistant and susceptible maize hybrids. *Plant Disease* 82: 281-284.

83. Windham GL, William WP, Davis FM. 1999. Effects of the Southwestern corn borer on *Aspergillus flavus* kernel infection and aflatoxin accumulation in maize hybrids. *Plant Disease* 83: 535-540.

84. Woloshuk CP, Cavaletto JR, Cleveland TE. 1997. Inducers of aflatoxin biosynthesis from colonized maize kernels are generated by an amylase activity from *Aspergillus flavus. Phytopathology.* 87: 164-169.

85. www.aspencancerconferrence.amc.org/3m2001goopman.htm(Molecular Epidemiology: Aflatoxin P53 mutations and liver cancer as a paradigm)

*Rev. Plant Pathol. Vol. 4, 2008*
*Indian Society of Mycology and Plant Pathology*
*Scientific Publishers (India), Jodhpur, pp. 301-334*

# MOLECULAR GENETICS OF 'MOB-ATTACK' BY PHYTOPATHOGENIC BACTERIA

***Appa Rao Podile, Debashish Dey and Burla Sashidhar***

Department of Plant Sciences, School of Life Sciences, University of Hyderabad, Hyderabad- 500 046, India. (E-mail correspondence: arpsl@uohyd.ernet.in)

**KEY WORDS:** Quorum sensing (QS), phytopathogenic bacteria, chemical signaling, cell-to-cell communication, acyl homoserine lactones (AHLs), cell density-dependent, mob-attack, autoinducer.

## ABSTRACT

Quorum sensing (QS) governs bacterial cell-to-cell communication *via* a cell density-dependent gene regulation that allows bacterial cells to express specific genes when they reach high cell density, through small diffusible chemical signaling molecules. The QS signaling molecules differ from species to species: Gram-negative bacteria use acyl homoserine lactones, where as Gram-positive bacteria use oligopeptides. Signaling molecules modulate both intra- and interspecies cell-to-cell communication, and play a major role in enabling bacteria behave like multicellular organisms. In recent years, researchers have shown that population of bacterial virulence is a very promising target for preventing infectious disease, since threshold cell population of pathogen is required to establish successful infection. By understanding and taking advantage of QS mechanisms, researchers target for new ways to control diseases. The present review dissects the molecular genetics, and highlights major differences and similarities in the mechanisms employed for QS in some important phytopathogenic bacteria, with an introduction that explains the QS mechanism involved in expression of bioluminescence by the marine symbiotic bacterium *Vibrio fischeri* and it's free-living relative *Vibrio harveyi*.

## 1. BACTERIA ALTER BEHAVIOUR IN A CELL-DENSITY-DEPENDENT MANNER

Bacteria cells with each other by chemical signaling to execute several important functions in accordance with the external surrounding environment as a cell-density-dependent behavior. Quorum sensing (QS) allows bacteria to monitor the environment and to alter behavior on a population-wide scale involving production and release of small diffusible signal molecules, signal detection and signal transduction mechanism. The earliest described QS system is the biolumin-escent marine symbiotic bacterium *Vibrio fischeri,* the paradigm for QS in many Gram-negative bacteria. Several bacterial pathogens (causing diseases in animals and plants) evolved to express their virulence factors through the QS system for the "mob-attack" of the host. By QS mechanism, bacteria sense their surroundings and determine when, and, equally important, or not to attack the susceptible host. Excellent reviews described this extraordinary cell-to-cell communication by bacteria, involving chemicals influencing the cell density-dependent behaviour of bacteria (1, 7, 65, 67, 68, 69, 71, 124, 127). The developments in the past decade, more particularly in understanding the significance of QS for the plant pathogenic bacteria, form the focus of this review.

QS is a well-understood mechanism of bacterial cell-cell communication that allows organisms to control physiological responses and express certain traits when bacteria are crowded together. This allows bacterial cells to act in a coordinate manner and reinforces the notion that individual bacteria benefit from cooperative group behavior to survive, compete and persist in nature, even invade a particular host. The QS involves the exchange of low molecular weight diffusible signal molecules within and between members of a localized population. The signal molecules, synthesized and when reached a threshold level, activate cognate receptor protein, in turn may trigger wide spread changes in the gene expression, are popularly referred as auto inducers (AI). As in higher organisms, the information supplied by these molecules is critical for synchronizing the activities of large group of cells, a step towards multicellularity in otherwise unicellular organisms.

## 2. BACTERIAL CELL-TO-CELL COMMUNICATION DEPENDS ON CHEMICAL SIGNALING SYSTEMS

Two major classes of communications systems have been described based on the nature of the signal. Various processed oligopeptides serve as signals in most of the Gram-positive bacteria (62, 65), while Gram-negative bacteria use basically acyl homoserine lactones (AHLs) as signaling molecules (124), and some species-specific systems including opine signaling in *Agrobacterium tumefaciens*, 3-hydroxy palmitic acid methyl ester signals in *Ralstonia solanacerum* and less-characterized diffusible molecules of *Xanthomonas campestris*. A third signaling system, as in cell-cell signaling in *Escherichia coli* and *Salmonella enterica* (1), through interspecies communication has been linked to auto-inducer-2 (AI-2), a furanosyl borate diester (24). A single bacterium produces more than one signaling molecules. The multiple cell-cell signaling molecules, for example two different AHLs, stimulate the expression of virulence factors as in *Pseudomonas aeruginosa*. The QS of bioluminescence in *Vibrio harveyi* and *V. fischeri* is also mediated by two extra cellular signaling molecules. In *Bacillus subtilis*, under nutrient depletion conditions, cells are stimulated by the accumulation of two oligopeptides in the growth medium (ComX pheromone for competence and CSF for sporulation factors), which lead to formation of spores (62).

## 3. QUORUM SENSING REGULATES VIRULENCE FUNCTIONS OF PATHOGENIC BACTERIA AND DRIVES THE "MOB-ATTACK" ON TO THE HOST

Pathogens must identify a susceptible host and then attack when they are in enough numbers to cause the disease. When bacteria with a small number attack the host, minimal damage caused to the host can activate host defense mechanisms. For example *Erwinia carotovora* subsp. *carotovora* (*Ecc*), a maceration pathogen, by producing extracellular enzymes releases plant cell contents (some oligosaccharides) into the surrounding environment. The plant detects specific oligosaccharides, to mount defense responses and eliminate the pathogen if they are present in sufficiently small number. Another advantage of QS is production of multiple virulence factors with one external signal. *Ecc* produces macerating enzymes and the antibiotic carbapenem, this means, by degrading plant tissue, releasing the vital nutrients into the environment may attract the neighboring bacteria and the antibiotic plays a role in eliminating competitors, leaving available nutrients for use of *Ecc* cells (82).

In the remaining part of this review, the available information is discussed under two major categories viz. AHL-based, and non-AHL-based regulation of virulence factors as cell density-dependent phenomenon in bacteria with special reference to plant pathogenic bacteria. A brief summary of the QS regulatory genes, signals and the regulated property in phytopathogenic bacteria is given in table 1.

## 4. ACYL HOMOSERINE LACTONES (AHL)-BASED REGULATION OF VIRULENCE FACTORS

The first discovered and well-studied AHL-based QS system is the *lux* system operating in *Vibrio fischeri*, which lies as a symbiotic organism in the light organs of a squid, *Euprymna scolopes* (40). The luciferase enzyme encoded by the *lux* gene cluster is responsible for imparting blue green light to the squid, which is exploited by the squid to perform counter illumination thereby protecting it from nocturnal predators (116).

Two divergently transcribed operons exist in *lux* gene cluster (Fig. 1). In one operon the *luxR* gene encodes the transcriptional regulator LuxR in one direction, while the rest of the *lux* genes are encoded by the other operon transcribed in opposite direction. The second operon consists of *luxI* encoding acyl HSL synthase (LuxI), *luxAB* encoding subunits of luciferase enzyme, *luxCDE* encoding the components of a multienzyme complex and *luxG* encoding a flavin reductase. The LuxI protein synthesizes small molecular weight acyl HSL signal molecules like N-(3-oxo-hexanoyl)-L-homoserine lactone (OHHL), a small diffusible signal molecule of fatty acid origin in case of *V. fischeri* (41, 44, 58). When the population size or cell density of *V. fischeri* is low, *luxI* gene is expressed at its basal level leading to low concentration of OHHL molecules. As the population size increases, the OHHL concentration also increases. Once its concentration crosses the threshold level, it binds with LuxR (Fig. 1), which undergoes conformational change mediated activation and binds 20 bp upstream promoter sequences of *luxI* (34, 35, 42). This specific sequence is having a dyad symmetry known as *lux* box. LuxR binding to *lux* box switches on the *lux* operon leading to bioluminescence and increased concentration of OHHL.

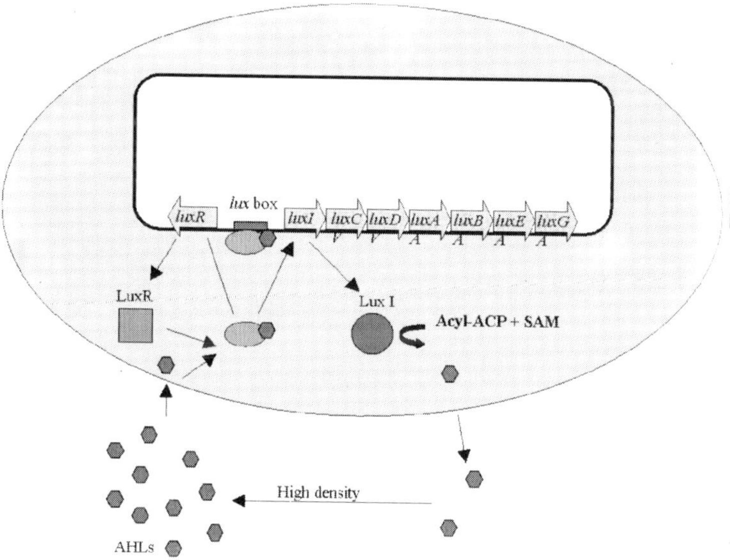


**Fig. 1.** Regulation of *lux* operon in *Vibrio fischeri*. At low cell density the *luxI* and *luxR* are expressed at their basal level. *luxI* encodes LuxII ( ⬢ ), which synthesizes AHLs ( ● ) from acylated acyl carrier protein (acyl ACP) and S-adenosylmethionine (SAM). AHL concentration goes on increasing in a cell density dependent manner. At high cell density, high concentration of AHL leads to activation of LuxR through a conformational change (note that square becomes oval upon binding to the AHL), which represses its own synthesis shown by ( ——| ) and switches on the other *lux* operon (Indicated as arrows) by binding to the *lux* box present in the upstream promoter regions.

At low cell density, the concentration of OHHL molecules is at its basal level of expression, which is enough to bind and activate few LuxR molecules. Once bound with OHHL, LuxR represses its own synthesis by an unknown mechanism (40), however, it is assumed that a *lux* box within *luxD* may be involved in this mechanism (104).

**Table 1:** Regulatory genes/molecules governing quorum sensing in different phytopathogenic bacterial:

| Bacterium | LuxR/I homologues | Major AHL/ Non AHL | Regulated property | Ref. |
|---|---|---|---|---|
| *Agrobacterium tumefaciens* | TraR, TraI | 3-oxo-C8-HSL | Conjugation | 47 |
| *Agrobacterium vitis* | AvhR | Unknown | Conjugation | 53 |
| *Erwinia carotovora* subsp. *carotovora* | CarR, ExpR ExpI (CarI) | 3-oxo-C6-HSL | Carbapenem antibiotic, exoenzymes | 4 |
| *Erwinia chrysanthemi* | ExpR, ExpI (EchR, EchI) | 3-oxo-C6-HSL | Pectinases | 80 |
| *Erwinia carotovora* subsp. *betavasculorum* | EcbR, EcbI | 3-oxo-C6-HSL | Pectate lyase | 29 |
| *Erwinia carotovora* subsp. *atroseptica* | VirR | 3-oxo-C6-HSL | Exoenzymes | 15 |
| *Pantoea stewartii* | EsaR, EsaI | 3-oxo-C6-HSL | Exopolysaccharide | 8 |
| *Ralstonia solanacearum* | SolR, SolI | C8-HSL | Unknown | 45 |
| *Xanthomonas campestris* pv. *campestris* | *PigB* | DSF and DF | Exopolysaccharide | 95 |

### 4.1 Expression of pathogenicity factors in plant pathogenic *Erwinia* is population density-dependent

The Gram-negative phytopathogenic *Erwinia amylovora* causes soft rot disease in several economically important crops (87). It produces several pathogenicity factors in a population density-dependent manner, such as production of extracellular enzymes and carbapenem antibiotic as well as the Hrp secretion system. However, different subspecies differ in host specificity and cause different disease types e.g. *E. carotovora* subsp. *atroseptica* (*Eca*) causes blackleg disease (stem rotting) of potatoes, whereas *E. carotovora* subsp. *carotovora* (*Ecc*) has a broader host range and typically causes soft rots (88).

The QS signaling systems of *E. carotovora* subspecies are of two classes defined by the nature of the major AHL analog produced (Fig. 2) as well as structural and functional characteristics of AHL synthase (AhlI) and AHL receptor (ExpR) (19). The class I strains produce 3-oxo-C8-HL (N-3-oxooctanoyl-l-homoserine lactone) as the major AHL analog and detectable levels of 3-oxo-C6-HL (N-3-oxohexanoyl-l-homoserine lactone) while class II strains produce 3-oxo-C6-HL as the major analog. *Ecc* produces many exoenzymes such as cellulases, proteases and pectinases. The production of these exoenzymes is controlled by QS mechanism (57, 91). At high cell density, the concentration of acyl-HSL, OHHL increases leading to up regulation of exoenzyme synthesis. CarI synthase, a homologue of LuxI (4, 20, 57, 91), directs OHHL production, the primary AHL in *Ecc*. The *carI* deletion strains do not produce carbapenem antibiotic, pectolytic enzymes, endoglucanases, and proteases and fail to secrete harpin and these are nonpathogenic to plants (111). The cognate LuxR homologue of LuxI needed for exoenzyme regulation has not been reported yet in *Ecc*. The production of β-lactam antibiotic, carbapenem in certain strains of *Ecc* (84) is mediated by QS mechanism consisting of *carI* and *carR* genes present in two different bi-directionally transcribed operons (57, 20, 91). The former synthesizes OHHL that activates CarR and latter activates the expression of *carA-H* genes needed for carbapenem production (73). Among these, *carABCDE* encode biosynthetic enzymes; *carFG* encode a resistance mechanism against the antibiotic effects (75,74), while *carH* function is still unknown. The *car* cluster is positioned almost 150 bp downstream of *carR*, with the intergenic region containing the transcriptional start of *carA*. The *carI* gene is not linked with *car* cluster and is found in a separate location. Mutations either in *carR* or in *carI* lead to no antibiotic production, while mutations only in *carI* affect exoenzyme production. Freely available CarR exists as dimer and is able to bind DNA in the absence of OHHL, but with low affinity (121).

Upon activation by cognate AHLs, it forms multimeric complex, which is able to bind promoter region very efficiently leading to transcriptional activation of *car* genes.

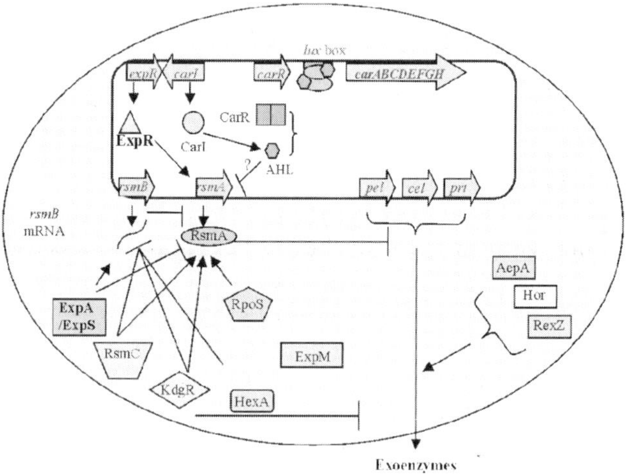


Figure 2. The regulation of exoenzyme production and carbapenem synthesis in *Ecc*. CarI is required for AHL ( ⬢ ) synthesis, while CarR, which undergoes a conformational change-mediated activation in presence of AHL and leads to activation of *car* cluster genes required for carbapenem synthesis.

The high concentration of AHL also leads to increased expression of exoenzymes through an unknown mechanism involving another unidentified LuxR homologue. Several regulators of exoenzyme production exist in *Ecc* such as RsmA/*rsmB* system, KdgR, ExpA/ExpS, RsmC, HexA, ExpM and RpoS. *rsmB* prevents target RNA degradation by RsmA by binding and sequestering it in the cell. ExpA/ExpS activates *rsmB* expression whereas represses RsmA. Conversely RsmC, KdgR and HexA repress *rsmB* expression and activate RsmA-mediated repression of virulence factor production. HexA activates RsmA through ExpM mediated repression of RpoS. However there are also other regulators like AepA, Hor and RexZ reported to regulate exoenzyme production. Positive regulation is shown by arrows while negative regulation is shown by ( ——| ).

The carbapenem synthesis normally occurs during the late log or early stationary phases of growth but it can be induced at early stages by exogenous addition of OHHL. When the concentration of OHHL is above 0.1 μg/ml, it activates CarR, which exists as a dimer. One dimer binds with 2 molecules of OHHL and then only it can bind the *carR-carA* intergenic region and induce carbapenem synthesis. The necessity of OHHL can be circumvented by overexpressing *carR*, suggesting a role of OHHL merely protecting CarR from proteolytic degradation (85).

The same mechanism of exoenzyme production may also be involved in *E. chrysanthemi* (*Ech*), the causal organism of soft rots and vascular wilts in several plants. There are two convergently transcribed genes *expI* and *expR* and three different AHLs – OHHL, N-hexanoyl-L-homoserine lactone (HHL) and N-decanoyl-L-homoserine lactone (DHL) of which OHHL and HHL are synthesized by *expI* gene product (80). Deletion of either *expI* or *expR* has no effect on the exoenzyme production even though ExpR has been reported to bind promoter regions of several pectinase genes in absence of AHL. The ability of even the mutant regulators to bind promoter regions could be a mechanism to avoid damaging effects of *expR* mutations on exoenzyme production. Several regulators such as KdgR, CRP, Pir and PecS (72) control the production of pectinases, which is the major virulence determinant. Among these KdgR is the most important positive or negative regulator of expression of most pectinase genes (101) and genes involved in rhamnose catabolism (55). Two genes in *Ech*, *paeY* and *paeX* encode pectin acetylesterase needed for enzymatic degradation of the pectin in plant cell walls, and its expression is regulated by KdgR and PecS regulators (107).

There is another system (EcbRI) existing in *E. carotovora* subsp. *betavasculorum* (*Ecb*), the causal agent of soft rot in sugar beet, for production of another unidentified antibiotic (29). This system consists of EcbI, responsible for AHL production and EcbR regulator. Deletion strains of *ecbI* gene are impaired in pectate lyase enzyme production and are unable to produce antibiotic.

The presence of a novel LuxR homologue in *E. carotovora* subsp. *atroseptica (Eca)* genome (11) and recent discovery of VirR (for Virulence Repressor) as a master regulator of QS regulation and repressor of the production of QS-dependent secreted proteins in the absence of OHHL (15) reveal the diversity and complexity of QS system in *Erwinia*. It also acts as a regulator of two virulence factors (Svx and Nip). The wide distribution of *virR* in *Eca*, *Ecc* and *Ecb* strains strongly suggests that VirR may also be important in regulating virulence factor synthesis in these and other soft-rotting *Erwiniae.* The virulence determinant Svx, secreted out by the type II machinery (28), has an N-terminal signal sequence and zinc binding-region signature, and the mature protein is post-translationally modified. Its transcription is under AHL-mediated QS control. The mutants of *svx* are impaired in virulence in *Eca*.

There is yet another system involved in QS regulation of virulence factors in *Ecc* known as RsmA/rsmB (regulator of secondary metabolism) system (20, 32), which is conserved in *E. coli* and many *Erwinia* and *Pseudomonas* spp. This system involves two components – an RNA binding protein, RsmA and an untranslated RNA molecule *rsmB* (Fig. 2). These together form a post transcriptional control system and act as an interface between quorum sensing signals and intracellular control network (66). RsmA degrades the target gene mRNA, thereby acting as a negative regulator of virulence factors including exoenzymes. *rsmB* negates its effect by binding and sequestering it in the cell. Deletion strains of *rsmA* overproduce exoenzymes and other pathogenicity determinants independent of presence of OHHL (20, 31) and induce HR in tobacco. OHHL regulates exoenzyme production *via* repression of *rsmA* transcript although the absence of *lux* box in *rsmA* promoter suggests that this might occur through another regulator. This is supported by recent reports where ExpR upregulates *rsmA* transcript activation by binding to *rsmA* promoter sequences thereby inhibiting virulence factor production in absence of OHHL. At high cell density, high concentrations of OHHL inhibit this ExpR effects (30).

There are other regulators, which control this RsmA/*rsmB* system such as RsmC, KdgR, RpoS and HexA. The RsmA expression is induced by RsmC, while it represses *rsmB*. Conversely, ExpA/ExpS represses RsmA and activates *rsmB*. RpoS activates RsmA whereas HexA leads to RsmA induction through inactivating RpoS and rsmB. KdgR is also reported to repress the exoenzyme production (66), In addition to these, several other regulators like AepA, Hor and RexZ also regulate secretion of exoenzymes in *Erwinia* spp. New virulence factors (Nip) involved in causing necrosis in host plants have been identified in *Erwinia* spp. and these are subject to regulation by a LuxR homologue, EccR and negative global regulator RsmA (86).

### 4.2 EsaR and EsaI, associated with QS in *Pantoea stewartii*, are homologues of LuxR and LuxI

*Pantoea stewartii* subsp. *stewartii* (*Pnss*; synon. *Erwinia stewartii*), the causal organism of Stewart's wilt and leaf blight of sweet corn and maize, produces large amounts of capsular polysaccharides, which blocks xylem vessels and limits the plant water transport. It grows in the apoplastic regions of the leaf and produces water-soaked lesions, blocking of xylem vessels and subsequent wilting (13). The *cps* locus encodes polysaccharides required for stewartan EPS synthesis and membrane trans-location (27). *cps* gene cluster is related in organization and function to typical group I capsule (M-antigen) biosynthetic gene clusters in other enteric bacteria. Mutations in the *cpsA-M* locus lead to loss of wilting and systemic movement. Mutants of *cps* locus are unable to cause wilting and do not show systemic movement and are, therefore, avirulent.

There are two different levels of controls for the *cps* genes. At primary level, the control occurs through EsaR and EsaI, a homologue of LuxR and Lux I, respectively (8) in a cell density-dependent manner. Secondary level of control is driven by RcsAB protein, which alters capsular synthesis in response to environmental factors in a number of different bacteria including *E.coli* (51). EsaR is an unusual LuxR homologue, as in the presence of OHHL, there is no induction of *cps* gene expression rather the effects are relieved (10). So, it acts as a negative regulator of *cps* gene cluster (6). The gene *esaR* encodes the regulator (EsaR), which can bind to a *lux* box in its own promoter and repress its synthesis, but it cannot regulate *esaI* gene as it lacks a *lux* box in its promoter. This is in accordance with *esaR* deletion mutants, which produce same levels of OHHL as in wild type. QS-dependent EPS expression is essential for Stewart's wilt disease development as the strains deficient in synthesis of a diffusible AHL inducer are avirulent in nature. Disruption of *esaR* gene leads to hypermucoidy and wilting.

At low cell density, when OHHL concentration is low or absent, EsaR dimerise into active DNA binding form and represses *esaR* and *cps* genes (98). Once OHHL concentrations increase at high cell density, the AHL binds and inactivates available EsaR, relieving *cps* repression. EsaR binds one molecule of OHHL per protein monomer (79). This binding makes EsaR unable to bind DNA or leads to conformational changes rendering the protein prone to proteolytic degradation. OHHL synthesis by EsaI is a constitutive process. At low cell density, EsaR sequesters the cellular OHHL. At high cell density, concentration of OHHL exceeds the levels of EsaR leading to its inactivation and *cps* gene expression.

The *hrp* type III secretion system is essential for *Pnss* to cause disease in host plants. The *hrp* regulon is also under control of EsaI/EsaR system, which consists of four *hrp* regulatory genes: *hepX* and *hrpY* encoding a sensor kinase and response regulator, together these form the HrpX/HrpY two-component signal transduction system, *hrpS* encoding NtrC like transcriptional enhancer (HrpS) and *hrpL* encoding a sigma factor (HrpL). These genes are arranged in the order *hrpL, hrpXY* and *hrpS* and constitute three operons. The expression of *hrpXY* is constitutive and autoregulated (76). HrpY activates *hrpS*, which in turn activates *hrpL,* which finally controls expression of *hrp* and *wts* genes. The mutants of *esaI* were lacking the water soaking ability in corn and failed to induce HR in tobacco. These mutations were restored by overexpression of *hrpS*, but not *hrpY* suggesting the role of EsaR as a direct/indirect repressor of *hrpS* (7).

When the susceptible corn seedlings are tested with mutant *esaI* strains, only water soaked lesions are observed without any wilting symptom. Whereas the *esaR* and *esaI/esaR* double deletion mutants are less virulent even though they produce sufficiently high amounts of EPS and *hrp* gene products. The possible reason behind reduced virulence of the mutants could be ascribed to the early expression of virulence factors. This observation suggests the role of QS-dependent expression of EPS, *hrp* genes and other virulence factors at proper time and location during infections (10).

EsaR directly binds EsaR DNA binding site within the *rcsA* promoter and represses the transcription of the *rcsA* gene (78). However, RcsA positively controls its own expression. It is also reported that the promoter proximal genes in the cps cluster are much more responsive to AHLs those present in the middle or 3' end of the gene cluster (78). The presence of HrpS-mediated novel autoregulatory loop which functions to rapidly induce *hrp* genes during infection is reported. The regulation is due to activation of *hrpS* by increased levels of HrpY, resulting from read-through transcription of *hrpXY* from the *hrpL* promoter (76).

Several regulatory proteins such as RcsC/YojN/RcsB/A are involved in the QS regulation as a multicomponent phosphorelay signal transduction system (51, 59, 112). RcsC is known to be reported as a transmembrane sensor kinase, which senses desiccation, changes in osmolarity or other membrane changes (84, 81, 109). Phosphate transfer takes place from histidine kinase domain to response regulator domain of RcsC. YojN (now renamed as RcsD) (70) is an inner membrane protein that is involved in phosphorelay transfer from RcsC to RcsB regulator (112, 102). The phosphorylated RcsB forms an activation complex with RcsA, which binds to RcsAB-specific binding sequences in the target gene promoters (120). RcsB also induces expression of RprA, a small RNA; ftsZ, a cell division gene and other genes including osmotically inducible *osmB* and *osmC* genes (70).

### 4.3 QS in *Agrobacterium tumefaciens* is under the control of opines

*Agrobacterium tumefaciens*, the causal agent of crown gall disease, harbors tumor inducing plasmid (Ti plasmid), which is responsible for, transfer of T-DNA into the chromosome of plant nucleus *via* a system of Vir proteins (129). This T-DNA possess the genes for plant hormone synthesis and opine production. The synthesis of plant growth hormones auxin and cytokinin results in deregulated cell division resulting in crown gall tumour. The resulting tumor produces opines in large quantity that serves as a food for the bacteria.

Both the plant- and bacterium-produced opines act to induce their own catabolic systems and transfer of Ti plasmid within the

*A. tumefaciens* population. QS plays a prominent role here to ensure that all the progeny receives the plasmid so that they can catabolise the opines available and also to ensure that the donor cells population is sufficiently high to achieve maximum conjugation to receive the Ti plasmid (90). The genes encoding for opine catabolism and uptake are induced by their respective cognate opines which fall into two classes. Based on these substrates, Ti plasmid has been divided into two types–octopine-type induced by octopine and nopaline-type induced by agropine A and B (43, 61). There are two different QS systems proposed for control of transfer of these different plasmids.

The presence of octopine activates the LysR type regulator, OccR in octopine inducible Ti plasmid (52, 118). The OccR induces transcription of *traR* gene, a member of 14 gene *occ* operon, encoding products for uptake and catabolism of octopine (46, 89). *traI* gene encodes OOHL synthase, which synthesizes acyl-HSL molecule OOHL. When the population density of bacterium increases, a threshold concentration of OOHL is reached resulting in dimerization and activation of TraR (98), a LuxR homologue, which is responsible for the transcriptional activation of *tra* and *trb* genes (Fig. 3) needed for conjugal Ti plasmid transfer (47) by binding the *tra* box present in the upstream promoter sequences. The *traI* gene is present in the *tra* operon whose transcription is induced by TraR, thereby suggesting the presence of a negative autoregulatory system. The N-terminal region of TraR forms ternary complexes with C-terminal segments of alpha subunit of *A. tumefaciens* RNA polymerase leading to transcription of its target genes (99).

OHHL serves to protect TraR from proteolysis and the mutants failing to detect OHHL undergo TraR proteolysis (17). A study on TraR-OOHL-DNA complexes showed that one molecule of OOHL is completely engulfed within the N-terminal domain of each TraR subunit and TraR binds OOHL *via* four hydrogen bonds. The native amino terminus of TraR may be involved in signaling proteolysis, however fusing it with other proteins protect from such degradation (16). TraR-OOHL complexes activate transcription of at least seven promoters on the Ti plasmid. Mutational analysis of TraR resulted in identification of a patch of amino acid residues that are critical for activation at both a class I and a class II promoter by contacting the C-terminal domain of the RNA polymerase α-subunit (122). TraR has been reported to increase the plasmid copy number at the level of transcription leading to three to four-fold increase in plant tumorigenesis (83).

In nopaline type plasmids, there is a FucR type regulator, AccR that represses *traR* transcription (Fig. 3). The presence of agropine molecules derepresses the AccR mediated blockage resulting in TraR production (9, 89, 60). Further mechanisms of regulation of QS by TraR are similar to octopine type regulator. Two proteins regulate TraR activation. One of these is an 11 kDa protein known as TraM, which is found, on both the types of Ti plasmids (56) and it inhibits the function of TraR activator by sequestering the protein (110). In *traM* mutants, TraR is active and there is constitutive transfer of Ti plasmids even in unfavorable conditions. In normal conditions, the concentration of TraR should cross that of TraM for the activation of *tra* and *trb* genes leading to transfer of Ti plasmid (90). The crystal structure elucidations reveal that TraM exists as dimer, with each monomer consisting mainly of two antiparallel alpha helices. Biochemical studies confirm that TraM exists as a homodimer in solution in equilibrium with the monomeric form, and the TraM homodimer undergoes dissociation in forming the antiactivation complex (23). TraM momomer forms a stable oligomeric complex with TraR, which likely accounts for two TraR and two TraM dimers (115). Deletion mutational studies of TraM suggests the presence of determinants at the far C-terminus and initial residues between 10-20 at the N-terminus play a role in TraR binding and antiactivator function (99). The second regulatory protein is TrlR (TraS) (130) which is induced by mannopine. A frameshift mutation in the C-terminal DNA binding domain of TraR results in origination of TrlR. Therefore, it shares much homology with TraR and forms stable heterodimers rendering it inactive and inhibits binding to target gene promoters. (18). There also exist other inhibitors of TraR like succinate and tryptone (131).

There is yet another regulatory system operating in *A. tumefaciens* strain 58, designated as AvhB (*Agrobacterium virulence homologue*VirB), based on homology of seven genes of this operon to genes encoding VirB proteins of Ti plasmid (24). This system mediates conjugal transfer and it is believed to be expressed in environments different from the Ti based *virB* system. The *luxR* homologue *avhR* in *Agrobacterium vitis* strain F2/5 is unique among all reported functional LuxR homologues in having substitutions at a very conserved region predicted to play important roles in AI binding in TraR (53).

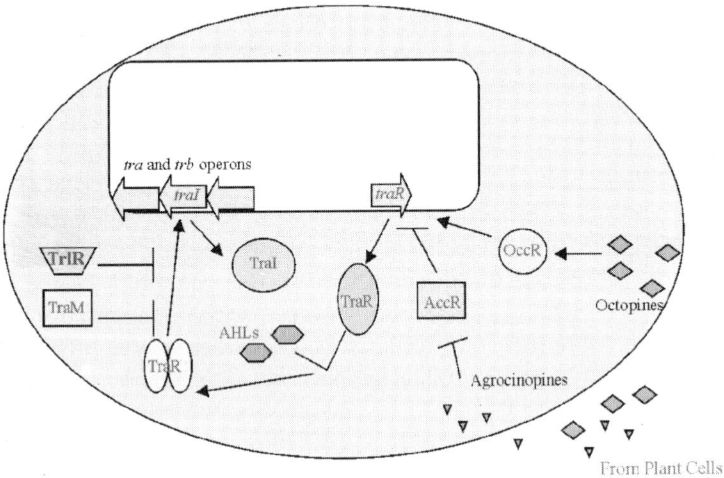


Figure 3. Quorum sensing in *Agrobacterium tumefaciens*. The plant cells having bacterial T DNA incorporated in nucleus encode for opines, which serve to induce their own catabolic system and transfer of T DNA among *Agrobacterium* progeny cells enabling them to use opines.

Upregulation of *traR* expression is met by induction of OccR by octopines ( ◆ ) in octopine-inducible plasmids whereas this is reverse in nopaline-inducible plasmids where this is met by repression of AccR by agrocinopines ( ▼ ). AHLs are synthesized by TraI, which is encoded by *traI* gene present in *tra* and *trb* operons. TraR dimerizes in the presence of AHLs and activates *tra* and *trb* operons whereas TraM and TrlR inhibits this TraR mediated activation. Arrows indicate the positive regulation whereas negative regulation is shown by ( ——| ).

## 5. NON AHL-BASED REGULATION OF VIRULENCE FACTORS

### 5.1 QS in *Ralstonia solanacearum* involves a typical Inducer/ Regulator path besides other systems

*Ralstonia solanacearum* causes vascular wilt in a wide variety of plant species throughout the world. They gain entry through the root system and remain in the xylem vessels. They secrete their primary virulence factor, exopolysachharide I (EPS I), an acidic polymer that contributes to wilting by blocking the xylem vessels and plant water transport (106). Besides EPS I, there are other factors, which contribute to pathogenicity like cell wall degrading enzymes, siderophores for iron acquisition, and flagella. All these factors together make a complex network, which is regulated in a bacterial cell density- dependent manner.

PhcA, a LysR type regulator, plays a prominent role in this network (14). It activates production of EPS, endoglucanases, and pectin methyl esterases but reduces expression of polygalacturonase and motility *via* PehRS two-component system (Fig. 4). Mutants of PheA are unable to cause wilting and are therefore avirulent in nature. The QS signal molecule in case of *R. solanacearum*, is 3-hydroxy-palmitic acid methyl ester (3-OH PAME) (45), a diffusible volatile compound synthesized by *phcB* of *phcBSR* operon (26). PhcB is having motif similar to SAM-dependent methyl transferases and is able to convert a fatty acid to its volatile methyl ester generating 3-OH PAME. The latter regulates PhcA activity.

PhcA expression is controlled by a two-component system consisting of regulators PhcR and PhcS. When the 3-OH PAME concentration is low at low cell density, PhcS phosphorylates PhcR, which represses phcA expression. However, no DNA binding domain is reported in phcR and, therefore, the actual method of phcA repression is still unknown. When 3-OH PAME concentration increases upto 5nM at higher cell densities, it relieves PhcRS-mediated repression of phcA leading to expression of virulence factors.

Towards initial stages of infection when the population density is low, low levels of 3-OH PAME results in repression of *phcA* expression thereby upregulation of motility factors (3, 48), however, the bacterium remains non-motile in rest of the infection stages (114). At higher cell densities, high levels of 3-OH PAME relieves PhcRS-mediated repression of PhcA, which activates the expression of EPS I and exoenzymes resulting in progression of disease after successful colonization is over.

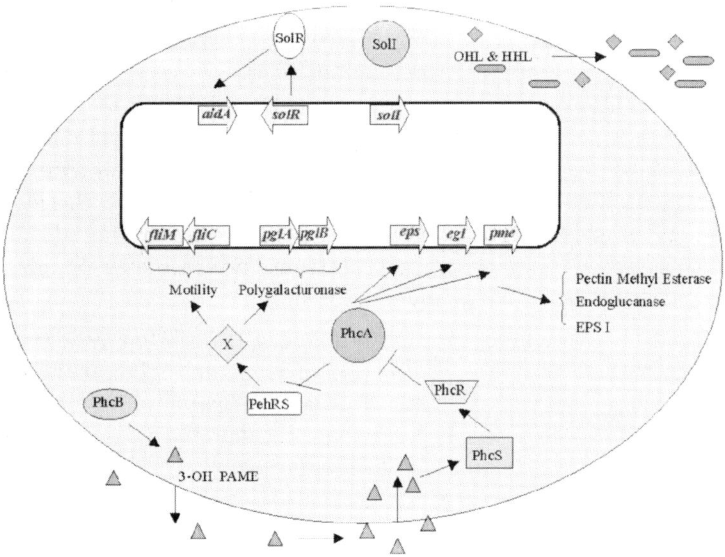


**Figure 4.** Quorum sensing in *Ralstonia solanacearum*. PhcB synthesizes 3-OH PAME signaling molecules, which are expressed at basal level at low cell density. PhcS phosphorylates PhcR, which in turn represse PhcA leading to PehRS regulated expression of polygalacturonase and motility factors. When cell density increases, high concentration of 3-OH-PAME ( ▲ ) relieves PhcRS-mediated repression of PhcA resulting in expression of virulence factors such as EPS I, Endoglucanase and Pectin methyl esterase.

There is another system of regulation involving *solI* and *solR* genes. *solI* encodes for SolI which synthesizes HHL (◆) and OHL ( ▬ ) whereas *solR* encodes for SolR which is reported to activate *aidA* gene of unknown function. Arrows show positive regulation while negative regulation is shown by (——┤).

Other QS regulatory mechanism operating in *R. solanacearum* (45) involves a typical AHL-based system. This consists of *solI* and *solR*, which are homologues of LuxI and LuxR, respectively. Expression of these *solIR* genes requires RpoS and PhcA regulator activation, which forms a hierarchy within *R. solanacearum* QS networks. This system consists of two AHL signal molecules – N-octanoyl-L-homoserine lactone (OHL) and HHL. The mutants of SolI are unable to produce these signal molecules. There is no distinct effect of *solIR* mutation on virulence and the target genes of *solR* are largely unknown. It is assumed that this system is involved in the regulation of virulence factors towards the terminal stages of wilt disease.

A third putative QS locus was shown in the genome of *R. solanacearum* (49, 105). This locus consists of a pair of open reading frame having homology to *solI* and *solR*, respectively. The complete understanding of these systems will unveil the complex regulation of QS systems in *R. solanacearum*. Transposon mutagenesis of *R. solanacearum* strain AW1-PC (phcA1) resulted in isolation of new gene *ssd*, whose expression is repressed by iron (III) and PhcA (12). Wild type strains having inactive *ssd* gene are incapable of scavenging iron but virulent on the host. PhcA represses the expression of *hrp* genes that code for type III secretion system, a major pathogenicity determinant *R. solanacearum*, triggered by a QS signal and complex nitrogen sources (50).

### 5.2 Two separate diffusible factors independently regulate virulence factors through QS in *Xanthomonas*

*Xanthomonas campestris* pv. *campestris* (*Xcc*) is a vascular pathogen of crucifers causing black rot disease. It gains entry into the leaves through the hydathodes and multiplies in the xylem vessels. Towards the initial stages of disease it produces V-shaped watery soaked lesions whereas in the later stages it causes severe wilt. Several virulence factors are employed by this bacterium for a successful pathogenesis, for example - a range of plant cell wall degrading enzymes and exo-polysachharide (EPS) composed of xantham gums (25, 38) and a *hrp* secretion pathway, together all these contribute to colonization and multiplication in the host tissue. The xanthan gums are valuable industrial product and the functions of several genes involved in its synthetic pathway are being elucidated for its commercial value.

The regulation of these virulence factors is governed by two separate QS systems: DF (Diffusible factor) and DSF (Diffusible extracellular factor) (5, 93). These two systems are functioning independent of each other as over expression of either factor is unable to complement a mutation in the other (95).

The first system involving DF, a butyrolactone produced by *Xcc* strain B24, regulates xanthomonadin pigment production (25, 92) whereas the second system involving DSF, a fatty acid derivative, regulates exoenzyme and cyclic glucan production in *Xcc* strain 8004 (117). However both of these contribute to EPS production in *Xanthomonas* (33, 124). DF induces a yellow pigment xanthomonadin production and expression of *gum* operon, the products of which help in synthesis and export of EPS (25). DF is synthesized by a *pigB* gene product and the presence of this gene in *Xcc* genome in essential for epiphytic survival of this bacterium (95). The *pigB* mutants are nonpigmented and are reduced in EPS production and impaired in epiphytic survival (94, 96). The structural analysis reveals that DF may be 2-(2- methyl-3- oxo bytyl) –butero lactone (25). Although DF and DSF are structurally similar and both induce virulence factor production, yet, they differ in biological activity and chromatographic mobility (95). *pigB* and DF are responsible for normal xanthomonadin production, which serves to protect it against UV light (97)

The DSF factor also helps in EPS expression through the *rpf* (Regulator of Pathogenicity Factor) locus (39, 113). There are nine *rpf* genes (*rpfA-I*) that are linked within a 22.1kbp region. Transposon mutagenesis within this region inhibits the production of virulence factors (5, 113). The gene products of *rpfB* and *rpfF* are homologous to fatty acyl CoA ligase and enoyl CoA hydratase, respectively and function to divert lipid metabolism towards DSF synthesis. The signaling molecule involved in DSF synthesis is a fatty acid derivative other than an acyl-HSL molecule. The deletion strains of both these genes are unable to produce DSF.

The rest of the genes of *rpf* operon encode for the regulators. The products of *rpfC* and *rpfG* encode a two-component regulatory system which senses and controls the DSF levels (108). The RpfC regulator is a hybrid protein formed from two different proteins as it contains both a sensor kinase and response-regulator like domains (113). The mutants of *rpfC* are repressed in DSF production while virulence factor production is activated. Upon autophosphorylaton, RpfC sensor kinase phorphorylates its cognate response regulator RpfG. The latter possess a typical input domain attached to an HP-GYP domain, the proteins containing this domain are mainly phosphodiesterases involved in cyclic diguanylate signaling. RpfG is supposed to regulate EPS and exoenzyme synthesis by governing the synthesis or turnover of cyclic di-GMP through probably a secondary regulator (6, 7). RpfC is stabilized in the cell by RpfH, which is homologous to transmembrane sensor domain of RpfC. The *rpfH* seems to be non-essential for virulence, since the mutants of *rpfH* are slightly reduced in virulence (108).

In *Xanthomonas oryzae* pv. *oryzae* (*Xoo*) *rpfF* mutants, the virulence and DSF levels are decreased, siderophore levels are increased while EPS and exoenzyme levels are unaffected. It is supposed that RpfF may be involved in controlling iron uptake (22). Until now the only *rpf* gene product involved in controlling iron uptake in *Xcc* is RpfA. The *rpfA* gene product is a homologue of bacterial aconitase enzyme (126). The mutants of *rpfA* are having low levels of intracellular iron and may regulate virulence factor expression in response to intracellular iron fluxes.

The composition of EPS produced by virulent and virulence-deficient strains in *Xoo* is found to be different as is evident from the NMR and FT-IR studies (63). The transposon mutagenesis studies of *Xcc* revealed the involvement of *wxcA* gene in both lipopolysaccharide (LPS) biosynthesis and also in EPS yield (68). The presence of *rpf* genes in several *Xanthomonas* spp. suggests their prime importance in virulence factor production. The *rpfI* present in *Xcc* is supposed to be responsible for massive tissue maceration during the infections. However, the addition of extracellular DSF to *Xcc* culture was insufficient for precocious induction of enzymes suggesting the presence of other regulatory factors of virulence such as nutrient availability and a protein homologous to cAMP receptor protein (54).

## 6. CONCLUDING REMARKS

QS systems demonstrated that bacteria have evolved multiple chemical language for communicating within and between species. Intra and interspecies cell-cell communication allows bacteria to coordinate various physiological activities and to behave like multicellular organisms. QS involved in regulating gene expression of diverse array of genes, some of which are making interaction between microbes and/or between microbes and plants for example symbiotic microbes will increase the symbiotic association of the plant-microbe interaction. Because quorum sensing is required for virulence of several bacteria, the QS system has become the target for development of new types of therapeutics, antipathogenic agents, agents that do not kill bacterial pathogens but that do interfere with their ability to cause infection. The challenge is to identify inhibitors of QS and test their effectiveness in the treatment of infection. Many microbes are associated with plants in different forms like symbiosis, pathogenesis or commensal and maintain relationship-utilizing QS.

Competing bacteria and susceptible hosts have evolved natural strategies for obstructing QS bacteria by destroying the chemical signals molecules or producing antagonistic signal that interferes with recognition of the bonafied AI molecule. Similarly in symbiotic association, bacteria and host have evolved tactics that enhance the QS ability of the beneficial bacteria. In the plant-beneficial rhizosphere bacterium *Pseudomonas fluorescens* CHA0, the GacS/GacA system is essential for the production of antibiotic compounds and hence for biological control of root-pathogenic fungi. These natural strategies that inhibit or enhance QS are being used as the models in the design of analogue synthetic therapies intended to manipulate QS system. Finally, biotechnological approaches aimed at exploring useful bacterial QS process have potential uses in improvement of industrial scale production of natural products such as antibiotics and antimicrobial molecules.

Plants which are susceptible to bacterial pathogens can be made resistant exploring QS mechanisms. Phytopathogenic bacteria depend quite significantly on QS regulation to coordinate their colonization and infection of plant host. Plants take advantage of this dependency. For example plant might be able to disturb QS in pathogens by enzymatically destroying their signals or by synthesizing molecules that mimic these QS signal, or alternatively it might have some early QS signal detection mechanism and trigger its defense responses at the appropriate site. Plants are not reported to have capabilities to produce compounds that specifically inhibit the synthesis of bacterial QS signal. But, some soil bacteria like *Bacillus subtilis* have the capability to produce a lactonase (*aii*A), that opens the lactone ring of AHLs' and was responsible for QS inhibition activity (36). The opened ring lactone has no detection ability by pathogen and a 1000-fold reduction in the QS signal. Transgenic potato and tomato expressing AiiA are highly resistant to infection by *Erwinia carotovora* subsp. *carotovora*, which depend heavily on 3-oxo-C6–AHL-mediated QS for pathogenicity. Tissue maceration, which occurs at a later stage of infection, was completely abolished in the AHL-deficient mutants. Both the incidence of disease and *in planta* growth of *P. syringae* pv. *tabaci* were greatly reduced in transgenic tobacco plants that produced AHL, compared to wild-type plants (37). Thus, destruction or inactivation of AHL signal *in situ* appears to be an alternative strategy for engineering disease resistance in plants. Perhaps non-specific destruction of QS signal by plant hosts would adversely affect not only pathogens, but valuable symbiotic bacteria as well. So, more specific mechanisms of plant disease resistance would be favored by natural selection.

## *Literature Cited*

1. Ahmer BMM. 2004. Cell-to-Cell signaling in *Escherichia coli* and *Salmonella enterica*. *Mol. Microbiol.* 52: 933-945.
2. Aizawa S, Hardwood CS, Kadner RJ. 2000. Signaling components in bacterial locomotion and sensory reception. *J. Bacteriol.* 182: 1459-1471.
3. Allen C, Gay J, Simon-Buela L. 1997. A regulatory locus, *pehSR*, controls polygalacturonase production and other virulence factors in *Ralstonia solanacearum*. *Mol. Plant-Microbe Interact.* 10: 1054-1064.
4. Bainton NJ, *et al.* 1992. *N*-(3-oxo-hexanoyl)-L-homoserine lactone regulates carbapenem antibiotic production in *Erwinia carotovora*. *Biochem. J.* 288: 997-1004.
5. Barber CE, *et al.* 1997. A novel regulatory system required for pathogenicity of *Xanthomonas campestris* is mediated by a small diffusible signal molecule. *Mol. Microbiol.* 24: 555-566.
6. Beck von Bodman S, *et al.* 2003. The quorum sensing negative regulators EsaR and ExpR(Ecc), homolo-gues within

the LuxR family, retain the ability to function as activators of transcription. *J. Bacteriol.* 185(23): 7001-7007.

7. Beck von Bodman S, Bauer WD, Coplin DL. 2003. Quorum sensing in plant-pathogenic bacteria. *Annu. Rev. Phytopathol.* 41: 455-482.
8. Beck von Bodman S, Farrand SK. 1995. Capsular polysaccharide biosynthesis and pathogenicity in *Erwinia stewartii* require induction by an *N*-acylhomoserine lactone autoinducer. *J. Bacteriol.* 177: 5000-5008.
9. Beck von Bodman S, Hayman GT, Farrand SK. 1992. Opine catabolism and conjugal transfer of nopaline Ti plasmid pTiC58 are coordinately regulated by a single repressor. *Proc. Natl. Acad. Sci. USA* 89:643-647.
10. Beck von Bodman S, Majerczak DR, Coplin DL. 1998. A negative regulator mediates quorum -sensing control of exopolysaccharide production in *Pantoea stewartii* subsp. *stewartii. Proc. Natl. Acad. Sci. USA* 95: 7687-7692.
11. Bell KS, *et al.* 2004. Genome sequence of the enterobacterial phytopathogen *Erwinia carotovora* subsp. *atroseptica* and characterization of virulence factors. *Proc. Natl. Acad. Sci. USA* 101: 11105-11110.
12. Bhatt G, Denny TP. 2004. *Ralstonia solanacearum* iron scavenging by the siderophore staphyloferrin B is controlled by PhcA, the global virulence regulator. *J. Bacteriol.* 186: 7896-7904.
13. Braun EJ. 1982. Ultrastructural investigation of resistant and susceptible maize inbreds infected with *Erwinia stewartii. Phytopathology* 72: 159-166.
14. Brumbley SM, Carney BF, Denny TP. 1993. Phenotype conversion in *Pseudomonas solanacearum* due to spontaneous inactivation of PhcA, a putative LysR transcriptional regulator. *J. Bacteriol.* 175: 5477-5487.
15. Burr T, *et al.* 2005. Identification of the central quorum sensing regulator of virulence in the enteric phytopathogen, *Erwinia carotovora:* the VirR repressor. *Mol. Microbiol* 59: 113-125.
16. Chai Y, Winans SC. 2004. Site-directed mutagenesis of a LuxR-type quorum-sensing transcription factor: alteration of autoinducer specificity. *Mol. Microbiol.* 51: 765-776.
17. Chai Y, Winans SC. 2005. Amino-terminal protein fusions to the TraR quorum-sensing transcription factor enhance protein stability and autoinducer-independent activity. *J. Bacteriol.* 187: 1219-1226.

18. Chai Y, Zhu J, Winans SC. 2001. TrlR, a defective TraR-like protein of *Agrobacterium tumefaciens*, blocks TraR function in vitro by forming inactive TrlR:TraR dimers. *Mol. Microbiol.* 40: 414-421.

19. Chatterjee A, *et al.* 2005. Comparative analysis of two classes of quorum-sensing signaling systems that control production of extracellular proteins and secondary metabolites in *Erwinia carotovora* subspecies. *J. Bacteriol.* 187: 8026-8038.

20. Chatterjee A, Cui Y, Liu Y, Dumenyo CK, Chatterjee AK. 1995. Inactivation of *rsmA* leads to overproduction of extracellular pectinases, cellulases, proteases in *Erwinia carotovora* subsp. *carotovora* in the absence of the starvation/cell density-sensing singal, *N*-(3-oxohexanoyl)-L-homoserine lactone. *Appl. Environ. Microbiol.* 61: 1959-1967.

21. Chatterjee A, Cui YY, Chatterjee AK. 2002. RsmA and the quorum-sensing signal, N-[3-oxohexanoyl]-L-homoserine lactone, control the levels of *rsmB* RNA in *Erwinia carotovora* subsp. *Carotovora* by affecting its stability. *J. Bacteriol.* 184: 4089-4095.

22. Chatterjee S, Sonti RV. 2002. *rpfF* mutants of *Xanthomonas oryzae* pv. *oryzae* are deficient for virulence and growth under low iron conditions. *Mol. Plant-Microbe Interact.* 15: 463-471.

23. Chen G, Malenkos JW, Cha MR, Fuqua C, Chen L. 2004. Quorum-sensing antiactivator TraM forms a dimer that dissociates to inhibit TraR. *Mol. Microbiol.* 52: 1641-1651.

24. Chen X, *et al.* 2002. Structural identification of a bacterial quorum-sensing signal containing boron. *Nature* 415: 545-549.

25. Chun W, Cui J, Poplawsky AR. 1997. Purification, characterization and biological role of a pheromone produced by *Xanthomonas campestris* pv. *campestris*. *Physiol. Mol. Plant Pathol.* 51: 1-14.

26. Clough SJ, Flavier AB, Schell MA, Denny TP. 1997. Differential expression of virulence genes and motility in *Ralstonia (Pseudomonas) solanacearum* during exponential growth. *Appl. Environ. Microbiol.* 63: 844-850.

27. Coplin DL, Majerczak DR. 1990. Extracellular polysaccharide genes in *Erwinia stewartii*: directed mutagenesis and complementation analysis. *Mol. Plant-Microbe Interact.* 3: 286-292.

28. Corbett M, *et al.* 2005. Identification of a new quorum-sensing-controlled virulence factor in *Erwinia carotovora* subsp. *atroseptica* secreted *via*

the type II targeting pathway. *Mol. Plant-Microbe Interact.* 18: 334-342.

29. Costa JM, Loper JE. 1997. EcbI and EcbR: homologs of LuxI and LuxR affecting antibiotic and exoenzyme production by *Erwinia carotovora* subsp. *betavasculorum. Can. J. Microbiol.* 43: 1164-1171.

30. Cui Y, *et al.* 2005. ExpR, a LuxR homolog of *Erwinia carotovora* subsp. *carotovora*, activates transcription of *rsmA*, which specifies a global regulatory RNA-binding protein. *J. Bacteriol.* 187: 4792-4803.

31. Cui Y, Madi L, Mukherjee A, Dumenyo K, Chatterjee AK. 1996. The RsmA-mutants of *Erwinia carotovora* subsp. *carotovora* strain Ecc71 over-express *hrpNEcc* and elicit a hypersensitive reaction-like response in tobacco leaves. *Mol. Plant-Microbe Interact.* 9: 565-573.

32. Cui YY, *et al.* 1995. Identification of a global repressor gene, *rsmA*, of *Erwinia carotovora* subsp. *Carotovora* that controls extracellular enzymes, *N*-(3-oxohexanoyl)-L-homoserine lactone and pathogenicity of soft-rotting in *Erwinia* spp. *J. Bacteriol.* 177: 5108-5115.

33. Denny TP. 1995. Involvement of bacterial exopolysaccharides in plant pathogenesis. *Annu. Rev. Phytopathol.* 33: 173-197.

34. Devine JH, Countryman C, Baldwin TO. 1988. Nucleotide sequence of the *luxR* and *luxI* genes and structure of the primary regulatory region of the *lux* regulon of *Vibrio fischeri* ATCC 7744. *Biochemistry* 27: 837-842.

35. Devine JH, Shadel GS, Baldwin TO. 1989. Identification of the operator of the *lux* regulon from the *Vibrio fischeri* strain ATCC 7744. *Proc. Natl. Acad. Sci. USA* 86: 5688-5692.

36. Dong YH, Xu JL, Li XZ, Zhang LH. 2000. AiiA an enzyme that inactivates the acylhomoserine lactone quorum sensing signal and attenuates the virulence of *Erwinia carotovora. Proc. Natl. Acad. Sci. USA* 97: 3526-3531.

37. Dong YH, *et al.* 2001. Quenching quorum sensing-dependent bacterial infection by an N-acyl homose-rine lactone. *Nature* 411: 813-817.

38. Dow JM, Daniels MJ. 1994. Pathogenicity determinants & global regulation of pathogenicity of *Xanthomonas campestris* pv. *campestris. Curr. Top. Microbiol. Immunol.* 192: 29-41.

39. Dow JM, Feng JX, Barber CE, Tang JL, Daniels MJ. 2000. Novel genes involved in the regulation of pathogenicity factor production within the *rpf* gene cluster of *Xanthomonas campestris. Microbiology* 146: 885-891.

40. Dunlap PV, Ray JM. 1989. Requirement for the autoinducer in transcriptional negative autoregulation of the *Vibrio fischeri luxR* gene in *Escherichia coli. J. Bacteriol.* 171: 3549-3552.

41. Eberhard A, Longin T, Widrig CA, Stranick SJ. 1991. Synthesis of the *lux* gene autoinducer in *Vibrio fischeri* is positively autoregulated. *Arch. Microbiol.* 155: 294-297.

42. Egland KA, Greenberg EP. 1999. Quorum sensing in *Vibrio fischeri*: elements of the *luxI* promoter. *Mol. Microbiol.* 31: 1197-1204.

43. Ellis JG, Kerr A, Petit A,Tempe J. 1982. Conjugal transfer of nopaline and agropine Ti-plasmids–the role of agropines. *Mol. Gen. Genet.* 186: 269-273.

44. Engebrecht J, Nealson K, Silverman M. 1983. Bacterial bioluminessence: isolation and genetic analysis of functions fron *Vibrio fischeri. Cell* 32: 773-781.

45. Flavier AB, Clough SJ, Schell MA, Denny TP. 1997. Identification of 3-hydroxypalmitic acid methyl ester as a novel autoregulator controlling virulence in *Ralstonia solanacearum. Mol. Microbiol.* 26:251-259.

46. Fuqua C, Winans SC. 1996. Localization of OccR-activated and TraR-activated promoters that express two ABC-type permeases and the *tra* R gene of Ti plasmid pTiR10. *Mol. Microbiol.* 20: 1199-1210.

47. Fuqua WC, Winans SC. 1994. A LuxR-LuxI type regulatory system activates *Agrobacterium* Ti plasmid conjugal transfer in the presence of a plant tumor metabolite. *J. Bacteriol.* 176: 2796–2806.

48. Garg RP, *et al.* 2000. Evidence that *Ralstonia eutropha* (*Alcaligenes eutrophus*) contains a functional homologue of the *Ralstonia solanacearum* Phc cell density sensing system. *Mol. Microbiol.* 38: 359-367.

49. Genin S, Boucher C. 2002. *Ralstonia solanacearum*: secrets of a major pathogen unveiled by analysis of its genome. *Mol. Plant Pathol.* 3: 111-118.

50. Genin S, Brito B, Denny TP, Boucher C. 2005. Control of the *Ralstonia solanacearum* Type III secretion system (Hrp) genes by the global virulence regulator PhcA. *FEBS Lett.* 11579: 2077-2081.

51. Gottesman S, Stout V (1991) Regulation of capsular polysachharide synthesis in *Escherichia coli* K12. *Mol. Microbiol.* 5: 1599-1606.

52. Habeeb LF, Wang L, Winans SC. 1991. Transcription of the octopine catabolism operon of

the *Agrobacterium* tumor-inducing plasmid pTiA6 is activated by a LysR-type regulatory protein. *Mol. Plant-Microbe Interact.* 4: 379-385.

53. Hao G, Zhang H, Zheng D, Burr TJ. 2005. *luxR* homolog *avhR* in *Agrobacterium vitis* affects the development of a grape-specific necrosis and a tobacco hypersensitive response. *J. Bacteriol.* 187: 185-192.

54. Hsaio YM, Tseng YH. 2002. Transcription of *Xanthomonas campestris prt1* gene encoding protease1 increases during stationary phase and requires global transcription factor Clp. *Biochim. Biophys. Res. Commun.* 295: 43-49.

55. Hugouvieux-Cotte-Pattat N. 2004. The RhaS activator controls the *Erwinia chrysanthemi* 3937 genes *rhiN*, *rhiT* and *rhiE* involved in rhamnogalacturonan catabolism. *Mol. Microbiol.* 51: 1361-1374.

56. Hwang I, Smyth AJ, Luo ZQ, Farrand SK. 1999. Modulating quorum sensing by antiactivation: TraM interacts with TraR to inhibit activation of Ti plasmid conjugal transfer genes. *Mol. Microbiol.* 34: 282-294.

57. Jones S, *et al.* 1993. The *lux* autoinducer regulates the production of exoenzyme virulence determinants in *Erwinia carotovora* and *Pseudomonas aeruginosa. EMBO J.* 12:2477-2482.

58. Kaplan HB, Greenberg EP. 1985. Diffusion of autoinducer is involved in regulation of the *Vibrio fischeri* luminescence system. *J. Bacteriol.* 163: 1210-1214.

59. Kelm O, Kiecker C, Geider K, Bernhard F. 1997. Interaction of the regulator proteins RcsA and RcsB with the promoter of the operon for amylovoran biosynthesis in *Erwinia amylovora. Mol. Gen. Genet.* 256:72-83.

60. Kim H, Farrand SK. 1997. Characterization of the *acc* operon from the nopaline-type Ti plasmid pTiC58, which encodes utilization of agropines A and B and susceptibility to agrocin 84. *J. Bacteriol.* 136: 775-785.

61. Klapwijk PM, Scheulderman T, Schilperoort RA. 1978. Coordinated regulation of octopine degradation and conjugative transfer of Ti plasmids in *Agrobacterium tumefaciens*: evidence for a common regulatory gene and separate operons. *J. Bacteriol.* 136: 775-785.

62. Kleerebezem, M., *et al.* 1997. Quorum sensing by peptide pheromones & two-component signal-transduction system in Gram-positive bacteria. *Mol. Microbiol.* 24: 895-904.

63. Kumar A, Sunish Kumar R, Sakthivel N. Compositional difference of the exopolysaccharides produced by the virulent

and virulence-deficient strains of *Xanthomonas oryzae* pv. *oryzae*. *Curr. Microbiol.* 46:251-255.

64. Latifi A, Foglino M, Tanaka K, Williams P, Lazdunski A. 1996. A hierarchical quorum-sensing cascade in *Pseudom-onas aeruginosa* links the transcriptional activators LasR and RhlR (VsmR) to expression of the stationary-phase sigma factor RpoS. *Mol. Microbiol.* 21: 1137-1146.

65. Lazazzera, B. A., Grossman, A. D. 1998. The inns and outs of peptide signaling *Trends Microbiol.* 7: 288-294.

66. Liu Y, Cui Y, Mukherjec A, Chatterjee AK. 1998. Characterization of a novel RNA regulator of *Erwinia carotovora* subsp. *carotovora* that controls production of extracellular enzymes and secondary metabolites. *Mol. Microbiol.* 29: 219-234.

67. Loh J, Pierson EA, Pierson LS III, Stacey G, Chatterjee A. 2002. Quorum sensing in plant-associated bacteria. *Curr. Opin. Plant Biol.* 5: 1-6.

68. Lu GT, JL, Wei GN, He YQ, Chen BS. 2004. The gene *wxcA* of *Xanthomonas campestris* pv. *campestris* 8004 strain involved in EPS yield. *Sheng Wu Gong Cheng Xue Bao.* 20: 477-483.

69. Shell MA. 2000. Control of virulence and pathogenicity genes of *Ralstonia solanacearum* by an elaborate sensory array. *Annu. Rev. Phytopathol.* 38: 263-292.

70. Majdalani N, Gottesman S. 2005. The Rcs Phosphorelay: A Complex Signal Transduction System. *Annu. Rev. Microbiol.* 59: 379-405.

71. March JC, Bentley WE. 2004. Quorum sensing and bacterial cross talk in biotechnology. *Curr. Opin. Biotechnol.* 15: 495-502.

72. Matsumoto H, Jitareerat P, Baba Y, Tsuyumu S. 2003. Comparative study of regulatory mechanisms for pectinase production by *Erwinia carotovora* subsp. *carotovora* and *Erwinia chrysanthemi*. *Mol. Plant-Microbe Interact.* 16: 226-237.

73. McGowan S, *et al.* 1995. Carbapenem antibiotic production in *Erwinia carotovora* is regulated by CarR, a homologue of the LuxR transcriptional activator. *Microbiology* 141: 541-550.

74. McGowan SJ, *et al.* 1997. Analysis of the carbapenem gene cluster of *Erwinia carotovora*: definition of the antibiotic biosynthetic genes and evidence for a novel ß -lactam resistance mechanism. *Mol. Microbiol.* 26: 545-556.

75. McGowan SJ, *et al.* 1996. Analysis of bacterial carbapenem antibiotic production genes reveals a novel ß-lactam biosynthesis pathway. *Mol. Microbiol.* 22: 415-426.

76. Merighi M, Majerczak DR, Coplin DL. 2005. A novel transcriptional autoregulatory loop enhances expression of the *Pantoea stewartii* subsp. *stewartii* Hrp type III secretion system. *FEMS Microbiol Lett.* 243: 479-487.

77. Merighi M, Majerczak DR, Stover EH, Coplin DL. 2003. The HrpX/HrpY two-component system activates *hrpS* expression, the first step in the regulatory cascade controlling the Hrp regulon in *Pantoea stewartii* subsp. *stewartii*. *Mol. Plant-Microbe Interact.* 16: 238-248.

78. Minogue TD, Carlier AL, Koutsoudis MD, von Bodman SB. 2005. The cell density-dependent expression of stewartan exopolysaccharide in *Pantoea stewartii* ssp. *stewartii* is a function of EsaR-mediated repression of the *rcsA* gene. *Mol. Microbiol.* 56: 189-203.

79. Minogue TD, Wehland M, Berhard F, von Bodman SB. 2002. The autoregulatory role of EsaR, a quorum sensing regulator in *Pantoea stewartii* subsp. *stewartii*: evidence for a repressor function. *Mol. Microbiol.* 44: 1625-1635.

80. Nasser W, Bouillant ML, Salmond G, Reverchon S. 1998. Characterization of the *Erwinia chrysanthemi expI-expR* locus directing the synthesis of two *N*-acylhomoserine lactone signal molecules. *Mol. Microbiol.* 29: 1391-1405.

81. Ophir T, Gutnick DL. 1994. A role for exopolysaccharides in the protection of microorganisms from desiccation. *Appl. Environ. Microbiol.* 60: 740-745.

82. Palva, T.K., Holmstrom, K.O., Heino, P., Palva, E.T. 1993. Induction of palnt defense responses by exoenzyme of *Erwinia carotovora* subsp. *carotovora*. *Mol. Plant-Microbe Interact.* 6: 190-196.

83. Pappas KM, Winans SC. 2003. A LuxR-type regulator from *Agrobacterium tumefaciens* elevates Ti plasmid copy number by activating transcription of plasmid replication genes. *Mol. Microbiol.* 48: 1059-1073.

84. Parker CT, Kloser AW, Schnaitman CA, Stein MA, Gottesman S, Gibson BW. 1992. Role of the *rfaG* and *rfaP* genes in determining the lipopolysaccharide core structure and cell surface properties of *Escherichia coli* K-12. *J. Bacteriol.* 174: 2525-2238.

85. Pemberton CL, Slater H, Salmond GPC. 2004. Chemical signaling by bacterial plant pathogens. In *Plant Microbio-*

*logy*, ed. M Gillings, A Holmes. BIOS Scientific Publishers, London, New York. pp. 129-151.

86. Pemberton CL, *et al.* 2005. Novel quorum-sensing-controlled genes in *Erwinia carotovora* subsp. *carotovora*: identification of a fungal elicitor homologue in a soft-rotting bacterium. *Mol. Plant-Microbe Interact.* 18: 343-353.

87. Perombelon MCM. 2002. Potato diseases caused by soft rot erwinias: an overview of pathogenesis. *Plant Pathol.* 51: 1-12.

88. Pérombelon M, Kelman A. 1980. Ecology of the soft rot *Ewrinias*. *Annu. Rev. Phytopathol.* 18: 361-387.

89. Piper KR, Beck von Bodman S, Hwang I, Farrand SK. 1999. Hierarchical gene regulatory systems arising from fortuitous gene associations: controlling quorum sensing by the opine regulon in *Agrobacterium*. *Mol. Microbiol.* 32: 1077-1089.

90. Piper KR, Farrand SK. 2000. Quorum sensing but not autoinduction of Ti plasmid conjugal transfer requires control by the opine regulon and the antiactivator TraM. *J. Bacteriol.* 182: 1080-1088.

91. Pirhonen M, Flego D, Heikinheimo R, Palva ET. 1993. A small diffusible signal molecule is responsible for the global control of virulence and exoenzyme production in the plant pathogen *Erwinia caroto-vora*. *EMBO J.* 12: 2467-2476.

92. Poetter K, Coplin DL. 1991. Structural and functional analysis of the *rcsA* gene from *Erwinia stewartii*. *Mol. Gen. Genet.* 229: 155-160.

93. Poplawsky, AR, Chun W. 1997. *pigB* determines a diffusible factor needed for extracellular polysaccharide slime and anthomonadin production in *Xanthomonas campestris* pv. *campestris*. *J. Bacteriol.* 179: 439-444.

94. Poplawsky AR, Chun W. 2002. *Xanthomonas campestris* pv. *campestris* extracellular polysaccharide is required for natural host infection as well as virulence. *Phytopathology.* 92: S66 (Abstr.).

95. Poplawsky AR, Chun W, Slater H, Daniels MJ, Dow JM. 1998. Synthesis of extracellular polysaccharide, extracellular enzymes, and xanthomonadin in *Xanthomonas campestris*: evidence for the involvement of two intercellular regulatory signals. *Mol. Plant-Microbe Interact.* 11: 68-70.

96. Poplawsky, A. R., Chun, W. 1998. *Xanthomonas campestris* pv. *campestris* requires a functional *pigB* for epiphytic survival and host infection.

*Mol. Plant-Microbe Interact.* 11: 466-475.

97. Poplawsky AR, Urban SC, Chun W. 2000. Biological role of xanthomonadin pigments in *Xanthomonas campestris* pv. *campestris*. *Appl. Environ. Microbiol.* 66: 5123-5127.

98. Qin Y, *et al.* 2000. Quorum-sensing signal binding results in dimerization of TraM and its release from membranes into the cytoplasm. *EMBO J.* 19: 5212-5221.

99. Qin Y, Luo ZQ, Farrand SK. 2004. Domains formed within the N-terminal region of the quorum-sensing activator TraR are required for transcriptional activation and direct interaction with RpoA from *Agrobacterium*. *J. Biol. Chem.* 279: 40844-40851.

100. Qin Y, Smyth AJ, Su S, Farrand SK. 2004. Dimerization properties of TraM, the antiactivator that modulates TraR-mediated quorum-dependent expression of the Ti plasmid *tra* genes. *Mol. Microbiol.* 53: 1471-1485.

101. Rodionov DA, Gelfand MS, Hugouvieux-Cotte-Pattat N. 2004. Comparative genomics of the KdgR regulon in *Erwinia chrysanthemi* 3937 and other gamma-proteobacteria. *Microbiology*. 150: 3571-3590.

102. Rogov VV, Bernhard F, Lohr F, Dotsch V. 2004. Solution structure of the *Escherichia coli* YojN histidine-phosphotransferase domain and its interaction with cognate phosphoryl receiver domains. *J. Mol. Biol.* 343: 1035-1048.

103. Ruby EG.1999. The *Euprymna scopoles-Vibrio fischeri* symbiosis: a biomedical mode; for the study of bacterial coloni-zation of animal tissue. *J. Mol. Microbiol. Biotechnol.* 1: 13-21.

104. Shadel GS, Baldwin TO. 1992. Identification of a distantly located regulatory element in the *luxCD* gene required for negative autoregulation of the *Vibrio fischeri luxR* gene. *J. Biol. Chem.* 267: 7690-7695.

105. Salanoubat M, Genin S, Artiguenave F 2002. Genome sequence of the plant pathogen *Ralstonia solanace-arum*. *Nature.* 415: 497-502.

106. Schell MA. 1996. To be or not to be: how *Pseudomonas solanacearum* decides whether or not to express virulence genes. *Eur. J. Plant Pathol.* 102: 459-4 69.

107. Shevchik VE, Hugouvieux-Cotte-Pattat N. 2003. PaeX, a second pectin acetylesterase of *Erwinia chrysanthemi* 3937. *J. Bacteriol.* 185: 3091-3100.

108. Slater H, *et al.* 2000. A two-component system involving an HD-GYP domain protein links

cell-cell signaling to pathogenicity gene expression in *Xanthomonas campestris. Mol. Microbiol.* 38: 986-1003.

109. Sledjeski DD, Gottesman S. 1996. Osmotic shock induction of capsule synthesis in *Escherichia coli* K-12. *J. Bacteriol.* 178: 1204-1206.

110. Swiderska A, *et al.* 2001. Inhibition of the *Agrobacterium tumefaciens* TraR quorum-sensing regula- tor: interactions with the TraM anti-activator. *J. Biol. Chem.* 276: 49449-49458.

111. Swift S, Winson MK, Chan PE (1993) A novel strategy for the isolation of *luxI* homologues: evidence for the widespread distribution of a LuxR:LuxI superfamily in enteric bacteria. *Mol. Microbiol.* 10: 511-520.

112. Takeda S, *et al.* 2001. A novel feature of the multistep phosphorelay in *Escherichia coli*: a revised model of the RcsC. YojN. RcsB signalling pathway implicated in capsular synthesis and swarming behavior. *Mol. Microbiol.* 40: 440-450.

113. Tang JL, *et al.* 1991. Genetic and molecular analysis of a cluster of *rpf* genes involved in positive regulation of synthesis of extracellular enzymes and polysaccharide in *Xanthomonas campestris* pv. *campestris. Mol. Gen. Genet.* 226: 409-417.

114. Tans-Kersten J, Huang H, Allen C. 2001. *Ralstonia solanacearum* needs motility for invasive virulence on tomato. *J. Bacteriol.* 183: 3597-3605.

115. Vannini A, Volpari C, Di Marco S. 2004. Crystal structure of the quorum-sensing protein TraM and its interaction with the transcriptional regulator TraR. *J. Biol. Chem.* 279: 24291-24296.

116. Visick KL, McFall-Nagi MJ. 2000. An exclusive contract: specificity in the *Vibrio fischeri-Euprymna scolopes* partnership. *J. Bacteriol.* 182: 1779-1787.

117. Vojnov AA, *et al.* 2001. Regulation of the synthesis of cyclic glucan in *Xanthomonas campestris* by a diffusible signal molecule. *Arch. Microbiol.* 176: 415-420.

118. Wang L, Helmann JD, Winans SC. 1992. The *Agrobacterium tumefaciens* transcriptional activator OccR causes a bend at a target promoter, which is partially relaxed by a plant tumor metabolite. *Cell* 69: 659-667.

119. Watson WT, *et al.* 2002. Structural basis and specificity of acyl-homoserine lactone signal production in bacterial quorum sensing. *Mol. Cell* 9: 685-694.

120. Wehland M, *et al.* 1999 Identification of an RcsA/RcsB recogn-

ition motif in the promoters of exopolysachharide biosynthetic operons from *Erwinia amylovora* and *Pantoea stewartii* subsp. *stewartii*. *J. Biol. Chem.* 274: 3300-3307.

121. Welch M, *et al.* 2000. *N*-acyl homoserine lactone binding to the CarR receptor determines quorum-sensing specificity in *Erwinia*. *EMBO J.* 19: 631-641.

122. White CE, Winans SC. 2005. Identification of amino acid residues of the *Agrobacterium tumefaciens* quorum-sensing regulator TraR that are critical for positive control of transcription. *Mol. Microbiol.* 55: 1473-1486.

123. Whitefield C, Roberts IS. 1999. Structure, assembly and regulation of expression of capsules in *Escherichia coli*. *Mol. Microbiol.* 31: 1307-1319.

124. Whitehead NA, *et al.* 2001. Quorum-sensing in Gram-negative bacteria. *FEMS Microbiol. Rev.* 25: 365-404.

125. Whitehead NA, *et al.* 2002. The regulation of virulence in phyto-pathogenic *Erwinia* species: quorum sensing, antibiotics and ecological considerations. *Antonie van Leeuwenhoek.* 81: 223-231.

126. Wilson TJG, *et al.* 1998. The *rpfA* gene of *Xanthomonas campestris* pv. *campestris*, which is involved in the regulation of pathogenicity factor production, encodes an aconitase. *Mol. Microbiol.* 28: 961-970.

127. Winans SC, Bassler BL. 2002. Mob psychology. *J. Bacteriol.* 184: 873-883.

128. Zenno S, Saigo K.1994. Identification of the genes encoding NAD(P)H-flavin oxidoreductases that are similar in sequence to *Escherichia coli* Fre in four species of luminous bacteria: *Photorhabdus luminescens*, *Vibrio fischeri*, *Vibrio harveyi* and *Vibrio orientalis*. *J. Bacteriol.* 176: 3544-3551.

129. Zhu J, *et al.* 2000. The bases of crown gall tumorigenesis. *J. Bacteriol.* 182: 3885-3895.

130. Zhu J, Winans SC. 1998. Activity of the quorum-sensing regulator TraR of *Agrobacterium tumefaciens* is inhibited by a truncated, dominant defective TraR-like protein. *Mol. Microbiol.* 27: 289-297.

131. Zhu J, Winans SC. 1999. Autoinducer binding by the quorum-sensing regulator TraR increases affinity for target promoters *in vitro* and decreases TraR turnover rates in whole cells. *Proc. Natl. Acad. Sci. USA.* 96: 4832-4837.

*Rev. Plant Pathol. Vol. 4, 2008*
*Indian Society of Mycology and Plant Pathology*
*Scientific Publishers (India), Jodhpur pp. 335-366*

# MOLECULAR BASIS OF PLANT'S INTERACTION WITH THE PATHOGEN

***A. Snehalatharani, M. Shashikanth, Mubarak Begum, J. Madhusudhana Reddy and K. Ulaganathan***

Centre for Plant Molecular Biology, Osmania University,
Hyderabad 500 007, Andhra Pradesh, INDIA

## ABSTRACT

Plants are mostly immobile and live in an environment full of pathogens which constantly challenge them, even then disease is a rare phenomenon. Plants are, in fact, resistant to most micro organisms by means of constitutive chemical or physical barriers and by induced systemic acquired resistance. Disease is therefore the exception rather than the rule when microbes and plants meet. In spite of the lack of specialized defense cells, plants do have a basic defense mechanism which is the frame work upon which the defense against specific pathogens is built. This basic defense mechanism consists of the following steps: 1. recognition of the pathogen by the plant, 2. localization of the pathogen (hypersensitivity reaction), 3. systemic signaling and 4. systemic acquired resistance. The first step in the basic defense is the process of recognition of the avirulence gene products of pathogens by the R proteins within the plant. Interaction between R gene product and avr gene product are the key determinants of whether a plant is susceptible or resistant to a particular pathogen. The most noticeable feature of this R gene dependent resistance response is the rapid cell death that is well defined as the hypersensitive reaction (HR). The HR cell death is often preceded by changes in ion fluxes, oxidative burst, and crosslinking of cell wall proteins.

Finally this results in the increase in salicylic acid (SA) biosynthesis and systemic signalling, transcriptional activation of various pathogenesis-related (PR) genes, and the establishment of a long-lasting systemic response known as systemic acquired resistance

## INTRODUCTION

Plants are mostly immobile and live in an environment full of pathogens which constantly challenge them; even then disease is a rare phenomenon. Broadly there are three reasons for the pathogen failure: 1. The plant is unable to support the niche requirements of a potential pathogen and is thus a non-host 2. The plant possesses preformed structural barriers or toxic compounds that confine successful infections to specialized pathogen species 3. Upon recognition of the attacking pathogen, defense mechanisms are elaborated and the invasion remains localized. Among these three, the last one *i.e* the elaboration of the defense mechanism upon recognition of a pathogen is the most important and common type of defense responsible for relatively disease free conditions in most plants.

## THE BASIC DEFENSE MECHANISM

There are thousands of species of plants in the environment which encounter a number of pathogenic microorganisms daily. Plants, unlike animals, do not have specialized cells dealing with defense and potentially all the plant cells are capable of defense. In spite of the lack of specialized defense cells, like those present in animals, plants do have a basic defense mechanism which is the frame work upon which the defense against specific pathogens is built. The basic defense mechanism, unlike the preformed defense, is present both in resistant and susceptible plants. This basic defense mechanism consists of the following steps: 1. recognition of the pathogen by the plant, 2. localization of the pathogen (hypersensitivity reaction), 3. systemic signaling and 4. systemic acquired resistance. Of these, the components involved in localization of the pathogen, systemic signaling and systemic acquired resistance are same both in susceptible and resistant variety of a plant, but there is a difference in the resistance gene(s) which is the major component of the pathogen recognition process. Consequently, recognition is the most important part of the defense mechanism determining the resistance of the plant to the particular pathogen. If there is perfect recognition of the pathogen then the signal is transmitted to the genome and the defense pathway is activated. But if recognition fails the pathogen is able to invade the plant since the defense mechanisms are not activated. The process of recognition gives specificity to the resistance, but for this, other subsequent steps (hypersensitivity, systemic signaling and systemic acquired resistance) do not provide any specificity.

## RECOGNITION OF PATHOGEN BY THE PLANT

The first step in the basic defense is the process of recognition of the foreign organism in contact with the plant as the pathogen. This was not understood for many decades. Recently, due to the availability of the sequence of a number of R genes, this process has been understood much better. What explanation we give today is actually based on a hypothesis proposed in 1947 by Flor (41) based on his work on the interaction between Flax plant and the rust pathogen *Melampsora lini.* His hypothesis, called the Gene for gene hypothesis, explains that "for every gene of resistance (R) (in plant) there is a corresponding avirulence gene (avr) in the pathogen and the interaction between the R genes and *avr* genes determine the resistance or susceptibility".

## PLANT RECOGNIZES THE PATHOGEN BASED ON AVR GENE PRODUCT

The plant recognizes an organism as pathogen based on the product of the avirulence gene. Avirulence genes have been identified from viruses, bacteria, fungi, nematodes and insects (65, 85). Generally avirulence gene products of pathogen elicit resistance response in the host plant which is host specific (44). As the plant recognizes the pathogen based on the avr gene product and many avr genes have already been cloned (127), the structure of the avr protein will reveal about the pathogen recognition process. Sequence and structural analysis of the avr proteins showed that they do not belong to any particular structural or functional group (73).

If the pathogen is identified based on a product of avr gene, then the questioin that can be asked is why the pathogen should have a gene whose product is responsible for revealing to the plant that the organism possessing it as a pathogen. The answer to the question is that the pathogen does not have a special category of genes (avirulence genes) that is the pathogen does not specify a gene to be an avirulence gene which is reveled by the lack of sequence or structure relationship among the avirulence genes. Thereby which gene is avirulence gene of a pathogen is determined by the plant and not by the pathogen.

The bacterial avirulence genes share little or no homology with the exception of avrBs3 family. All avrBs3 family genes contain a central domain, a repeat domain, a nuclear localization signal and an acidic transcriptional activation domain. Repeat domain of avrBs3 family determines the recognition specificity. The other bacterial avirulence genes that belong to avrRxv/yopJ family are from varied organisms ranging from mammalian pathogens *Yersinia pseudotuberculosis*, *Salmonella* to plant pathogens *X. compestris pv. vesicatoria*, *Pseudomonas* and *Erwinia amylovora*. They possess similarity only in the region of protease catalytic site (95).

Majority of fungal avirulence genes were cloned from the fungi that colonize intercellular spaces in plant tissues. Avr9, a fungal avirulence gene of *Cladosporium fulvum* was found to be a pre protein of 63 amino acid peptide (124), whereas Avr4 encodes a 135 amino acid pre protein with a characteristic signal peptide for extra cellular targeting (63). Both Avr9 and Avr4 from *C. fulvum* do not share homology with each other or with sequences in the databases. The two proteins are low molecular weight cysteine rich proteins secreted by *C. fulvum* during infection of tomato. Another avirulence gene Pi-ta from *Magnaporthe grisea* encodes a 223 amino acid protein. The sequence of the gene does not share significant homology with other known proteins except for a small domain. The viral avr genes are nothing but, essential virus components such as coat protein, replicase and the movement protein.

As the plant has to sense the pathogen based on biochemical interaction with the pathogen or its gene product, it selects a gene product from pathogen which can be sensed by its sensory mechanism. To distinguish different kinds of pathogens the plant would have selected different gene products of pathogens. If this is correct, the evolutionary pressure on the pathogen would have made the pathogen to modify or make the gene ineffective. Though several mutations and variations have been observed in the avr gene products, the pathogen is unable to get rid of the gene probably due to the vital nature of the gene of the organism. Plant would have chosen the gene product which is vital for the pathogen so that its sensing process is durable.

A number of bacterial avirulence genes were found to play a role in pathogenicity on susceptible plants. In *Pseudomonas syringae* pathovars several avr genes are located in the regions flanking the pathogenicity clusters (HRP) (73). Many avirulence genes have been shown to play a role in virulence (68, 131). The avrBs2 gene product has a sequence related to agrocinopine synthase and glycerol phosphodiestarase. Genome wide analysis of the hrp regulon in *Pseudomonas syringae* pv *tomato* DC3000 revealed that several of the 48 hrp promoters with highly significant HMM E-values expressed homologues of known effector proteins Avr or Hop (Hrp dependent outer proteins) (97). The viral avirulence genes are nothing but the essential viral components, coat protein, replicase and movement protein.

## RESISTANCE GENES ARE USED FOR SENSING THE PATHOGEN

Resistance genes (R genes) are the most important components of the sensing mechanism of the plant. Plant uses the R gene product directly or indirectly with the help of any other proteins to sense the *avr* gene product directly or indirectly. The specificity of sensing by an R gene depends on the presence of corresponding *avr* gene product in the pathogen. In contrast to the diverse group of *avr* encoded proteins R gene products are structurally related proteins with extracellular or intracellular leucine rich repeats (LRR) (7, 27, 57).

As the *avr* gene product has been found to undergo frequent variation to escape sensing by the plant, the R gene product is unable to sense the variants. To enable sensing of avr gene product variants, R gene product undergoes evolution resulting in new R genes which can sense the modified *avr* gene products. Due to this there are a number of R genes for sensing a particular pathogen. Each of these R gene variants can only sense the corresponding *avr* gene product variants and this process is called race specificity. Due to race specificity the total number of R genes in a plant is substantial. For example, in *Arabidopsis thaliana*, more than 150 R genes have been identified and in rice and other crop plants the number of R genes is two to three times higher than that of *Arabidopsis* (119).

## STRUCTURE OF RESISTANCE GENE PRODUCTS

Proteins that are encoded by R genes have to fulfil two tasks: first to recognize the pathogen derived signal and second to initiate a coordinated plant defense reaction. The third requirement of R gene is the capacity for rapid evolution of specificity to match the new virulent races of the pathogens that are rapidly evolving.

Studies of resistance proteins have indicated that the highly variable LRR domains determine recognition of the pathogen Avr products (32, 59) whereas the more conserved TIR-NBS or CC-NBS regions are believed to propagate the perception signal. LRR motif of R gene contains Leucine rich repeats with the consensus sequence XXLXLXX. However, there are some R genes which do not contain LRR region. *eg:* Pto(tomato) (*Pseudomonas syringae pv tomato* resistance) and *Arabidopsis* RPW8 (Resistance to Powdery mildew 8), which encodes kinase and CC domains respectively but not LRRs (133). Still non LRR and LRR containing R proteins might utilize common pathways. *eg*: Pto resistance depends on the NB-LRR protein Prf (*Pseudomonas* resistance to fenthion sensitivity). Similarly Arabidopsis RPS5 (resistance to *Pseudomonas syringae* 5), which is a typical NB-LRR protein requires the Pto like PBS1 (avrPphB susceptible 1) kinase (84, 115).

Comparative sequence analysis for both NB-LRR and extra cellular LRR classes of R proteins revealed that recognition specificity resides largely in the LRRs (35). However studies of the Arabidopsis NB-LRR gene RPS5 uncovered a mutation in the LRR that suppressed the function of multiple R genes, indicating that LRRs are not only involved in signal perception but also in downstream signaling.

Some R proteins, recognize more than one pathogen signal. RPM1 and Pto each recognize different pairs of structurally distinct effectors from the same or related bacteria. For example RPM1 can recognize AvrB and AvrRpm1 (84), Pto recognizes AvrPto and a newly discovered protein from the same pathogen, AvrPtoB (67), Mi-1 confers resistance to a nematode and to an aphid pest, indicating that it recognizes a signal from each. Some R loci show dual specificity accounted for by tightly linked nearly identical genes. Examples include *HRT* and *RPP8* of *Arabidopsis*, and *Gpa2* and *Rx1* of potato.

## INTERACTION BETWEEN R AND AVR PROTEINS:

The ability to discriminate between pathogens from non-pathogenic microbes is a key feature of plants that forms the basis for the activation of defense mechanisms against attempted microbial infection (53). The structural diversity of avr proteins and structural similarity between R proteins rises the question about how structurally similar R proteins recognize structurally diverse avr proteins. Though many models have been proposed to explain this interaction, they generally fall under two categories: Direct interaction between R and avr proteins or indirect interaction between the two. Direct interaction has been observed in a number of plant pathogen interactions.

Several lines of evidence indicate that the LRR domains of R proteins may act as determinants of specificity and ligand binding. Recently, the LRR-like domain of the rice resistance protein Pita was shown to be required for interaction with Avr-Pita (59.) Ellis *et al.* (36) found that LRRs from the flax L proteins might determine specificity combinatorially with an amino-terminal domain. Further Ren *et al.* (100) identified an Arabidopsis protein (TIP) that specifically interacts with the Turnip crinkle virus coat protein. The tobacco N gene is involved in localization of TMV infection and the elicitation of hypersensitive response. Levy *et al.* (79) showed that a basal level of transcription occurs even in uninoculated tobacco plants. Accumulation of N-mRNA in infected plants indicates that transcription is stimulated by TMV infection. It supports the possible interaction of preexisting N protein and an incoming viral directed protein activator. In the interaction of tomato with *Cladosporium fulvum*, Avr proteins are secreted into the extracellular space and the R proteins were found to be localized to the plasma membrane (98). Four oomycetes avr genes, Avr1b-1 from the soybean pathogen *Phytophthora sojae*, ATR 13 and ATR1NdWsB from the *Arabidopsis* pathogen *Hyaloperonospora parasitica* and Avr 3a are detected by cognate R proteins in the host cytoplasm (4).

The guard hypothesis predicted that the cellular location of the Avr protein might determine the location of not only the R protein, but also the virulence target of the Avr protein and this has been found in the interaction between AvrB of *Pseudomonas syringae* and RPM1 of Arabidopsis and the interaction between AvrB of *Pseudomonas syringae* and Rpg1-b of *Arabidopsis thaliana*.

## HYPERSENSITIVITY RESPONSE

Once the plant senses the pathogen (*i.e* identifies the microbe in contact with it as pathogen using the R gene based sensing), localization of the pathogen takes place. Localization of the pathogen is done by killing the cells in contact with the pathogen. This process commonly referred to as "hypersensitivity reaction" (HR) (51, 113) occurs only in resistant plants in response to pathogenic viruses, bacteria, fungi or nematodes, and is associated with a multitude of biochemical processes that make these dead cells, and the adjacent living cells, an inhospitable environment for microbes (70). This hypersensitive cell death delimits the infected zone and avoids the multiplication and spread of the pathogen.

One of the earliest events in the HR is the rapid accumulation of reactive oxygen species (ROS) and nitric oxide (NO) through the activation of enzyme systems similar to neutrophil NADPH oxidase (66, 74) and nitric oxide synthase (NOS) (17). Both NO and ROS are necessary to trigger host cell death.

### *Oxidative burst*

The oxidative burst in plant cells was found to be a biphasic response that extends over many hours after exposure to a pathogen. There was a primary peak 1–2 h after infection, followed by a secondary peak of greater magnitude 3–6 h after infection. The second phase was found to be triggered by avirulent pathogens (*i.e* the plant is able to sense the pathogen), which do not infect the plant. The second phase was not observed with virulent pathogens, which are able to infect the plant (*i.e.* the plant is unable to sense the pathogen) (74, 106).

Before it was thought that cell death during HR is due to ROS only (29). But recent evidence indicates that it is more complex than observed (56). Cell death is not always found to be associated ROS and there are elicitors that cause cell death, but do not evoke a strong oxidative burst (33, 138). The Arabidopsis dnd1 mutation in a gene that encodes a cyclic nucleotide-gated ion channel, in some measure, uncouples cell death from the defense response (23,138). Moreover, double mutants in both AtrbohD and AtrbohF genes encoding the Arabidopsis homologs of gp91phox subunit of the NADPH oxidase do not produce ROS, but exhibit a residual HR (122).

Recent studies by Costet *et al.* (26) suggest that ROS induce production of intercellular messenger, but are not themselves the second messengers. The second messengers are salicylic acid (SA), jasmonic acid (JA), systemin as well as ABA and ethylene (72, 105).

### *Nitric oxide*

In addition to reactive oxygen species, plant cells also accumulate high levels of nitric oxide in response to infection with bacterial, viral and fungal pathogens. Elicitation of epidermal tobacco cells induces overproduction of NO within minutes. Similarly, high levels of NO have been detected in tobacco plants 24 h after infection with tobacco mosaic virus.

How NO and ROS actually kill is still unclear. But there is evidence that NO induces cell death by activating proteases (111) and over expression of AtCYS1, a cysteine protease inhibitor, was found to block cell death activated by either avirulent pathogens or by nitrosative stress in *A. thaliana* suspension cultured cells and in transgenic tobacco plants (2).

NO function together with ROS in triggering hypersensitive cell death and is also involved in other defense functions that are independent of ROS. It functions independently of ROS in the induction of various defense genes including pathogenesis-related proteins and enzymes of phenylpropanoid metabolism involved in the production of lignin, antibiotics and the secondary signal salicylic acid. Noritake *et al.* (93) showed that NO stimulates a signal transduction pathway by the production of phytoalexin in higher plant cells. Moreover, the first enzyme of phenylpropanoid biosynthesis pathway, phenylalanine ammonia-lyase (PAL), is induced after infiltration of tobacco leaves and cells with commercially available NOS or NO donors (28).

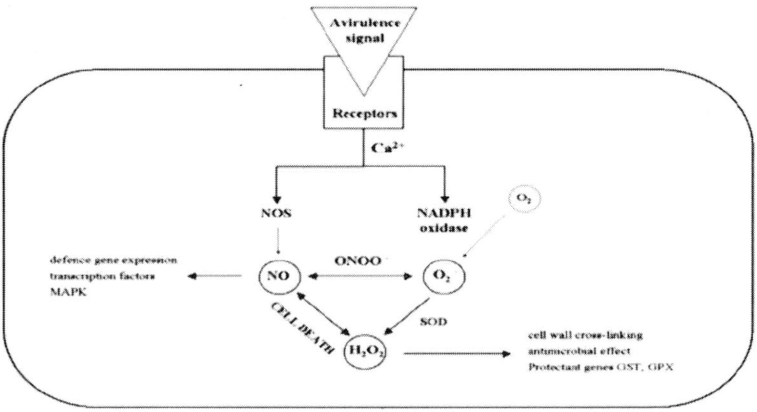


NO- Nitric oxide; NOS- Nitric oxide synthase; ONOO- Peroxynitrite; $H_2O_2$ - Hydrogen peroxide; GPX- Glutathione peroxidase; GST-Glutathione S transferase; SOD-Super oxide dismutase; MAPK-Mitogen activated protein kinase.

**Fig. 1.** Nitric oxide and Hydrogen peroxide mediated plant cell death

The regulation of plant defense responses operating downstream of NO and ROS have also been attributed to mitogen-activated protein kinase (MAPK) cascades. A MAPK was recently found to be activated by NO in Arabidopsis (21), but its role in the induction of defense genes has not yet been investigated. SIPK is typically induced by SA and by $H_2O_2$ (134) and shows SA-mediated NO inducibility (71), while WIPK is not induced by either SA or NO. A number of other kinases are being identified that might constitute a complex signaling network leading to resistance.

## SYSTEMIC SIGNALING

Immediately after the pathogen is localized by hypersensitivity reaction, signal is transmitted systemically to activate the defense genes leading to systemic acquired resistance. Salicylic acid, jasmonic acid and ethylene are the three important signaling molecules in the plant which operate in inducing defense. Among these, salicylic acid is considered to be the signal acting in inducing defense genes in response to the R gene based race specific resistance while the jasmonic acid and ethylene are considered to mediate non-pathogen (abiotic) stress mediated induction of the defense genes.

Several mutation studies emphasized the role of Salicylic acid in plant defense responses and in establishment of SAR. These mutants include

(a) *eds (enhanced disease susceptibility)* mutant: *eds5–1* accumulates *PR-1* mRNA to only around 10% of wild type levels following virulent *P. syringae* infection, but is capable of mounting at least a partial SAR response with WT levels of *PR-2* and *PR-5* (102).

(b) *sid* (*salicylic acid induction deficient*) mutants: *sid1* and *sid2* (92) also showed reduced PR-1 gene expression and are unable to accumulate SA after pathogen infection.

(c) *pad* (phytoalexin deficient mutants): These mutants have defects in phytoalexin and SA accumulation, and reduced *PR* expression following pathogen attack (virulent *P. syringae*) (50). *PAD4* acts upstream of SA and encodes a protein with predicted similarity to triacyl glycerol lipases similar to EDS1 (40).

(d) Mutants with activated SA mediated response: A number of mutants with activated SA mediated defense responses showing spontaneous lesions resembling HR cell death were identified (81).

(e) cpr mutants (*constitutive expression of PR genes*): *cpr1* and *cpr6* have constitutively high levels of *PR-1* gene expression and show enhanced disease resistance that is blocked by *eds1, eds5*, or *pad4* (22, 60).

*(f) edr1* (*enhanced disease resistant 1*) is another mutant that is involved in SA signalling. It exhibits enhanced disease

resistance to *P. syringae* and the fungi *Erysiphe cichoracearum,* but does not constitutively express *PR-1, PR-2* or *PR-5* (42). *EDR1* encodes a MAPK kinase and is thought to act upstream of SA and the mutants *eds1* and *pad4* to negatively regulate SA signalling (43).

(g) *npr1* (*nonexpressor of PR genes1* and *sai1* (*SA-insensitivity1*)] acts downstream of SA and is impaired in its ability to express *PR* genes in the presence of SAR inducers or avirulent pathogens (50).

NPR1 interacts with TGA2 through its ankyrin repeat domain and TGA2 binds to an SA-responsive promoter element in the *PR-1* gene that resembles the *as-1*-like motif (139). By analyzing single or combined deletion knockout mutants of *TGA2, TGA5*, and *TGA6*, Zhang *et al.* (140) discovered that *PR* gene expression and pathogen resistance induced by the SA analogue INA was blocked in the triple *TGA* knockout mutant, but not in single or double *TGA* knockouts mutants. Additionally, this SAR compromised phenotype mimicked that of mutations in *NPR1* (99).

WRKYs are another important family of transcription factors involved in SA-mediated defense responses, which bind to the W-box motif containing a TGAC core (39). The *NPR1* promoter contains several W-box motifs and when they are mutated in promoter::reporter gene fusions, promoter activity is abolished in response to SA suggesting that WRKY transcription factors are important for *NPR1* expression (137). Additionally, analysis of *PR-1* co-regulated genes induced during SAR revealed an over-representation of W-box or W-box like motifs in their promoters (83).

## SYSTEMIC ACQUIRED RESISTANCE

After the hypersensitivity reaction, the systemic signals induce expression of a large number of genes collectively called as systemic acquired resistance (SAR) proteins or pathogenesis related proteins (PR Proteins) leading to a condition of resistance during which no pathogen can attack the plant. This phenomenon was initially observed in tobacco. It has been found that tobacco plants challenged with TMV subsequently developed increased resistance to secondary infection in distal tissues (103, 104). This spread of resistance through out the plant tissue was termed as Systemic Acquired Resistance (SAR).

Resistance through SAR is long lasting and effective against a broad spectrum of pathogens (including virus, bacteria and fungi), which may differ from the SAR-inducing organism (107). SAR confers quantitative protection against a broad spectrum of microorganisms. The time needed for the establishment of SAR depends on both the plant and the type of inducing organism. A very rapid induction was reported for cucumber, where SAR sets in as early as 7 h after a primary inoculation with *Pseudomonas syringae* (110). But in case of Tobacco, injection of spores of *Peronospora parasitica* pv *tabaci* under the epidermis of the stem leads to the expression of SAR in 2–3 weeks after the primary inoculation. The level of protection may vary depending on the organism used for the primary inoculation. Once established, SAR can last over weeks for example in cucumber, inoculation of the first leaf, followed 2–3 weeks later by a second booster inoculation, protected the plants up to flowering.

At the Molecular level, SAR is characterized by the increased expression of a large number of pathogenesis-related genes (*PR* genes), in both local and systemic tissues. PR proteins were first described in the 1970s by Van Loon, who observed accumulation of various novel proteins after infection of tobacco with TMV. It is generally thought that SAR results from the concerted effects of many PR proteins rather than a specific PR protein.

Lignification is one of the mechanisms observed in many plant species following attempted infection by pathogenic organisms such as fungi, bacteria, viruses, and nematodes (86). It is not restricted to incompatible interactions in gene-for-gene systems but is also observed in non host resistance and SAR. Lignin is formed by random dehydrogenative polymerization of precursors produced in the phenylpropanoid pathway. The first step in this pathway is the deamination of phenylalanine to cinnamic acid catalyzed by the enzyme phenylalanine ammonia-lyase (PAL). PAL provides precursors for lignin and for several phenylpropanoid-derived secondary plant products involved in resistance. Incorporation of lignin into a plant cell wall is bound to strengthen it mechanically and to make it more resistant to degradation by enzymes secreted by an invading pathogen.

Lignified cell walls could also constitute a barrier preventing free nutrient movement and therefore help to starve a pathogen. Lignin precursors themselves might exert a toxic effect on pathogens or, by binding to fungal cell walls, make them more rigid and impermeable, thus hindering further growth or uptake of water and nutrients (114).

Strengthening of the cell wall can also occur by peroxidase catalyzed cross-linking of hydroxyl proline-rich structural cell wall glycoproteins. Glycine-rich proteins, which are found in the cell wall of many plants, accumulate systemically in tobacco plants infected with tobacco mosaic virus (TMV) and in virus-infected rice plants. In Arabidopsis, the transcript level of several glycine-rich proteins is slightly elevated after SA treatment. These structural modifications of plant cell walls after infection may contribute to resistance, either by stopping pathogen ingress directly or by slowing down the penetration process, thus allowing the plant to activate further defense mechanisms.

## SAR PROTEINS

During systemic acquired resistance a spectrum of genes are expressed and the expressed products act together to provide a condition of resistance against the pathogens. Individually, these proteins possess antimicrobial activity which is exploited for developing pathogen resistance in plants. Homologues of these proteins are also present in other organisms like bacteria, fungi, insects, animals and humans. The proteins expressed during SAR include chitinases (PR3), chitin binding proteins, beta-1,3 glucanases (PR2), PR1 proteins, PR4 proteins, thaumatin like proteins (PR5), thionins, lipid transfer proteins, defensins, 2S storage albumins, trypsin inhibitors, ribosome inactivation proteins (RIPs) and polygalacturanase inhibitors (PGI) (123).

### *Chitinases (PR-3)*

Chitinases have been studied from various plants and many genes coding for chitinases have been cloned and characterized. It is one of the most studied enzymes showing antifungal activity. They have been purified from a number of plants. Chitinases are constitutively expressed at low levels in plants but are induced to high levels by pathogens, abiotic stress, fungal cell wall components and oligosaccharides. Chitinases show synergistic effect in combination with glucanases (77).

The antifungal activity of chitinases has been attributed to the hydrolyzing activity of the N-acetyl glucosamine polymer, chitin, which is a major component of fungal cell wall. These are encoded by small multi gene family. Chitinases have been shown to be toxic to a number of fungal pathogens and over expression of chitinases in a number of plants resulted in reduced fungal infection.

### *β-1, 3-glucanases (PR-2)*

They are glucan endo-1,3-beta-glucosidases. They catalyze endo type hydrolytic cleavage of the 1, 3-beta-D glucosidic linkages in beta-1, 3-glucans. They are constitutively produced at low levels under normal conditions and are over expressed during infection. They exist as multiple structural isoforms that differ in size, isoelectric point, primary structure, cellular localization and regulation. Four major classes of glucanases have been identified recently. The class I isoforms include basic isoforms of 33 kD size which accumulate in the vacuoles. The class II includes the acidic isoforms and they accumulate in the apoplast. The class III enzymes are also acidic with 35 kD size protein. Both class II and class III proteins are acidic and do not possess C-terminal extension present in the class I enzymes (108). Recently a new class (class IV) has been identified which is a PR-like protein (Tag 1). It is also acidic protein with 35 kD size and shares conserved sequences found in all classes of tobacco enzymes. The role of glucanases in defense reactions involves solubilization of cell wall of fungi.

### *Chitin-binding proteins*

Chitin binding proteins are a class of anti-microbial proteins present in plants. They have a domain that is specifically suited to bind to chitin. Though chitin binding domains are present in chitinases and other proteins there are proteins which have only chitin binding domains. They are non-enzymatic proteins showing antifungal activity. Included in this group are the nettle lectin, hevein and Ac -AMPs isolated from *Amaranthus* seeds. The nettle lectins inhibited several fungi including *Colletotrichum lindemuthianum, Phoma betae, Phycomyces blakesleeanus, Septoria nodorum* and *Trichoderma viridae*. The Ac-AMPs isolated from *Amaranthus caudatus* seeds showed a wide spectrum of antifungal/antimicrobial activity (10). Similar peptides were purified from *Mirabilis jalapa and* were designated as Mj-AMP1 and Mj-AMP2 (15). Four low molecular weight non-enzymatic proteins showing antifungal activity against *Aspergillus flavus, Fusarium moniliforme and Macrophomina phaseolina* were purified from sorghum (48, 49).

## PR-1 Proteins

These are the most abundant proteins of the PR family. They belong to small multi gene families. The acidic *N. tobaccum* PR1 a was the first to be purified and characterized. PR 1 proteins were shown to accumulate in tobacco after TMV infection. There are both acidic and basic PR 1 proteins. Three acidic PR 1 proteins (PR-1a, PR-1b and PR 1c) with molecular weights of 15-17kD and a basic PR 1 protein were identified from tobacco. In tomato a PR protein of 14 kD named as P14 was identified to accumulate extracellularly on infection with viroids and fungal pathogens. A similar protein was identified from *Arabidopsis thaliana* by Metzler et al (88).

## *PR-4 Proteins*

These are neutral, extracellular proteins with molecular weight of 13-14.5kD. The antifungal activity of the PR4 group of proteins was observed by Hejgaard *et al.* (55). No specific function or enzymatic activity has been attributed to the PR 4 proteins. They are coded by small gene families consisting of 2-4 genes. Their synergistic response with chitinases and glucanases has also been studied.

## *PR-5 Proteins*

It consists of a set of thaumatin like proteins showing homology to sweet protein, thaumatin, isolated from the fruits of *Thaumatococcus danielli*. This family includes osmotin, a basic polypeptide which was shown to accumulate in tobacco under salt stress. This family has been categorized in to three groups- acidic, neutral and basic (69). The regulation of osmotin gene expression has been studied most extensively. The anti fungal activity of this group of proteins is attributed to the inhibition of fungal hyphal growth and spore germination shown by permatins, a potent thaumatin like protein in maize, sorghum, oat and wheat seeds.

### *Thionins*

Thionins are basic and cysteine-rich low molecular weight (~5 KD) proteins identified in both monocots and dicots. Thionins show toxicity against many bacterial and fungal pathogens. They function by altering the membrane permeability, inhibiting macromolecule biosynthesis and by participating in redox reactions (45).

### *Lipid - transfer proteins [ns-LTPs]*

Lipid-transfer proteins (LTPs) another group of antifungal proteins found in plants (46). The antimicrobial activity of LTPs seems to be dependent on the target cell and the nature of each protein (15).

### *Plant defensins*

Plant defensins are small, cysteine-rich, basic, structurally distinct peptides that possess antimicrobial activity. Plant defensins are typically 45-54 amino acids in length (10). Most plant defensins are synthesized with a signal peptide (Expression of several plant defensins can be induced following fungal infection of vegetative tissues (9, 37). Plant defensins inhibit the growth of a wide range of fungi at micromolar concentrations (120).

### *Inhibitor proteins*

Ribosome inactivation proteins are a group of proteins induced during systemic acquired resistance which are potent catalytic inactivators of eukaryotic protein synthesis. These proteins have been grouped into two types. The type I RIPs are 26- 32 KD monomeric proteins and has been purified from more than 30 plants. The type II RIPs includes the lectins containing galactose binding domains facilitating membrane recognition and transport across cell membranes. Plant RIPs have great specificity for non-host ribosome. The antifungal activity of the RIPs was elucidated first in barley against *Trichoderma* and *Fusarium sporotrichioides*

(78). Inhibitors of pathogen proeases and polygal-acturanases are also induced during systemic acquired resistance.

## PROSPECTS AND CONCLUSION

The molecular basis of plant pathogen interaction could not be understood for a long time. But by the cloning of a number of R genes and avr genes, the mystery of interaction is being solved. Production of mutants in *Arabidopsis thaliana* has helped in elucidation of signaling and regulatory genes associated with this interaction. Though some gaps are there to be understood, the overall interaction process is now clear. The basic interaction between plant and pathogen starts with recognition of each other i.e the plant has to sense the pathogen to activate the defense and the pathogen has to sense the plant to activate its offensive mechanisms. The resistance or susceptibility of a plant to a particular pathogen is determined by the recognition process where the R genes are used as sensors to sense the pathogen based on certain gene products of the pathogen which are collectively called the avr proteins. This sensing may be due to direct or indirect interaction between the proteins. Once the pathogen is sensed it is localized and killed by the hypersensitive cell death which leads to systemic signaling and activation of a battery of genes involved in systemic acquired resistance. Though this is the basic interaction, based on the pathogens offensive mechanisms a number of other defenses are incorporated into this basic defense. For example if the pathogen produces a toxin, the plant produces an enzyme which detoxifies this toxin. In this interaction, this review has focused only from the plant angle of the interaction and the pathogen angle is not considered here.

**Table 1.** Plant disease resistance (R) genes.

| Class | R gene | Plant | Pathogen(s) | Disease | Avr gene | Race specificity | References |
|---|---|---|---|---|---|---|---|
| 1 | Pto | Tomato | *Pseudomonas syringae* | Bacterial rot | AvrPto, AvrPtoB | Yes | 67 |
| 2 | Bs2 | Pepper | *Xanthomonas campestris* | Bacterial leaf spot | AvrBs2 | Yes | 117 |

| | | | | | | |
|---|---|---|---|---|---|---|
| Dm3 | Lettuce | *Bremia lactucae* | Powdery mildew of lettuce | | Yes | 89 |
| Gpa2a | Potato | *Globodera pallida* | Cyst Nematode | | Yes | 126 |
| Hero | Potato | *G. rostochiensis, G. pallida* | Cyst Nematode | | Yes | 38 |
| HRTb | *Arabidopsis* | Turnip Crinkle Virus | Viral disease | Coat Protein | Yes | 25 |
| I2 | Tomato | *Fusarium oxysporum* | Root rot | | Yes | 109 |
| Mi | Tomato | *Meloidogyne incognita* | Cyst Nematode | | Yes | 90 |
| Mi | Tomato | *Macrosiphum euphorbiae* | Insect | | Yes | 128 |
| Mla | Barley | *Blumeria graminis* | Fungus | | Yes | 141 |
| Pib | Rice | *Magnaporthe grisea* | Blast disease | | Yes | 129 |
| Pi-ta | Rice | *M. grisea* | Blast disease | AVR-Pita | Yes | 94 |
| R1 | Potato | *Phytophthora infestans* | Late blight | | Yes | 1 |
| Rp1 | Maize | *Puccinia sorghi* | Rust | | Yes | 24 |
| RPM1 | *Arabidopsis* | *P. syringae* | Bacterial disease | AvrRpm1 AvrB | Yes | 52 |
| RPP8b | *Arabidopsis* | *Peronospora parasitica* | Downy mildew | | Yes | 87 |
| RPP13 | *Arabidopsis* | *P. parasitica* | Blight | | Yes | 5 |
| RPS2 | *Arabidopsis* | *P. syringae* | Bacterial disease | AvrRpt2 | Yes | 91 |
| RPS5 | *Arabidopsis* | *P. syringae* | Bacterial disease | AvrPphB | Yes | 130 |
| Rx1a | Potato | Potato Virus X | Viral disease | Coat Protein | Yes | 3 |
| Rx2 | Potato | Potato Virus X | Viral disease | Coat Protein | Yes | 3 |

| | | | | | | | |
|---|---|---|---|---|---|---|---|
| | Tm-2 | Tomato | Tomato Mosaic Virus | Viral disease | Movemen t protein | Yes | 75 |
| | RCY1 | *Arabi-dopsis* | Turnip crinkle virus | Viral disease | Coat Protein | Yes | 118 |
| | Sw-5 | Tomato | Tomato Spotted Wilt Virus | Viral disease | | Yes | 11 |
| | Xa1 | Rice | *X. oryzae* | Bacterial leaf blight | | Yes | 136 |
| 3 | L | Flax | *Melampsora lini* | Rust | | Yes | 76 |
| | M | Flax | *M. lini* | Rust | | Yes | |
| | N | Tobacco | Tobacco Mosaic Virus | Viral disease | Helicase | Yes | 79 |
| | P | Flax | *M. lini* | Rust | | Yes | 32 |
| | RPP1 | *Arabi-dopsis* | *P. parasitica* | Blight | | Yes | 8 |
| | RPP4 | *Arabi-dopsis* | *P. parasitica* | Blight | | Yes | 125 |
| | RPP5 | *Arabi-dopsis* | *P. parasitica* | Blight | | Yes | 96 |
| | RPS4 | *Arabi-dopsis* | *P. syringae* | Bacterial disease | AvrRps4 | Yes | 47 |
| 4 | Cf-2c | Tomato | *Cladosporium fulvum* | Fungus | Avr2 | Yes | 82 |
| | Cf-4d | Tomato | *C. fulvum* | Fungus | Avr4 | Yes | 121 |
| | Cf-5c | Tomato | *C. fulvum* | Fungus | | Yes | 31 |
| | Cf-9d | Tomato | *C. fulvum* | Fungus | Avr9 | Yes | 62 |
| 5 | Xa21 | Rice | *Xanthomonas oryzae* | Bacterial leaf blight | | Yes | 112 |
| 6 | Hm1 | Maize | *Cochliobolus carbonum* | Fungus | | Yes | 61 |
| | HS1 pro-1 | Beet | *Heterodera schachtii* | Cyst nematode | | Yes | 14 |
| | mlo | Barley | *B. graminis* | Fungus | | | 13 |

| | | | | | | |
|---|---|---|---|---|---|---|
| Rpg1 | Barley | *Puccinia graminis* | Rust Fungus | | No | 12 |
| RPW8 | *Arabidopsis* | *Erisyphe chicoracearum* | Powdery mildew | | No | 133 |
| RRS1-R | *Arabidopsis* | *Ralstonia solanacearum* | Bacteria | | Yes | 30 |
| RTM1 | *Arabidopsis* | Tobacco Etch Virus | Virus | | Yes | 19 |
| RTM2 | *Arabidopsis* | Tobacco Etch Virus | Virus | | Yes | 132 |
| Vele, Ve2e | Tomato | *Verticillium alboatrum* | Fungus | | Yes | 64 |
| Rsv1 | Soybean | *Soybean Mosaic Virus* | Virus | P3(NBS-LRR multi allelic gene) | Yes | 54 |

**Table 2.** Avirulence genes with virulence or virulence associated functions

| S.N. | Gene | Pathogen | Possible function and localization in the host | References |
|---|---|---|---|---|
| 1 | *avrBs2* | *X. c.* pv. *vesicatoria* | Cytoplasmic | 116 |
| 2 | *avrXa7* | *X. o.* pv. *oryzae* | NLS, AD DNA binding, nucleus | 142 |
| 3 | *avrb6* | *X. c.* pv. *malvacearum* | Nucleus | 135 |
| 4 | *pthN* | *X. c.* pv. *malvacearum* | Nucleus | 16 |
| 5 | *pthA* | *X. citri* | Cell hyperplasia, nucleus | 34 |
| 6 | *avrRxv* | *X. c.* pv. *malvacearum* | MAP kinase binding? | 20 |
| 7 | *avrRpm1* | *P. s.* pv. *maculicula* | MYM, inner membrane? | 101 |
| 8 | *avrPto* | *P. s.* pv. *tomato* (T1) | MYM, kinase binding, inner membrane | 18 |
| 9 | *avrE* | *P. s.* pv. *tomato* (PT23) *E. amylovora* | Unknown | 6 |
| 10 | *avrF* | *P. s. pv. syringae E. amylovora* | *Chaperone* | 6 |
| 11 | *avrA* | *P. s. pv. tomato (PT23)* | *Cytoplasmic* | 80 |

| | | | | |
|---|---|---|---|---|
| 12 | *avrRpt2* | *P. s. pv. tomato* | *Cytoplasmic* | - |
| 13 | *avrPphF* | *P. s. pv. phaseolicola* | *Unknown* | 58 |
| 14 | *virPphA* | *P. s. pv. phaseolicola* | *Unknown, inhibits HR* | 58 |
| 15 | AvrPitA | *Magnaporthe grisea* | Protease | 59 |
| 16 | Avr9 | *Cladosporium fulvum* | NR | 124 |
| 17 | Nla | Potato Virus Y | Proteinase | |
| 18 | Coat protein | Turnip Crinkle Virus | coat protein | 100 |

## *Literature cited*


1. Ballvora A, *et al.* 2002. The R1 gene for potato resistance to late blight (*Phytophthora infestans*) belongs to the leucine zipper/ NBS/ LRR class of plant resistance genes. *Plant J.* 30:361–71.
2. Belenghi B, *et al.* 2003. AtCYS1, a cystatin from *Arabidopsis thaliana*, suppresses hypersensitive cell death. *Eur J Biochem* 270:1-12.
3. Bendahmane A, Kohn BA, Dedi C, Baulcombe DC. 1995. The coat protein of potato virus X is a strain-specific elicitor of Rx1-mediated virus resistance in potato. *Plant J.* 8:933–41.
4. Birch PRJ, *et al.* 2006. Trafficking arms: oomycete effectors enter host plant cells. *Trends in Microbiol.* 14:8-11.
5. Bittner-Eddy PD, Crute IR, Holub EB, Beynon JL. 2000. RPP13 is a simple locus in *Arabidopsis thaliana* for alleles that specify downy mildew resistance to different avirulence determinants in *Peronospora parasitica. Plant J.* 21:177–88.
6. Bogdanove AJ, *et al.* 1998. Homology and functional similarity of an hrp-linked pathogenicity locus, dspEF, of *Erwinia amylovora* and the avirulence locus avrE of *Pseudomonas syringae patho-var tomato. Proc. Natl. Acad. Sci.* USA 95:1325-1330.
7. Bonas U, Lahaye T. 2002. Plant disease resistance triggered by pathogen derived molecules: refined models of specific recognition. *Curr. Opin. Microbiol.* 5: 44–50.
8. Botella MA, *et al.* 1998. Three genes of the Arabidopsis RPP1 complex resistance locus recognize distinct *Peronospora parasitica* avirulence determinants. *Plant Cell* 10:1847–60.
9. Broekaert WF, *et al.* 1997. Antimicrobial peptides from plants. *Crit Rev Plant Sci* 16: 297-323.
10. Broekaert, WF, Terras, FRG, Cammue, BPA, Osborn, RW.

1995. Plant defensisns: novel antimicrobial peptides as components of the host defense system. *Plant Physiol.* 108: 1353-58.

11. Brommonschenkel SH, Frary A, Tanksley SD. 2000. The broad-spectrum tospovirus resistance gene Sw-5 of tomato is a homolog of the root-knot nematode resistance gene Mi. *Mol. Plant-Microbe Interact.* 13: 1130 - 38.
12. Brueggeman R, *et al.* 2002. The barley stem rust-resistance gene Rpgl is a novel disease-resistance gene with homology to receptor kinases. *Proc. Natl. Acad. Sci.* USA 99:9328–33.
13. Buschges R, *et al.* 1997. The barley Mlo gene: a novel control element of plant pathogen resistance. *Cell* 88:695–705.
14. Cai D, *et al.* 1997. Positional cloning of a gene for nematode resistance in sugar beet. *Science* 275:832–34.
15. Cammue BPA, *et al.* 1995. A Potent Antimicrobial Protein from Onion Seeds Showing Sequence Homology to Plant Lipid Transfer Proteins. *Plant Physiol.*109: 445-455.
16. Chakrabarty PK, Duan YP, Gabriel DW. 1997. Cloning and characterization of a member of the Xanthomonas avr/pth gene family that evades all commercially utilized cotton R genes in the United States. *Phytopathology* 87:1160-1167.
17. Chandok MR, Ytterberg AJ, van Wijk KJ, Klessig DF. 2003. The pathogen-inducible nitric oxide synthase (iNOS) in plants is a variant of the P protein of the glycine decarboxilase complex, *Cell* 113, 469–482.
18. Chang JH, Rathjen JP, Bernal AJ, Staskawicz BJ, Michelmore RW. 2000. AvrPto enhances growth and necrosis caused by *Pseudomonas syringae pv. tomato* in tomato lines lacking either Pto or Prf. *Mol Plant Microbe Interact.* 13:569-571.
19. Chisholm ST, *et al.* 2000. cloning of the Arabidopsis RTM1 gene, which controls restriction of long-distance movement of tobacco etch virus. *Proc. Natl. Acad. Sci.* USA 97:489–94.
20. Ciesiolka LD, *et al.* 1999. Regulation of expression of avirulence gene avrRxv and identification of a family of host interaction factors by sequence analysis of avrBsT. *Mol Plant Microbe Interact.* 12:35-44.
21. Clarke A, Desikan R, Hurst RD, Hancock JT, Neill SJ. 2000. No way back: nitric oxide and programmed cell death in *Arabidopsis thaliana* suspension cultures. *Plant J.* 24:667-677.
22. Clarke JD, Aarts N, Feys BJ, Dong X, Parker JE. 2001. Constitutive disease resistance requires EDS1 in the Arabidopsis mutants cpr1 and cpr6 and is partially EDS1-dependent in cpr5. *Plant J.* 26: 409–420.

23. Clough SJ, Fengler KA, Yu IC, Lippok B, Smith RK, Jr., Bent AF. 2000. The Arabidopsis dnd1 "defense, no death" gene encodes a mutated cyclic nucleotide gated ion channel. *Proc. Natl. Acad. Sci.* USA 97: 9323–9328.
24. Collins N, *et al.* 1999. Molecular characterization of the maize Rp1-D rust resistance haplotype and its mutants. *Plant Cell* 11:1365–76.
25. Cooley MB, Pathirana S, Wu HJ, Kachroo P, Klessig DF. 2000. Members of the Arabidopsis HRT/RPP8 family of resistance genes confer resistance to both viral and oomycete pathogens. *Plant Cell* 12:663–76.
26. Costet L, Dorey S, Fritig B, Kauffmann S. 2002. A pharmacological approach to test the diffusible signal activity of reactive oxygen intermediates in elicitor-treated tobacco leaves. *Plant Cell Physiol* 43: 91–98.
27. Dangl JL, Jones JDG. 2001. Plant pathogens and integrated defense responses to infection. *Nature*. 411:826-833.
28. de Pinto MC, Tommasi F, De Gara L. 2002. Changes in the antioxidant systems as part of the signaling pathway responsible for the programmed cell death activated by nitric oxide and reactive oxygen species in tobacco bright-yellow 2 cells. *Plant Physiol*. 130: 698 – 708.
29. Desikan R, Reynolds A, Hancock JT, Neill SJ. 1998. Harpin and hydrogen peroxide both initiate programmed cell death but have differential effects on defence gene expression in Arabidopsis suspension cultures. *Biochem. J.* 330: 115–120.
30. Deslandes L, *et al.* 2002. Resistance to *Ralstonia solanacearum* in *Ara-bidopsis thaliana* is conferred by the recessive RRS1-R gene, a member of a novel family of resistance genes. *Proc. Natl. Acad. Sci.* USA 99:2404–2409.
31. Dixon MS, *et al.* 1998. The tomato Cf-5 disease resistance gene and six homologs show pronounced allelic variation in leucine-rich repeat copy number. *Plant Cell* 10:1915–25.
32. Dodds P, Lawrence G, Ellis J. 2001. Six amino acid changes confined to the leucine-rich repeat beta-strand/beta-turn motif determine the difference between the P and P2 rust resistance specificities in flax. *Plant Cell* 13:163–78.
33. Dorey S, Kopp M, Geoffroy P, Fritig B, Kauffmann S. 1999. Hydrogen peroxide from the oxidative burst is neither necessary nor sufficient for hypersensitive cell death induction, phenylalanine ammonia lyase stimulation, salicylic acid accumulation, or scopoletin consumption in cultured tobacco cells treated with elicitin. *Plant Physiol* 121: 163–172.

34. Duan YP, Castañeda A, Zhao G, Erdos G, Gabriel DW. 1999. Expression of a single, host-specific, bacterial pathogenicity gene in plant cells elicits division, enlargement, and cell death. *Mol. Plant Microbe Interact.* 12: 556-560.

35. Ellis J, Dodds P, Pryor T. 2000. Structure, function and evolution of plant disease resistance genes. *Curr. Opin. Plant Biol.* 3:278–84.

36. Ellis JG, Lawrence GJ, Luck JE, Dodds PN. 1999. Identification of regions in alleles of the flax rust resistance gene L that determine differences in gene-for-gene specificity. *Plant Cell* 11:495–506.

37. Epple P, Apel K, Bohlmann H. 1997. ESTs reveal a multigene family for plant defensins in *Arabidopsis thaliana*. *FEBS Letters* 400:168-172.

38. Ernst K, Kumar A, Kriseleit D, Kloos DU, Phillips MS, Ganal MW. 2002. The broad-spectrum potato cyst nematode resistance gene (Hero) from tomato is the only member of a large gene family of NBS-LRR genes with an unusual amino acid repeat in the LRR region. *Plant J.* 31:127–36.

39. Eulgem T, Rushton PJ, Robatzek S, Sommssich IE. 2000. The WRKY superfamily of plant transcription factors. *Trends Plant Sci.* 5: 199–206.

40. Falk A, *et al.* 1999. EDS1, an essential component of R gene-mediated disease resistance in Arabidopsis has homology to eukaryotic lipases. *Proc. Natl. Acad. Sci.* USA 96, 3292–3297.

41. Flor HH. 1947. Host-parasite interactions in flax rust—its genetics and other implications. *Phytopathology* 45:680–85.

42. Frye CA, Innes RW. 1998. An Arabidopsis mutant with enhanced resistance to powdery mildew. *Plant Cell* 10: 947–956.

43. Frye CA, Tang D, Innes RW. 2001. Negative regulation of defense responses in plants by a conserved MAPKK kinase. Proc. Natl. Acad. Sci. USA 98: 373–378.

44. Gabriel DW, Rolfe BG.1990. Working models of specific recognition in plant–microbe interactions. *Annu. Rev. Phytopathol.* 28:365-391.

45. Garcia-Olmedo F, *et al.* 1992. Charecterization and anlysis of thionin genes. In: Genes involved in plant defense, (eds Boller, T, Meins, F) pp 283-302. Springer-Verlag, New York, U.S.A.

46. Garcia-Olmedo F, Molina A, Segura A, Moreno M. 1995. The defensive role of nonspecific lipid-transfer proteins in plants. *Trends Microbiol* 3:72-74.

47. Gassmann W, Hinsch ME, Staskawicz BJ. 1999. The Arabidopsis RPS4 bacterial-resistance gene is a member of the TIR-NBS-LRR family of

disease-resistance genes. *Plant J.* 20:265–77.

48. Ghosh M, Ulaganathan K. 1996. Mature seeds of sorghum contain proteins toxic to aflatoxigenic *Aspergillus flavus*. *J. stored prod. Res.* 32: 339-343.
49. Ghosh M. 2000. In Identification and isolation of antifungal proteins and cloning a gene encoding for them in Sorghum, Ph D thesis, Osmania University, Hyderabad.
50. Glazebrook J, Rogers EE, Ausubel FM. 1996. Isolation of Arabidopsis mutants with enhanced disease susceptibility by direct screening. *Genetics* 143: 973–982.
51. Goodman RN,. Novacky AJ. 1994. The Hypersensitive Reaction in Plants to Pathogens. APS Press, St. Paul, MN.
52. Grant MR, *et al.* 1995. Structure of the Arabi-dopsis RPM1 gene enabling dual specificity disease resist-ance. *Science* 269:843–46.
53. Hammond-Kosack KE, Jones JDG. 1997. Plant disease resistance genes. *Annu. Rev. Plant Phys. Plant Mol. Biol.* 48:575-607.
54. Hayes AJ, *et al.* 2004. Recombination With in a Nucleotide-Binding-Site/ Leucine-Rich-Repeat Gene Cluster Produces New Variants Conditioning Resistance to Soybean Mosaic Virus in Soybeans. *Genetics.* 166: 493-503.
55. Hejgaard J, Jacobson S, Bjorn SE, Kragh KM.1992. Antifungal activity of chitin binding PR-4 type proteins from barley grain and stressed leaf. *FEBS Letters* 307:389-92.
56. Hoeberichts FA, Woltering EJ. 2003. Multiple mediators of plant programmed cell death: Interplay of conserved cell death mechanisms and plant-specific regulators. *Bioessays* 25: 47–57.
57. Holub EB. 2001. The arms race is ancient history in Arabidopsis, the wildflower. *Nat. Rev. Genet.* 2:516-527.
58. Jackson RW, *et al.* 1999. Identification of a pathogenicity island, which contains genes for virulence and avirulence, on a large native plasmid in the bean pathogen *Pseudomonas syringae pathovar phaseolicola*. *Proc Natl Acad Sci* USA 96: 10875-10880.
59. Jia Y, McAdams SA, Bryan GT, Hershey HP, Valent B. 2000. Direct interaction of resistance gene and avirulence gene products confers rice blast resistance. *EMBO J.* 19:4004–14.
60. Jirage D, *et al.* 2001. Constitutive salicylic acid-dependent signalling in cpr1 and cpr6 mutants requires PAD4. *Plant J.* 26: 395–407.
61. Johal GS, Briggs SP. 1992. Reductase activity encoded by the HM1 disease resistance gene in maize. *Science* 258:958–87.

62. Jones DA, Thomas CM, Hammond-Kosack KE, Balint-Kurti PJ, Jones JDG. 1994. Isolation of the tomato Cf-9 gene for resistance to *Cladosporium fulvum* by transposon tagging. *Science* 266:789–93.
63. Joosten MHAJ, Cozijnsen TJ, de Wit PJGM. 1994. Host resistance to a fungal tomato pathogen lost by a single base-pair change in an avirulence gene. *Nature* 367: 384–86.
64. Kawchuk LM, *et al.* 2001. Tomato Ve disease resistance genes encode cell surface-like receptors. *Proc. Natl. Acad. Sci.* USA 98:6511–15.
65. Keen NT. 1990. Gene-for-gene complementarity in plant–pathogen interactions. *Annu. Rev. Genet.* 24:447-463.
66. Keller T, *et al.* 1998. A plant homolog of the neutrophil NADPH oxidase gp91phox subunit gene encodes a plasma membrane protein with Ca2+ binding motifs, *Plant Cell* 10:255 – 266.
67. Kim YJ, Lin NC, Martin GB. 2002. Two distinct Pseudomonas effector proteins interact with the Pto kinase and activate plant immunity. *Cell* 109:589–98.
68. Kjemtrup S, Nimchuk Z, Dangl JL. 2000. Effector proteins of phytopathogenic bacteria: bifunctional signals in virulence and host recognition. *Curr. Opin. Microbiol.* 3:73-78.
69. Koiwa H, Sato F, Yamada Y. 1994. Characterization and accumulation of tobacco PR-5 proteins by IEF-immunoblot analysis. *Plant Cell Physiol.* 35:821-27.
70. Kombrink E, Somssich IE. 1995. Defense responses of plants to pathogens. *Adv. Bot. Res.* 21: 2–33.
71. Kumar D, Klessig DF. 2000. Deferential induction of tobacco MAP kinases by the defense signals nitric oxide, salicylic acid, ethylene, and jasmonic acid. *Mol. Plant Microbe Interact.* 13: 347 –351.
72. Kunkel BN, Brooks DM. 2002. Cross talk between signaling pathways in pathogen defense. *Curr. Opin. Plant Biol.* 5: 325–331.
73. Lahaye T, Bonas U. 2001. Molecular secrets of bacterial type III effector proteins. *Trends Plant Sci.* 6:479-485.
74. Lamb C, Dixon RA. 1997. The oxidative burst in plant disease resistance. *Ann rev physiol plant mol biol.* 48:251-275.
75. Lanfermeijer FC, Dijkhuis J, Sturre MJG, Haan P, Hille J. 2003. Cloning and characterization of the durable Tomato mosaic virus resistance gene Tm-22 from *Lycopersicon esculentum*. *Plant Mol. Biol.* 52:1039-1051.
76. Lawrence GJ, Finnegan EJ, Ayliffe MA, Ellis JG. 1995. The L6 gene for flax rust resistance is related to the Arabidopsis

bacterial resistance gene RPS2 and tobacco viral resistance gene N. *Plant Cell* 7:1195–206.

77. Lawton K, Ward E, Payne G, Moyer M, Ryals J. 1992. Acidic and basic class III chitinase messenger RNA accumulation in response to TMV infection of tobacco. *Plant Mol. Biol.* 19: 735-43.
78. Leah R, Tommerup H, Svendsen I, Mundy J. 1991. Biochemical and molecular characterization of three barley seed proteins with antifungal properties. *J. Biol. Chem.* 266, 1564–1573.
79. Levy M, Edelbaum O, Sela I. 2004. Tobacco mosaic virus regulates the expression of its own resistance gene N1. *Plant Physiol.*135: 2392-2397.
80. Lorang JM, Keen NT. 1995. Characterization of avrE from *Pseudomonas syringae pv. tomato*: a hrp-linked avirulence locus consisting of at least two transcriptional units. *Mol Plant Microbe Interact.* 8:49-57.
81. Lorrain S, Vailleau F, Balagué C, Roby D. 2003. Lesion mimic mutants: keys for deciphering cell death and defense pathways in plants? *Trends Plant Sci.* 8: 263–271.
82. Luderer R, de Kock MJD, Dees RHL, de Wit PJGM, Joosten MHAJ. 2002. Functional anlaysis of cysteine residues of ECP elicitor proteins of the fungal tomato pathogen *Cladosporium fulvum*. *Mol. Plant Pathol.* 3:91–95.
83. Maleck K, *et al.* 2001. The transcriptome of *Arabidopsis thaliana* during systemic acquired resistance. *Nature Genet.* 26: 403–410.
84. Martin GB, Bogdanove AJ, Sessa G. 2003. Understanding the functions of plant disease resistance proteins. *Annu. Rev. plant Biol.* 54:23-61.
85. Martin GB. 1999. Functional analysis of plant disease resistance genes and their downstream effectors. *Curr. Opin. Plant Biol.* 1999. 2:273-279.
86. Mauch-Mani B, Slusarenko AJ. 1996. Production of salicylic acid precursors is a major function of phenylalanine ammonia-lyase in the resistance of Arabidopsis to *Peronospora parasitica. Plant Cell* 8:203–12.
87. McDowell JM, *et al.* 1998. Intragenic recombination and diversifying selection contribute to the evolution of downy mildew resistance at the RPP8 locus of Arabidopsis. *Plant Cell* 10:1861–74.
88. Metzler MC, Cutt JR, Klessig DF. 1991. Isolation and characterization of a gene encoding a PR-1 like protein from *Arabidopsis thaliana. Plant Physiol.* 96:346-48.
89. Meyers BC, Shen KA, Rohani P, Gaut BS, Michelmore RW. 1998. Receptor-like genes in the major resistance locus of lettuce

are subject to divergent selection. *Plant Cell* 10:1833–46.

90. Milligan SB, *et al.* 1998. The root knot nematode resistance gene Mi from tomato is a member of the leucine zipper, nucleotide binding, leucine-rich repeat family of plant genes. *Plant Cell* 10:1307–19.
91. Mindrinos M, Katagiri F, Yu GL, Ausubel FM. 1994. The *A. thaliana* disease resistance gene RPS2 encodes a protein containing a nucleotide-binding site and leucine-rich repeats. *Cell* 78:1089–99.
92. Nawrath C, Métraux JP. 1999. Salicylic acid induction-deficient mutants of Arabid-opsis express PR-2 and PR-5 and accumulate high levels of camalexin after pathogen inoculation. *Plant Cell* 11: 1393–1404.
93. Noritake T, Kawakita K, Doke N. 1996. Nitric oxide induces phytoalexin accumulation in potato tuber tissues. *Plant and Cell Physiol.* 37: 113-116.
94. Orbach MJ, Farrall L, Sweigard JA, Chumley FG, Valent B. 2000. A telomeric avirulence gene determines efficacy for the rice blast resistance gene Pi-ta. *Plant Cell* 12:2019–2023.
95. Orth K, *et al.* 2000. Disruption of signaling by Yersinia effector YopJ, a ubiquitin-like protein protease. *Science* 290:1594-1597.
96. Parker JE, Coleman MJ, Dean C, Jones JDG. 1997. The Arabidopsis downy mildew resistance gene RPP5 shares similarity to the Toll and Interleukin-1 receptors with N and L6. *Plant Cell* 9:879–94.
97. Petnicki-Ocwieja T, *et al.* 2002. Genome wide identification of proteins secreted by the Hrp type III protein secretion system of *Pseudomonas syringae pv. tomato* DC3000. *Proc. Natl. Acad. Sci.* USA 99:7652-7657.
98. Piedras P, Rivas S, Droge S, Hillmer S, Jones JD. 2000. Functional, c-myc-tagged Cf-9 resistance gene products are plasma-membrane localized and glycosylated. *Plant J.* 21:529–36.
99. Pieterse CM, Van Loon LC. 2004. NPR1: the spider in the web of induced resistance signaling pathways. *Curr. Opin. Plant Biol.* 7: 456–464.
100. Ren T, Qu F, Morris TJ. 2000. HRT gene function requires interaction between a NAC protein and viral capsid protein to confer resistance to turnip crinkle virus. *Plant Cell* 12:1917-1926.
101. Ritter C, Dangl JL. 1995. The avrRpm1 gene of *Pseudomonas syringae pv. maculicola* is required for virulence on Arabidopsis. *Mol Plant Microbe Interact.* 8: 444-453.
102. Rogers EE, Ausubel FM. 1997. Arabidopsis enhances disease susceptibility mutants exhibit enhanced susceptibility to

several bacterial pathogens and alterations in PR-1 gene expression. *Plant Cell* 9: 305–316.

103.Ross AF. 1961. Systemic acquired resistance induced by localized virus infections in plants. *Virology* 14:340–58.

104.Ryals JA, *et al.* 1996. Systemic acquired resistance. *Plant Cell* 8:1809–19.

105.Ryan CA, Pearce G. 2001. Polypeptide hormones. *Plant Physiol* 125: 65–68.

106.Scheel D. 2002. Oxidative burst and the role of reactive oxygen species in plant–pathogen interactions. In: Inze D, Van Montagu M (eds) Oxidative Stress in Plants. Taylor & Francis, New York, pp. 137–153.

107.Schneider M, *et al.* 1996. Systemic acquired resistance in plants. *Int. J. Cytol.* 168:303– 40.

108.Simmons CR, Litts JC, Huang N, Rodriguez RL. 1994. Structure of a rice beta-glucanase gene regulated by ethylene, cytokinin, wounding, salicylic acid and fungal elicitors. *Plant Mol. Biol.* 18:33-45.

109.Simons G, *et al.* 1998. Dissection of the Fusarium I2 gene cluster in tomato reveals six homologs and one active gene copy. *Plant Cell* 10:1055–68.

110.Smith JA, M´etraux JP. 1991. *Pseudomonas syringae pv. syringae* induces systemic resistance to *Pyricularia oryzae* in rice. *Physiol. Mol. Plant Pathol.* 39:451–61.

111.Solomon M, Belenghi B, Delledonne M, Menachem E, Levine A. 1999. The involvement of cysteine proteases and protease inhibitor genes in the regulation of programmed cell death in plants. *Plant Cell* 11: 431 – 444.

112.Song WY, *et al.* 1995. A receptor kinase-like protein encoded by the rice disease resistance gene, Xa21. *Science* 270:1804-6.

113.Stakman EC. 1915. Relation between *Puccinia graminis* and plants highly resistant to its attack. *J. Agric. Res.* 4: 193–200.

114.Stroemberg A, Brishammar S. 1993. A histological evaluation of induced resistance to *Phytophthora infestans (Mont.)* de Bary in potato leaves. *J. Phytopathol.* 137:15–25.

115.Swiderski MR, Innes RW. 2001. The Arabidopsis PBS1 resistance gene encodes a member of a novel protein kinase subfamily. *Plant J.* 26:101–12.

116.Swords KMM, *et al.* 1996. Spontaneous and induced mutations in a single open reading frame alter both virulence and avirulence in *Xanthomonas campestris pv. vesicatoria* avrBs2. *J Bacteriol.* 178: 4661-4669.

117.Tai TH, *et al.* 1999. Expression of the Bs2 pepper gene confers resistance to bacterial spot disease in tomato. *Proc. Natl. Acad. Sci.* USA 96:14153–58.

118. Takahashi H, *et al.* 2002. RCY1, an *Arabidopsis thaliana* RPP8/ HRT family resistance gene, conferring resistance to cucumber mosaic virus requires salicylic acid, ethylene and a novel signal transduction mechanism. *Plant J.* 32: 655–667.

119. The Arabidopsis Genome Initiative, Analysis of the genome of the flowering plant *Arabidopsis thaliana*. 2000. *Nature* 408: 796-815.

120. Thevissen K, *et al.* 1996. Fungal membrane respo-nses induced by plant defensins and thionins. *J. Biol. Chem.* 271: 15018-15025.

121. Thomas CM, *et al.* 1997. Characterization of the tomato Cf-4 gene for resistance to *Cladosporium fulvum* identifies sequences that determine recognitional specificity in Cf-4 and Cf-9. *Plant Cell* 9:2209–24.

122. Torres MA, Dangl JL, Jones JD. 2002. Arabidopsis gp91phox homologues AtrbohD and AtrbohF are required for accumulation of reactive oxygen intermediates in the plant defense response. *Proc Natl Acad Sci* USA 99: 517–522.

123. Ulaganathan K, Basha S, Daida P. 2003. SAR proteins and SAR protein homologues and their use in developing fungal resistance in plants. *Annu. Rev. Plant Pathol.* 2: 475-497.

124. van den Ackerveken GF, Vossen P, De Wit PJ. 1993. The Avr9 race specific elicitor of *Cladosporium fulvum* is processed by endogenous and plant proteases. *Plant Physiol.* 103:91–96.

125. van der Biezen EA, *et al.* 2002. Arabidopsis RPP4 is a member of the RPP5 multigene family of TIR-NB-LRR genes and confers downy mildew resistance through multiple signalling components. *Plant J.* 29:439–51.

126. Van der Vossen EA, *et al.* 2000. Homologues of a single resistance-gene cluster in potato confer resistance to distinct pathogens: a virus and a nematode. *Plant J.* 23:567–76.

127. Vivian A, Arnold DL. 2000. Bacterial effector genes and their role in host–pathogen interactions. *J. Plant Pathol.* 82:163-178.

128. Vos P, *et al.* 1998. The tomato Mi-1 gene confers resistance to both root-knot nematodes and potato aphids. *Nat. Biotechnol.* 16: 1365–69.

129. Wang ZX, *et al.* 1999. The Pib gene for rice blast resistance belongs to the nucleotide binding and leucine-rich repeat class of plant disease resistance genes. *Plant J.* 19:55–64.

130. Warren RF, Henk A, Mowery P, Holub E, Innes RW. 1998. A mutation within the leucine-rich repeat domain of the Arabidopsis disease resistance gene RPS5 partially suppresses multiple bacterial and downy

mildew resistance genes. *Plant Cell* 10:1439–52.

131. White FF, Yang B, Johnson LB. 2000. Prospects for understanding avirulence gene function. *Curr. Opin. Plant Biol.* 3:291-298.

132. Whitham SA, *et al.* 2000. Arabidopsis RTM2 gene is necessary for specific restric-tion of tobacco etch virus and encodes an unusual small heat shock-like protein. *Plant Cell* 12:569–82.

133. Xiao S, *et al.* 2001. Broad spectrum mildew resistance in *Arabidopsis thaliana* mediated by RPW8. *Science* 291:118-120.

134. Yang KY, Liu Y, Zhang S. 2001. Activation of a mitogen activated protein kinase pathway is involved in disease resistance in tobacco, *Proc. Natl. Acad. Sci.* USA 98: 741 – 746.

135. Yang Y, Yuan Q, Gabriel DW. 1996. Water soaking function(s) of XcmH1005 are redundantly encoded by members of the Xanthomonas avr/pth gene family. *Mol Plant Microbe Interact.* 9: 105-113.

136. Yoshimura S, *et al.* 1998. Expression of Xa1, a bacterial blight-resistance gene in rice, is induced by bacterial inoculation. *Proc. Natl. Acad. Sci.* USA 95:1663–68.

137. Yu D, Chen C, Chen Z. 2001 Evidence for an important role of WRKY DNA binding proteins in the regulation of NPR1 gene expression. *Plant Cell* 13: 1527–1540.

138. Yu IC, Parker J, Bent AF. 1998. Gene-for-gene disease resistance without the hypersensitive response in Arabidopsis dnd1 mutant. *Proc Natl. Acad. Sci. USA* 95: 7819–7824.

139. Zhang Y, Fan W, Kinkema M, Li X, Dong X. 1999. Interaction of NPR1 with basic leucine zipper protein transcription factors that bind sequences required for salicylic acid induction of the PR-1 gene. *Proc. Natl. Acad. Sci.* USA 96: 6523–6528.

140. Zhang Y, Tessaro MJ, Lassner M, Li X. 2003. Knockout analysis of Arabidopsis transcription factors TGA2, TGA5, and TGA6 reveals their redundant and essential roles in systemic acquired resistance. *Plant Cell.* 15: 2647–2653.

141. Zhou JM, *et al.* 2000. NPR1 differentially interacts with members of the TGA/OBF family of transcription factors that bind an element of the PR-1 gene required for induction by salicylic acid. *Mol. Plant-Microbe Interact.* 13:191–202.

142. Zhu W, *et al.* 1999. The C terminus of AvrXa10 can be replaced by the transcri-ptional activation domain of VP16 from the herpes simplex virus. *Plant Cell* 11:1665-1674.

*Rev. Plant Pathol. Vol. 4, 2008*
*Indian Society of Mycology and Plant Pathology*
*Scientific Publishers (India), Jodhpur pp. 367-391*

# PLANT QUARANTINE IN INDIA IN THE WAKE OF INTERNATIONAL AGREEMENTS: A REVIEW

***R.K. Khetarpal and Kavita Gupta***

Division of Plant Quarantine, National Bureau of Plant Genetic Resources, New Delhi 110 012, India

KEY WORDS: Plant quarantine, regulations, PQ Order 2003, SPS agreement

## ABSTRACT

Exchange of plants and planting material can be a potential source of introducing exotic pests. Quarantine measures thus act as filters against entry of such species by restricting their introduction and in case introduced, preventing their establishment and further spread. The International Plant Protection Convention (IPPC) of FAO encourages cooperation among various countries and ensures that each country establish a National Plant Protection Organization to discharge such functions. In India, the Directorate of Plant Protection, Quarantine and Storage (DPPQS) under the Ministry of Agriculture implements the plant quarantine regulations for bulk consignments and National Bureau of Plant Genetic Resources under Indian Council of Agricultural Research (ICAR) is the nodal agency for safe movement of germplasm including transgenics. They work under the Plant Quarantine (Regulation of Import into India) Order 2003 which came into force from January 1, 2004, under the Destructive Insects and Pests Act of 1914. Although the regulations are now in place, there are a number of issues related to quarantine as it has been drafted more for facilitating bulk imports than for exchange of germplasm. Under this order a pest risk analysis (PRA) has been made mandatory and the various schedules of the Order give lists of crops for which a generic PRA is given. A number of cultivated crops (and their germplasm including wild relatives/ land races) do not find mention in any of the schedules. Hence, a detailed PRA becomes obligatory for them prior to import. The size of the consignment/sample, technique to be used for detection of minute amounts of pest, availability of diagnostic reagents and reference collection for exotic pests, and post-entry quarantine testing are important technical issues in quarantine. Over the years, during quarantine processing, the pests intercepted include many that are not known to occur in India, have different races/ biotypes/ strains not known to occur in India; are present on a new host, are from a country from where they were never reported before or are an entirely new pest species hitherto unreported in science or are reported to be present widely in India. These interceptions, especially of pests not yet reported from India signify the importance of quarantine in preventing the introduction of destructive exotic pests. To make quarantine more effective, research needs to be intensified on the development of sensitive and non-destructive detection and salvaging techniques, finding alternatives to methyl bromide, developing molecular techniques for detection of races/ biotypes/ strains and also low levels of infections. Such preparedness is essential under the WTO regime for a country like India which needs to enhance its efforts in developing national strategy to identify areas of research, streamline the technical and policy issues and network for safe exchange and quarantine.

## BACKGROUND

The characteristics and distribution of plant species and their pests in a given area are in a state of biological equilibrium under natural conditions. In an entirely natural environment, new species do not appear or disappear suddenly except by human intervention nor is there any major fluctuation in the population density of native species other than temporary seasonal variations. But after agriculture started being practiced, the change in the native vegetation accelerated and new plants were introduced into new areas. The pests of the plants also moved along with the plants into foreign or new territories. Modern agriculture led to the realization of the importance of diseases and pests. At the same time the disastrous affects of certain introduced pests in various parts of the world highlighted the need for quarantine, which are well documented (50).

The plant diseases or damage to plants are caused by specific pests which get carried into new areas and find suitable environmental conditions as well as susceptible hosts to survive and complete their life cycles. These pests vary considerably in vitality, some being short lived and resistant to desiccation and extremes of temperature, while others are susceptible to these factors and are delicate and short lived. The scope of quarantine is influenced by biological characteristics of the pest organism; the kinds and part of plants affected; whether or not the part is used in commerce (such as root, tuber, cutting, seed or flower); alternate hosts; the nature of the inoculum in case of a plant disease and whether the host or other potential carrier can be treated to eliminate the possibility of its carrying any viable reproductive form of the pest organism.

**EXAMPLES OF PESTS INTRODUCED INTO NEW AREAS**

- The introduction of seed potatoes infected with the late blight pathogen *(Phytophthora infestans)* into Ireland from Peru caused the well- known Irish famine of 1845. This forced millions of people to migrate *en masse* from Europe, as a result of almost total failure of potato crop due to attack of this dreaded pathogen.
- Powdery mildew *(Uncinula necator),* root eating aphid (*Viteus vitifoliae)* and downy mildew (*Plasmopara viticola*) of grape were introduced in succession into France in mid nineteenth century from North America. These pests caused extensive damage to the grapevines and virtually annihilated the wine industry of the country.
- Coffee rust (*Hemileia vastatrix)* appeared in Sri Lanka in 1875, was so severe that the production went down from

400 million pounds in 1870 to just 5 million pounds in 1889. The disease entered India in 1876 from Sri Lanka and within a decade, the coffee industry of South India was badly affected.

- Germany put embargo in 1875 on the import of potatoes from USA to keep out the dreaded Colorado potato beetle (*Leptinotarsa decemlineata*) causing severe losses in USA.
- In 1898, most European countries put embargo on American import directed mainly to keep out the San Jose´ scale (*Quadraspidiotus perniciosus),* a highly destructive pest of pome fruits.
- The cotton pink boll worm (*Pectinophora gossipiella*) was first recorded in India in 1842 and has spread from India to all major cotton growing areas of the world, inflicting important economic losses.
- The movement of the devastating stored grain pest Khapra beetle (*Trogoderma granarium*) is a classical example of movement of pest in international trade. Its introduction was recorded in USA, Italy and Zimbabwe in 1940s and the pest is still spreading continuously in international trade.

India, too, has had several introductions of exotic pests causing severe economic and environmental losses.

- A noxious weed *Lantana camara* was introduced in 1809 from Central America, which spread to wastelands and pastures causing displacement of native species.
- The fluted scale *(Icerya purchasi*), a serious pest of citrus and native of Australia was introduced into India before 1928 from Sri Lanka probably on wattles (*Acacia* sp.) to later become a serious pest on citrus in south India.
- The San Jose scale (*Quadraspidiotus perniciosus*), a pest of apple was introduced into India in 1930s and causes enormous losses in apple orchards in Himachal Pradesh.
- Bunchy top of banana entered India from Sri Lanka and causes loss to banana of over Rs 4 Crores annually. Bunchy top was reported for the first time in 1943 in Kottayam area of Travancore (54).

- The dreaded Golden nematode (*Heterodera rostochiensis*) introduced from UK along with exotic seed material in 1960s has been causing severe infestation of potato tubers in the Nilgiris (23). Potato wart caused by *Synchytrium endobioticum* was first reported from the Darjeeling area of West Bengal by Ganguly and Paul (14). Both these diseases were introduced in India from Europe.
- Heavy losses in grain yield of *Cicer arietinum* crop in states of Haryana, Madhya Pradesh, Punjab and adjoining areas occurred during 1981-82 due to the introduction of virulent biotype of Aschochyta blight from the Middle East.

All these examples clearly demonstrate that imported seed/ planting material, especially in bulk, without proper quarantine may result in introduction and establishment of quarantine pests into new areas, which may severely damage the crop production and economy of a nation.

## IMPORTANCE OF PLANT QUARANTINE

Plant quarantine is a government endeavour enforced through legislative measures to regulate the introduction of planting materials, plant products, soil, living organisms, *etc.* in order to prevent inadvertent introduction of pests (including fungi, bacteria, viruses, nematodes, insects and weeds) harmful to the agriculture of a country/state/region, and if introduced, prevent their establishment and further spread (24). Plant quarantine measures act as filters against the entry of exotic species and check or delay the introduction of unwanted organisms.

The first plant quarantine law was promulgated in Rouen, France in 1860 to suppress and prevent the spread of common barberry, which is the alternate host for wheat rust. Other countries to establish the quarantine laws after France were Germany, USA, UK and Australia. In 1881, a California Pest Act was passed to prevent the spread of grape Phylloxera. Subsequently, a number of countries became aware of the necessity and importance of plant quarantine for the safety and health of their agriculture. The first effort towards an International Agreement on Plant Protection was made in 1914 under the auspices of the

International Institute of Agriculture, Rome. This was followed by an International Convention of Plant Protection in 1919 under which Agreements regarding the issue and acceptance of phytosanitary certificates were finalized. This got a set back due to Second World War and was later appreciated by Food and Agriculture Organization (FAO) and in 1951, an International Plant Protection Convention (IPPC) was convened by the FAO (19). India became a party to this agreement in the year 1956 along with Australia, Sri Lanka, UK, Netherlands, Indonesia, Portugal and Vietnam. At present, there are more than 160 signatory members of the IPPC.

The IPPC is aimed at developing international cooperation among various countries in preventing the introduction and spread of pests across national boundaries to secure common and effective measures to this end. This convention is applicable mainly to the regulated pests involved with international movement of plants and planting material. The IPPC through its International Standard on Phytosanitary Measures on the "*Principles of plant quarantine as related to international trade*" was formulated with an aim to facilitate the process of developing international standards for plant quarantine. The general principles indicate the process of development of phytosanitary measures as applicable to international commerce, while the specific principles are related to procedures in plant quarantine.

The IPPC requires that each country establish a national plant protection organization to discharge the functions specified by it. In India, the Directorate of Plant Protection, Quarantine and Storage (DPPQS) under the Ministry of Agriculture is the body for implementation of plant quarantine regulations as per the IPPC.

## NATIONAL PLANT QUARANTINE SET-UP

### *Legislation*

The quarantine law was enacted for the first time in India in 1914 as Destructive Insects and Pests (DIP) Act (10). A gazette notification entitled Rules for Regulating the Import of Plants *etc.* into India was published in 1936 to enforce the plant quarantine regulations. Over the years, the DIP Act was revised and amended several times, more so recently, to meet the emerging scenario of liberalized trade under the world trade organization (WTO).

In 1984, a notification was issued under DIP Act namely Plants, Fruits and Seeds (Regulation of Import into India) Order then popularly known as the PFS Order which was revised in 1989 after the announcement of the New Policy on Seed Development by the Government of India in 1988, proposing major modifications for smooth quarantine functioning. The new policy covered the import of seeds/ planting materials of wheat, paddy, coarse cereals, oil seeds, pulses, vegetables, flowers, ornamentals and fruit crops. So under the revised PFS Order, 1989, care was being taken that there was absolutely no compromise on plant quarantine requirements while liberalizing import.

The establishment of WTO Agreements in 1995 although offering greater economic opportunities, has resulted in fresh challenges of increased pest risk due to liberalized international trade in agricultural commodities. With varied agricultural produce and surplus food production, India is now in a position to expand its agricultural trade, but requires National Standards to be developed for all critical phytosanitary activities. There has been a pressing need for compliance with international phytosanitary regulations. Hence, to fill-up the gaps in existing PFS Order *viz.*, regulating import of germplasm/Genetically modified organisms (GMOs)/ transgenic plant material; live insects/ fungi including bio-control agents *etc.* and to fulfill India's legal obligations under the international Agreement, the **Plant Quarantine (Regulation of Import into India) Order 2003** (48) came into force from January 1, 2004, henceforth referred to as PQ Order 2003. Under this Order the need for incorporation of Additional/ Special declarations for freedom of import commodities from quarantine and alien pests, on the basis of standardized PRA, particularly for seed/ planting materials is also taken care of.

Under the PQ Order, apart from PRA being a precondition for all imports, the scope of plant quarantine activities has been widened to incorporate additional definitions. The other salient features of the Order are:

- Inclusion of agricultural imports classified as:
  a. Prohibited plant species (Schedule IV)

b. Restricted species where import is permitted only by authorized institutions (Schedule V)

c. Restricted species whose import is permitted only with additional declarations of freedoms from quarantine/ regulated pests and subject to specified treatment certifications (Schedule VI)

d. Plant material whose import for consumption/industrial processing is permitted with a normal Phytosanitary Certificate (Schedule VII)

- Prohibition of import of commodities with weed/ alien species contamination as per Schedule VIII
- Restriction on import of packaging material of plant origin unless treated.
- Provisions include regulating the import of:

a. Soil, peat and sphagnum moss

b. Germplasm/ GMOs/ transgenic material for research

c. Live insects/microbial cultures and biological control agents

d. Import of timber and wooden logs

- Permit requirement is enforced on imports of seeds including flower seeds, propagating material and mushroom spawn cultures
- Additional declarations are specified in the Order for import of more than 300 agricultural commodities, specifically listing as many as 600 quarantine pests and 61 weed species (now 31 in subsequent amendments of the PQ Order).
- Notified points of entry are increased to 130 from the previous 59.
- Certification fee and inspection charges are rationalized.

So far, i.e. till October 2006, ten amendments of the Plant Quarantine Order 2003 have been notified to the WTO and two draft amendments have been sent to experts for comments. These amendments have revised lists of quarantine pests; incorporated ISPM No. 15 compliance; recognised efficacy of irradiation treatment; pest free areas and cold treatments for fruit flies to allow import of fresh fruits; and revised the lists of crops under Schedules VI and VII to include 411 and 284 crops and commodities, respectively.

**Infrastructure**

For the proper implementation of the regulations described above it is essential to have well developed infrastructure in place. The Directorate of Plant Protection, Quarantine and Storage (DPPQS) of Ministry of Agriculture is the apex body for implementation of plant quarantine regulations. It has a national network of 29 plant quarantine stations at different airports (10), seaports (10) and land frontiers (9). In all, two categories of materials being imported under the PQ Order, 2003: (a) bulk consignments for consumption and sowing/planting, and (b) samples of germplasm in small quantities for research purposes. The Plant Quarantine Stations under the DPPQS undertake quarantine processing and clearance of consignments of the first category (21). The five major Plant Quarantine Stations have been modernized recently under an FAO-UNDP funded projcct and undertake the quarantine processing of huge amount of samples received annually in close collaboration with the plant quarantine stations located at various places in the country.

NBPGR on the other hand undertakes the quarantine processing of all plant germplasm and transgenic planting material under exchange for which it has well-equipped laboratories and green house complex (44). Recently, a containment facility has also been established for processing transgenics (34). NBPGR also has a well equipped quarantine station at Hyderabad which mainly deals with the export samples of International Crop Research Institute for Semi-arid Tropics under a MoU (7).

Of the two categories of material being imported, bulk imports for sowing/ planting carry maximum risk as thorough examination and treatment becomes difficult and planting area required is also too large. Quarantine processing is often restricted to smaller samples derived from them and based on results of these samples the whole consignment is rejected/detained or released after giving fumigation treatment or fungicidal/ insecticidal treatments. The bulk consignments meant for consumption pose lesser hazards. However, certain small samples meant for research purposes are of immense quarantine importance. These samples usually comprise of germplasm material or wild relatives or landraces of a crop, and are thus more likely to carry diverse biotypes/races/ strains of the pest. Besides, in case of true seed, generally, risks are more due to deep-seated infections than with surface-borne contamination. However, import of vegetative propagules present a much higher order of risk than true seeds.

Almost all countries regulate the importation of plants and planting material because of the pest risk posed by such imports. The introduction of pests into a new locality is brought about in various ways: (a) the host may be the carrier; (b) inert materials such as packing material may carry resting stages of the organism; (c) insect vectors and birds may transport it; (d) air currents may carry the pest to long distances. The first two modes of distribution lend themselves to curtailment by quarantine measures. The last two, are by and large beyond human control and are a major limitation in the control of pest by exclusion.

However, the international developments in the form of trade related activities and the thrust of the WTO Agreements imply that countries need to update their quarantine or plant health services to facilitate pest free import/ export.

## WORLD TRADE ORGANIZATION

The establishment of the WTO in 1995 has provided unlimited opportunities for international trade of agricultural products. Under this legal standards have been formulated in the form of Sanitary and Phytosanitary (SPS) Measures for regulating the international trade. The WTO Agreement on the Application of SPS measures concerns the application of food safety and animal and plant health regulations. It recognizes government's rights to take SPS measures but stipulates that they must be based on science, should be applied only to the extent necessary to protect human, animal or plant life or health and should not arbitrarily or unjustifiably discriminate between members where identical or similar conditions prevail (22, 31).

The SPS Agreement aims to overcome health-related impediments of plants and animals to market access by encouraging the "establishment, recognition and application of common SPS measures by different members". The primary incentive for the use of common international norms is that these provide the necessary health protection based on scientific evidence and improve trade flow at the same time.

SPS measures are defined as any measure applied within the territory of the Member State:

- to protect animal or plant life or health from risks of entry, establishment or spread of pests, diseases, disease-carrying/ causing organisms;
- to protect human or animal life or health from risks arising from additives, contaminants, toxins or disease causing organisms in food, beverages or foodstuffs;
- to protect human life or health from the risk of diseases carried by animals, plants or their products, or from the entry, establishment /spread of pests; or
- to prevent or limit other damage from the entry, establishment or spread of pests.

The SPS Agreement explicitly refers to three standard-setting international organizations commonly called as the 'three sisters' whose activities are considered to be particularly relevant to its objectives: the FAO/ World Health Organization (WHO) Codex Alimentarius Commission, the Office International des Epizooties, and the international and regional organizations operating within the framework of the IPPC which develops the International Standards for Phytosanitary Measures (ISPMs) which provide guidelines on pest prevention, detection and eradication. To date, twenty four standards have been developed and several others are at different stages of development.

Prior to the establishment of WTO, governments on a voluntary basis could adopt international standards, guidelines, recommendations and other advisory texts. Although these norms remain voluntary, a new status has in effect been conferred upon them by the SPS Agreement. A WTO member adopting such norms is presumed to be in full compliance with the SPS Agreement.

## PHYTOSANITARY IMPLICATIONS OF SPS AGREEMENT ON INDIA

Compliance to the SPS Agreement requires that countries also incorporate the element of pest risk analysis (PRA) for making regulatory decisions during import and export. The IPPC has published international standards on general requirements of pest risk analysis and the procedure for undertaking it as ISPM-2, ISPM-11 and ISPM-21 (20). A PRA consists of three stages *viz.*, initiation, risk assessment and risk management, whose compilation should be fully documented in the event a review or dispute arises. There are two ways of initiating a PRA – pathway-based and pest-based. The entire procedure facilitates the identification and characterization of risk associated with the pathway. The databases such as the Crop Protection Compendium (6) on pests, their biology and ecology, experience and expert judgment are useful in interpreting the implications. Besides this, PRA also requires information on the management options including certification standards, and other control measures, which are considered in totality during trade negotiations. The seed-transmitted Karnal bunt of wheat is the example where a pest- based PRA was developed where as in case of soybean, a commodity-based PRA was developed by India for taking trade decisions with the US counterpart (47). Moreover, this procedure needs to be completely transparent so that the SPS measures taken or proposed to be taken do not become disguised barriers to trade (9). PRA is a procedure by which quarantine services can technically justify new regulations, phytosanitary guidelines or safeguards, operational procedures, entry status of imported articles and resource allocations (16, 17).

Identification of pest free areas (PFAs) is another area, which requires immediate attention and is a challenge to be faced for boosting our exports. To get a realistic picture of the country, extensive surveys and surveillance are a must for authentic mapping of endemic pests present in localized pockets to identify PFAs. Most of the developed countries have already equipped themselves with database on endemic and alien species/ pests and climatic models for analyzing spread and establishment of new pests in their territories. In this regard, a national database on endemic pests is being developed and updated at Regional Plant

Quarantine Station, Chennai. Further, a database on potential quarantine pests for India for various crops is being developed by NBPGR, New Delhi and the one on cereals has recently been published (11).

The *Article 3* of the SPS Agreement encourages members to establish, recognize and apply common SPS measures as prescribed by the ISPM or can have a higher standard provided they are scientifically justifiable. For effective compliance to this article of the SPS Agreement, India is faced with a daunting task of harmonizing/ developing the national standards on phytosanitary measures. The DPPQS has developed certain standards such as on PRA, certification of facilities for treatment of wood packaging material, quarantine procedures for methyl bromide fumigation, etc. Besides these other standards on post-entry quarantine, establishment of pest–free areas, etc. are in the pipeline. The development of national standards for PRA is a vital breakthrough in terms of our compliance but the administrative process needs to be put in place for their effective implementation to facilitate trade (18).

Besides technical challenges in terms of PRA and PFA, the regulatory mechanism having a direct bearing on quarantine and trade also needs to be revised and updated from time to time.

## ISSUES FOR PLANT QUARANTINE IN INDIA

There are a number of issues related to quarantine in exchange of plant/ planting material. The legislative issues are applicable to both material being exchanged in bulk as well as small samples meant for research, where as those of quarantine processing and methodology are specific.

### *Legislative Issues*

The national quarantine legislation needs to be in harmony with the international norms laid down by the IPPC and most of the countries are in the process of trying to gear up their activities to comply with it. India has attempted to do so by bringing out the new PQ Order 2003. Under this the imports have been classified as

1. Prohibited plant species (Schedule IV);

2. Restricted species where import is permitted only by authorized institutions (Schedule V);
3. Restricted species permitted only with additional declarations of freedom from quarantine/ regulated pests and subject to specified treatment certifications (Schedule VI) and;
4. Plant material imported for consumption/ industrial processing permitted with normal Phytosanitary Certificate (Schedule VII).

Under the PQ order, a PRA has been made mandatory for all material being imported into the country other than those present in Schedules V, VI and VII. However, the legislation has been drafted more for facilitating bulk imports than for exchange of germplasm.

*Issues specific to germplasm exchange*

- The various schedules V, VI and VII of the Order give lists of crops for which a generic PRA is given and detailed PRA is not required. In case of germplasm, a large number of species of cultivated crops (and their wild relatives/ land races) with useful traits are imported but do not find mention in any of the schedules. Hence, a detailed PRA becomes obligatory for them which, in fact, is elaborate and modalities of its preparation are still not clear. Also, the presence of wild relatives, land races of germplasm whose pest profile is not recorded is likely to hamper PRA preparation. There is a need to waive off the requirement of having a detailed PRA in case of germplasm of species not mentioned in any of the aforesaid schedules. This is more relevant in the present context when access to germplasm is becoming more and more difficult under the Convention on Biological Diversity, 1992. Also, many countries from where a pest is not reported find it difficult to certify in the phytosanitary certificate the freedom from those pests a requirement under Schedule VI. To resolve this, provision needs to be there in the PQ Order to waive off the requirements for certifying under Additional Declaration for freedom from pests not reported from that country. This would greatly help the indentors in procuring germplasm of their interest from varied sources.

- A section of the scientific community does not fully appreciate the concept of risks associated with imported germplasm who tend to circumvent the quarantine procedures or policies developed as safeguards. Circumvention of safeguards arises out of the misconception that plant quarantine is more an impediment than a safety mechanism. Unfortunately, at times, plant quarantine officials are too conservative and/ or plant breeders are too liberal towards the risks associated with the entry of germplasm. This point needs to be properly appreciated and it should be ensured that imported material is made available without compromising with quarantine procedures.

*Issues specific to bulk material in trade*

- In India, the quarantine officials at ports of entry do not have the authority to search and seize the personal baggage of the incoming travelers, and the powers are vested only with customs officials who are not professionally trained to understand the importance of pests and are primarily interested in intercepting gold, narcotics, weapons and other items of their interest.
- Despite the fact that the five major Plant Quarantine Stations have been modernized recently, they remain very poorly staffed. There is also a lack of trained personnel to deal effectively with the quarantine processing of huge amount of samples received annually. For example, there are very few trained virologists and not a single trained nematologist in any of the quarantine stations in India. Besides, the other stations especially the ones at the seaports dealing with timber log imports need a complete revamping in terms of space, expertise and facilities to ensure pest-free timber imports.
- Under the DIP Act, 1914 and its subsequent amendments, there is a provision of domestic quarantine to restrict the inter-state movement of nine invasive pests *viz.*, fluted scale, San José scale, coffee berry borer,

codling moth, banana bunchy top and mosaic viruses, potato cyst nematode, potato wart and apple scab. However, these pests are spreading to neighbouring states as there are no inter-state barriers or checks on the movement of seeds and planting materials of their host species. Besides, the spread of recent introductions like sunflower downy mildew (*Plasmopara halstedii*) in 1984, American serpentine leaf miner (*Liriomyza trifolii*) in 1991, spiraling white fly (*Aleurodicus dispersus*) in 1993 and biotype 'B' of white fly *Bemisia tabaci* in 1999, etc. needs to be reviewed, monitored and regulatory measures have to be taken up to check their further spread within the country.

## Technical Issues

The quarantine laboratories adopt various general and specific detection techniques for different pests based on the nature and the volume of the material imported. The imported material generally includes seeds (true seed), grains, vegetative propagules, seedlings, root stocks, stem cuttings and *in-vitro* samples. The quarantine procedures integrate the various methodologies of pathology, entomology and nematology for detection of pests and so also for salvaging the infected/infested samples (44,7).

Over the years a large number of pests have been intercepted in imported bulk consignments (21) and in germplasm and other research material (28,1,4) during quarantine processing. The intercepted pests can be divided into different categories (Table 1). The significant interceptions which are yet not reported from India include insects like *Acanthoscelides obtectus* in *Cajanus cajan*, *Anthonomus grandis* in *Gossypium* spp., *Ephestia elutella* in Macadamia nuts and *Vigna* spp., *Quadrastichodella eucalytii* in Eucalyptus, nematodes like *Heterodera schachtii, Ditylenchus dipsaci, D. destructor, Rhadinaphelenchus cocophilus, etc.* in soil clods and plant debris, fungi like *Claviceps purpurea* in seeds of wheat, barley, *etc., Peronospora manshurica* on soybean, *Fusarium nivale* on wheat, barley and *Aegilops, Uromyces betae* on sugarbeet, bacteria like *Xanthomonas campestris* pv. *campestris* on *Brassica* spp. and viruses like *Barley stripe mosaic virus* in wheat, *Pea seed borne mosaic virus* in pea and fababean, *Bean yellow mosaic virus* in french bean, *Cowpea mottle virus* in cowpea, etc.

(28). Such interceptions signify the success of quarantine as other wise these pests could have entered the country and played havoc with the plant biodiversity and agriculture if not detected in quarantine.

**Table 1.** Examples of different categories of pests intercepted in quarantine

| SNo | Category of pest intercepted | Pest/ Host/ Source country (Reference) |
|---|---|---|
| | Not known to occur in India | *Peronospora manschurica*/ Soybean/ USA (46) |
| | | *Uromyces betae*/ Sugarbeet/ USA and Italy (3) |
| | | *Fusarium nivale* / Wheat/ UK (12) |
| | | *Cowpea mottle virus*/ Cowpea/ Philippines (37) |
| | | *Tomato black ring virus* /French bean/ CIAT, Colombia (8) |
| | | *Heterodera schachtii* / Sugarbeet/ Denmark (39) |
| | | *Anthonomus grandis* / Cotton/ USA (43) |
| | | *Quadrastichodella eucalyptii*/ Eucalyptus/ Australia (42) |
| | Known to occur but the race/ biotype/strain intercepted is not known to occur | *Helminthosporium maydis*/ *race T* / Sorghum/ USA (49) |
| | | *Pea seed-borne mosaic virus* / Broadbean (37) |
| | | *Burkholderia solanacearum biovar 3/ Groundnut/* Australia (5) |
| | Intercepted on a host on which it was never reported before | *Alternaria zinniae*/ Tobacoo/ Japan (30) |
| | | *Pseudomonas syringae pv syringae* / *Hibiscus cannabinus*/ Bangladesh (52) |
| | | *Aphelenchoides besseyi* / *Stylosanthes hamata* / Australia (15) |
| | | *Merobruchus columbinus* / *Samanea saman* / UK (55) |
| | | *Bruchus ervi* / *Acacia brachustachva*/ Australia (56) |
| | | *Pachymerus lacerdae* / *Orbygnya phalerata*/ Italy (26) |
| | Intercepted from a country from where it was never reported before | *Peronospora manschurica*/ soybean/ Malaysia (2) |
| | | *Heterodera zeae* / *Vetiveria zizanioides*/ Tanzania (41) |
| | | *Bruchus ervi* / *Acacia brachustachva*/ Australia (56) |

| | |
|---|---|
| A new species hitherto unreported | *Drechslera pluriseptata*/ *Eleusine coracana*/ Zambia (29) |
| | *Tylenchorhynchus neoclavicaudatus* on potato tubers from USA (45) |
| | *Polenchus minutus*/ Palm plants/ UK (40) |
| Known to occur in India but possess a wide host range | *Collectotrichum graminicola, Drechslera turcica* and *Gloeocercospora sorghi* / Sorghum/Nigeria (35) |
| | *Drechslera siccans*/ Soybean/ USA (33) |
| | *Claviceps purpurea*/ *Avena sativa* / USA (4) |

The technical issues include issues pertaining to quarantine processing *i.e.* methodology for detection and salvaging of pests and are enumerated as follows:

- In case of bulk consignments, a proper sampling of the lot is to be done as per the norms for ensuring an effective processing of the material. However, due to stacking of containers of bulk material or inaccessibility, proper sampling procedures are not followed leading to oversight during quarantine processing. The opposite is the case in germplasm where the amount/ size of the sample is very crucial from processing point of view, as sampling procedures meant for bulk material cannot be adopted here. Besides, the technique used should enable detection of miniscule amounts of pest in the samples drawn and be non-destructive at the same time. The sample size may not be enough for direct testing of seed through the various techniques being applied to detect different types of pests. Hence, in case of small but precious samples of germplasm extreme precaution is needed to ensure that the result obtained in the tested part is not a false positive or a false negative (27).
- The seed samples of leguminous crops known to carry seed-transmitted viruses (38) are grown in post-entry quarantine glasshouses and the seedlings showing symptoms are tested for viruses by biological and serological tests, and only the harvest from the virus-free plants are released to the indentors (37, 8). Also, chemically treated material of international trials, which cannot be easily subjected to laboratory testing, are also subjected to post-entry quarantine growing in isolation (13). Thus, post-entry quarantine testing

and release from indexed disease free plants may take one crop season delaying the use of material. Singh and Khetarpal (53) describe the various methods of salvaging germplasm infected with plant pathogens.

- For detection, often the diagnostic reagents such as antisera for viruses/ bacteria are not easily available for exotic pests for which a repository of antisera needs to be established, as one has to deal with exotic pests in quarantine. For specialized detection and identification of viruses of exotic origin, it becomes mandatory to have the corresponding antisera for carrying out serological assays or the sequence of the exotic virus or a part of its genome in order to synthesize the corresponding primers for their detection by PCR. This requires an antisera bank of exotic viruses, a database on sequences of virus specific primers, and also a repository of seeds of indicator hosts. Expertise is also required to identify new pathogens or strains unknown to a region.
- The quarantine for *in-vitro* material under exchange needs special considerations (36). As per the recent amendments made under the PQ Order, the Advanced Centre of Plant Virology of IARI, New Delhi, Indian Institute of Horticultural Research, Bangalore and Institute of Himalayan Bioresource Technology, Palampur have been identified for ensuring virus-free status in the imported *in-vitro* material.
- The identity of the species, when intercepted in quarantine, needs to be established speedily to determine the quarantine risk posed by the pest and the suitable disinfestations procedure to be applied. Identification of quarantine insect pests is difficult because few institutions have an insect reference collection especially for exotic insect pests and identification keys are not readily available. Moreover, even if information on the species is available, it is scattered in various journals and quick retrieval is a problem. Therefore, attempts need to be made to consolidate the taxonomic information with the preparation of digitized keys for quick and reliable identification.
- Research prioritization in quarantine would involve
    - Generation of comprehensive epidemiological data on important pest/ diseases to determine their *tolerance*

*limits*. This would also help in developing the Pest Risk Analysis. The National preparedness in this regard has been outlined (51, 32, 17).

- Development of sensitive and non-destructive detection and salvaging techniques to detect low levels of infections as the quarantine samples need to be subjected to various techniques for detection of the variety of pests. This is more challenging in case of small samples of germplasm as often sample size is very small.
- Development of suitable alternatives to methyl bromide, a widely used quarantine fumigant being phased out due to its adverse environmental impacts (25). This is now designated as ozone-depleting substance and a potential health hazard to various organisms in the Montreal Protocol (1987) and is to be eventually phased out. India ratified Montreal Protocol and is legally committed to phase out the use of MB except for pre-shipment and quarantine purposes by 2015.
- Development of molecular techniques for detection of races/ biotypes/ strains also needs to be intensified as they are also considered as pests under the IPPC definition of pest, which should be sensitive enough to detect even low levels/ concentrations of pests.
- Studies on factors affecting likelihood of survival of pest under different conditions of transport, mode of dispersal, distribution of hosts/ alternate hosts at destination, potential for establishment, reproductive strategy and method of pest survival, potential vectors and natural enemies of the pest in the area etc need to be urgently undertaken to authentically prepare a PRA during exchange.

  - At present the bulk import of seeds/ grains of transgenic crops are not allowed. It is expected that they may be allowed on case-by-case basis in near future after biosafety issues are addressed satisfactorily. In that case, the DPPQS needs to develop infrastructure and expertise to handle the transgenics in quarantine (34).

## FUTURE THRUSTS

The liberalized trade under WTO has put a big responsibility on quarantine personnel for properly achieving their objectives of excluding the exotic pests or to carry out eradicative measures. The risk of pest introduction and means to stop the establishment of these pests into new areas continues to be a major concern. Unless proper phytosanitary measures are taken, pests could get transported all over the globe, become established in new areas and devastate agricultural production. On the other hand there are tremendous opportunities to the growers involved in export of agricultural commodities to boost the national economy if they meet the international quality standards and overcome phytosanitary constraints. Active participation by India in IPPC and Codex Alimentarius meetings to negotiate for reasonable standards would also give the necessary push to trade. All this is possible if the government agencies involved in SPS issues work in complete coordination with each other, to achieve international standards, which would ultimately boost exports of agricultural commodities while safeguarding our agriculture and environment during imports. This would indeed require a revamping of infrastructure and expertise in terms of quarantine for bulk material in trade.

### *Literature Cited*

1. Agarwal PC, *et al.* 2004a. Pathogenic fungi intercepted in introduced germplasm during the year 2001. *Indian J. Agril. Sci.* 74 (8): 451-454.
2. Agarwal PC, Khetarpal RK. 1985. Interception of *Peronospora manshurica* in imported soybean seeds. *FAO Pl. Prot. Bull.* 33 (1): 39.
3. Agarwal PC, Lambat AK, Majumdar A, Kaur P. 1982. Detection of seed-borne rust spores on safflower seeds by dry examination. *Seed Res.* 10: 180-182.
4. Agarwal PC, *et al.* 2004b. Pathogenic fungi intercepted in introduced germplasm during the year 2002. *Indian J. Microbiol.* 44 (1): 37-42.
5. Anitha K, *et al.* 2004. Interceptions of bacterial wilt of groundnut from introduced germplasm- a case study. *Indian J. Pl. Prot.* 32: 93-97.

6. CAB International. 2005. Crop Protection Compendium. CAB International, Wallingford: UK.
7. Chakrabarty SK, *et al.* 2005. *Germplasm Exchange and Plant Quarantine of ICRISAT Mandate Crops*, ICRISAT, Patencheru, Hyderabad. 72 p.
8. Chalam VC, *et al.* 2004. Interception of plant viruses in exotic germplasm imported during 2002-2004. In *Souvenir & Abstracts of Symposium on Crop Surveillance: Disease Forecasting and Management*, February 19-21, 2004, Indian Agricultural Research Instt., New Delhi, 48 pp.
9. Desai S. 2004. Research and policy issues in mycotoxin contamination management in agricultural commodities in India. *Ann. Rev. Pl. Pathol.* 3:441-470.
10. Destructive Insects and Pests (DIP) Act, 1914 and the Plants, Fruits and Seeds (Regulation of Import into India) Order, 1989 of Government of India. 106p.
11. Dev U, *et al.* 2005. *Potential Quarantine Pests for India: Cereals*. National Bureau of Plant Genetic Resources, New Delhi 124p.
12. Dev U, Majumdar A, Agarwal PC, Ram Nath. 1989. Interception of *Fusarium nivale* in exotic wheat (*Triticum* species) germplasm. *Indian J. Agril. Sci.* 59: 602-603.
13. Dev U, *et al.* 2003. Quarantine processing of chemically treated wheat (*Triticum* species), barley (*Hordeum vulgare*) and Triticale germplasm imported during 1992-2001. *Indian J. Agril. Sci.* 73: 462-465.
14. Ganguli A, Paul DK. 1973. Wart disease of potato in India. *Sci. & Cult.* 19: 605-606.
15. Gokte N, Mathur VK, Lal A, Rajan. 1987. *Fraximus americanum* a new host record for *Aphelenchoides besseyi*. *Indian J. of Nematol.* 19: 80.
16. Gupta Kavita, B M Pandey and R K Khetarpal. 2002. Pest risk analysis and its significance to India in WTO regime. *Seed Tech News.* 32 (2&3). pp 266-269.
17. Gupta Kavita, Khetarpal RK. 2004. Concept of regulated pests, their risk analysis and the Indian scenario. *Ann. Rev. Pl. Pathol* Vol 3: 409-441.
18. Gupta Kavita, Khetarpal RK. 2005. International Standards of Phytosanitary Measures. In: *Plant Protection in the New Millennium* Volume I (eds. A V Gadewar and B P Singh). Satish Serial Publishing House Delhi, pp 109-139.
19. http://www.ippc.org.
20. http://www. ISPMs\English Home- International Phytosanitary Portal (IPP).html accessed on April 25, 2006.

21. http://www.plantquarantineindia.org/docfiles/appendix-8.htm.

22. http://www.wto.org.

23. Jones FGW. 1961. The potato root eel worm, *Heterodera rostichiensis* in India. *Curr. Sci.* 30: 187.

24. Kahn RP. 1989. *Plant Protection and Quarantine: Selected Pests and Pathogens of Quarantine Significance*. Vol II. CRC Press Inc., Florida, 265 pp.

25. Kapur ML, Kavita Gupta, Shashi Bhalla, Khetarpal RK. 2004. Potential alternatives to methyl bromide for safe movement of horticultural material. In: *First Indian Horticulture Congress-2004,* Nov. 6-9, 2004, Horticultural Society of India, New Delhi, pp 68-69.

26. Kapur ML, Verma BR, Lal B. 1988. Interception of *Pachymerus lacerdae* (Coleoptera: Bruchidae) from imported nuts of *Orbygna phalerata* (Palmae) – a new host record. *J. Young Ent. Quart.* 6 (1): 31-32.

27. Khetarpal RK. 2004. A critical appraisal of seed health certification & transboundary movement of seeds under WTO regime. *Indian Phytopathol.* 57(4): 408-427.

28. Khetarpal RK, *et al.* 2001a. Plant quarantine of germplasm under exchange. In: *National Bureau of Plant Genetic Resources: A Compendium of Achievements* (eds. B S Dhillon *et al.*). NBPGR, New Delhi, pp.90-115.

29. Khetarpal RK, Ram Nath, Lal SP. 1984. A new species of *Drechslera* recorded on seeds of *Eleusine coracana. Indian Phytopathol.* 37 (2): 320 – 321.

30. Khetarpal RK. 1984. *Alternaria zinniae* on seeds of tobacco. *FAO Pl. Prot. Bull.* 32(3): 111.

31. Khetarpal RK, Kavita Gupta, 2002. Implications of Sanitary and Phytosanitary Agreement of WTO on plant protection in India. *Ann. Rev. Pl. Pathol.* 1: 1-26.

32. Khetarpal RK, and Kavita Gupta. 2006. Policy issues in plant protection. In: *One Hundred years of Plant Pathology in India: An Overview* (eds. S S Chahal, R K Khetarpal and T S Thind) Scientific Publishers, Jodhpur, 381- 401.

33. Khetarpal RK, Lambat AK. 1981. Fungi recorded on imported seeds of soybean germplasm. *Seed Res.* 9 (2): 199-201.

34. Khetarpal RK, Pandey BM. 2001. Plant quarantine and safe exchange of transgenics. In *Transgenic Crops and Biosafety Concerns* (eds. G J Randhawa, *et al.*). National Bureau of Plant Genetic Resources, New Delhi, pp 101-109.

35. Khetarpal RK and Ram Nath. 1982. Fungi recorded on seeds of sorghum germplasm

imported from Nigeria. *Seed Res.* 10 (2): 172-174.

36. Khetarpal RK, Parakh DB, Chalam VC. 2004. Virus indexing of plant germplasm. In *Conservation Biotechnology of Plant Germplasm* (eds. B B Mandal *et al.*). National Bureau of Plant Genetic Resources, New Delhi, India/ IPGRI, Rome, Italy/ FAO, Rome Italy, 99-104 pp.

37. Khetarpal RK, Singh S, Parakh, and Chalam VC. 2001b. Viruses intercepted in exotic germplasm during 1991-2000 in quarantine. *Indian J. Pl. Genet. Res.* 14: 127.

38. Kumar CA, *et al.* 1994. *Check-List of Seed-Transmitted Viruses: Leguminous Host.* Technical Bulletin, NBPGR, New Delhi, 14 p.

39. Lal A, Mathur VK. 1986. Spread of *Heterodera schachtii* cysts through sugarbeet seed. *Indian J. Nematol* 16: 122-123.

40. Lal A, Mathur VK. 1995. Nematodes intercepted from imported plant material: Description of *Polenchus minutus. Indian J. Nematol* 25: 98- 102.

41. Lal A, Mathur VK, Rajan. 1990. Identification of *Heterodera zeae* on vetiver plants imported from United Republic of Tanzania and variations observed. *FAO Plant Prot. Bull.* 39: 44.

42. Lal B, Verma BR. 1988. Seed infesting chalcid, *Quadrastichodella eucalypti* (Timbererlake) and its importance in plant quarantine. *J. Young Entomol.* 6 (12): 44-66.

43. Lal B, Verma BR, Wadhi SR. 1988. Quarantine risk in cotton seed import. *Indian J. Pl. Prot.* 16: 241- 243.

44. Mathur VK, Lal A. (Eds.). 1997. *Plant Quarantine Activities at NBPGR.* NBPGR Scientific Monograph No. 6, 96 pp.

45. Mathur VK, Sanwal KC, Lal A. 1978. *Tylenchorhynchus neoclavicaudatus* (Nematoda: Tylenchorhynchinae) in soil washings from imported tubers. *Indian J. Nematol.* 8: 148-150.

46. Mukewar PM, *et al.* 1980. Interception of *Peronospora manshurica* in imported seeds of soybean. *Seed Res.* 8(2):170-173.

47. Nagarajan S, Khetarpal RK. 2004. Seed and seed health care. In Souvenir: *National Conference on Seed: A Global Perspective*, March 26-28, 2004, Indian Society of Seed Technology and Indian Council of Agricultural Research, New Delhi, 22-23 pp.

48. Plant Quarantine (Regulation of Import into India) Order. 2003. The Gazette of India Extraordinary, Part II Section 3 Sub-section (ii) No. 1037 Published by Authority. 105p.

49. Ram Nath AK, Lambat, Payak MM, Lilaramani J, Rani I. 1973. Interception of race 'T' of *Helminthosporium maydis* on sorghum seed. *Curr. Sci.* 42: 872-874.

50. Ram Nath VK, Mathur, Verma BR. 1988. Plant quarantine activities in Indian context. In: *Plant Genetic Resources-Indian Perspective. NBPGR*_(eds. R S Paroda, R K Arora and K P S Chandel), New Delhi, pp. 470-480.

51. Reddy OR. 2002. Pest risk analysis: WTO-SPS Agreement/ IPPC. In: *Resources Management in Plant Protection, Vol. 1.* (eds. Sarath Babu, K, *et al.*). Plant Protection Association of India, National Plant Protection Training Institute Campus, Rajendranagar, Hyderabad, pp 114 -119.

52. Singh B, Anitha K, Agarwal PC, Ram Nath. 1995. Interception of *Pseudomonas syringae* pv. *syringae* on *Hibiscus cannabinus. Indian Phytopathol.* 48: 352.

53. Singh B, Khetarpal RK (Eds.) 2005. *Methods of Salvaging Germplasm Infected with Plant Pathogens.* NBPGR, New Delhi. 94p.

54. Verghese MK. 1945. Bunchy top disease of plantains. *Indian Farming.* 6:239-240.

55. Verma BR. 1991. Interception of *Merbruchus columbinus* (Sharp) (Bruchidae: Coleoptera) from *Samanea saman* seeds and new host record for this bruchid. *Indian J. Entomol.* 53 (3): 530.

56. Verma BR, Kapur ML, Lal B. 1991. A new host record for *Bruchus ervi* Froeiich (Coleoptera: Bruchidae). *Indian Jour. Entomology* 53 (1): 171-172

*Rev. Plant Pathol. Vol. 4, 2008*
*Indian Society of Mycology and Plant Pathology*
*Scientific Publishers (India), Jodhpur pp. 393-420*

# STATUS OF MICROBIAL PESTICIDES IN INDIA

***S. Chavan, M. Kulkarni and M.V. Deshpande***

Biochemical Sciences Division, National Chemical Laboratory,
Pune-411 008, India

KEY WORDS: Biological control, entomopathogen, fungal structures, hydrolytic enzymes microbial pesticides, mycoparasites.

## ABSTRACT

Virtually every new major outbreak of plant disease has led to different concepts concerning plant protection. The application of naturally occurring microorganisms with the ability to control or suppress a pest population has received modest research support over the past several decades, however, the public's concern of pesticide residues in food and in public spaces has motivated great interest in biological means of pest control. The viruses and bacteria, in particular are effective after ingestion while fungal agents have different modes of action. They act by contact and do not need ingestion, The understanding of the molecular aspects of biocontrol organism-pest interactions, the role of hydrolytic enzymes and their inhibitors which attack defence means of the pest are the prime areas of research. If micro-organism-based pesticide industry continues to be a chemical industry model, then the efforts are necessary to increase shelf-life of the preparation. The future studies regarding identification of the fast killing strain, use of consortium (one or two strains in combination), suitable formulation and application strategies are essential to broaden the scope of the microbe- based technology.

## INTRODUCTION

According to the Food and Agriculture Organization (FAO), the estimate of the world population for 2001 was $6.134 \times 10^9$ inhabitants which will be nearly $8.5 \times 10^9$ in 2025. It is indeed inevitable therefore to have an additional agricultural production of $2.4 \times 10^9$ tones/year. However, this additional production may not be by bringing more land under cultivation but it could be by improving the crop productivity. This can be achieved in part by suitable control of losses due to biotic agents (pests, diseases, weeds), which on an average are estimated to be more than 40% of the potential production. Currently, the control of plant pests, diseases and weeds is achieved mainly by spraying crops with a vast amount of synthetic chemical pesticides. In India, the pesticide consumption varies vastly across different states, depending on several factors, including cropping patterns, irrigation facilities, pest resurgence and resistance situations and so on. Table 1 describes the quantities of pesticides consumed by various states in India.

Table 1. Major Pesticide Consuming States

| States | Quantity (in tonnes) |
|---|---|
| Andhra Pradesh | 13000 |
| Uttar Pradesh | 11000 |
| Tamil Nadu | 9500 |
| Maharashtra | 6900 |
| Punjab | 6400 |
| West Bengal | 5800 |
| Haryana | 5200 |
| Gujarat | 5100 |
| Karnataka | 4400 |
| Rajasthan | 2900 |

From Agricultural Research Data Book (2001)

In the literature, it is mentioned that in developing countries every year around 10,000 people die and another 400,000 suffer from various effects of pesticides, and India accounts for one third of the pesticide poisoning cases in the third world. Biopesticides are quickly emerging as important tools in reducing pesticide use and risk. The characteristics, in general which make them potential alternative to chemicals are: a narrow target range and highly-specific mode of action; preventive in nature if timely applied; field persistence is generally not toxic; compatible nature with other components of Integrated Pest Management programme; and generally safe to humans and plants.

The biopesticides can be classified mainly in two groups as biochemicals and microbials. The molecules which are structurally and/or functionally related to the molecules of the pests can be used as biochemical pesticides which have a nontoxic mode of action. For instance, pheromones are naturally-occurring chemicals that insects use to find mates. Chemically synthesized pheromones can disrupt insect mating by creating confusion during the search for mates, or by attracting insects to traps (44).

Microbial pesticides are naturally-occurring or genetically altered bacteria, fungi, algae, viruses, or protozoans that are used to suppress pests by either producing a toxin specific to the pest, causing disease, preventing establishment of pest microorganisms through competition, or other modes of action. The further discussion will highlight the status of bacteria, viruses, fungi and their products as microbial pesticides. In view of the increasing awareness of the applications of fungi in biotechnology, the major emphasis will be on fungi and fungal products.

## 2. MICROORGANISMS IN BIOCONTROL OF INSECT PEST

### 2.1 Bacteria and viruses

The most widely accepted bacterial pesticide is Bt, a mixture of *Bacillus thuringiensis* spores and its toxin applied as a foliage spray. *B. thuringiensis* is a naturally-occurring soil bacterium that is toxic to the larvae of several species of insects. The total share of *B. thuringiensis* of today's biopesticide market accounts for over 90 (4, 35). One of the reasons is Bt preparations remain stable over years of storage and even in the presence of ultra violet of day light. However, major concern in the use of Bt preparation is its less effectiveness against III instar larvae, which damage crop extensively.

The other bacteria which are being studied extensively are, *Pseudomonas, Agrobacterium, Serratia* and others. The use of *Bacillus subtilis* as a fungicide has been demonstrated on a number of diseases, including cornstalk rot (*Fusarium roseum*), onion white rot (*Sclerotium cepivorum*), potato charcoal rot (*Macrophomina phaseolina*), bean rust (*Uromyces phaseoli*), so on and so forth.

The Table 2 describes some of the common bacterial biopesticides, which are commercially available. The genetically modified bacterial strains are also being used as biopesticides in USA on a large scale: *Agrobacterium radiobacter* k1026 used to treat fruit and vegetable plants to prevent crown gall tumors formation; *Pseudomonas fluorescens* has been modified with a number of different Cry delta-endotoxin genes from different subspecies of *B. thruingiensis* and the resulting biopestide is produced by killing modified *P. fluorescens*, The advantage of the modified strain is its longer persistence in the sunlight than *Bacillus* (http://www.isis.org.uk/).

Table 2. Some common bacterial biopesticides

| Name | Target pest/pathogen | Remark |
|---|---|---|
| *Agrobacterium radiobacter* | *Agrobacterium tumefaciens* (crown gall) | In USA and Australia. Trade names: Norbac 84-C™, Galltrol™, Nogall™, Diegall™ |
| *Bacillus popilliae* | Larvae of Japanese beetles, chafers, some May & June beetles | In USA. Trade names: Doom™ , Japademic™ |
| *Bacillus subtilis* | *Rhizoctonia, Fusarium, Alternaria* and *Aspergillus,* (root rots and seedling diseases) | In USA. Trade names: Kodiak™ and HB™, Epic™, Concentrate™, Quantum 4000, System 3™ |
| *Bacillus thuringiensis* var. *kurstaki* | Most lepidoptera larvae with high gut pH. e.g. Cabbage worm, European corn borer, diamondback moth, spruce budworm, gypsy moth caterpillars and other forest caterpillars, and Indian meal moth larvae in stored grain. | In USA.. 26 products such as Dipel™, Larvo-BT™, Thuricide™, JavelinWG™, Vault™, Raptor™ |

| | | |
|---|---|---|
| *Burkholderia cepacia* (formerly *Pseudomonas cepacia*) | Soil pathogens- *Fusarium, Phythium*, some nematodes | Deny™ , Blue Circle™ |
| *Pseudomonas fluorescens* | *Erwinia amylovora* on apple, cherry, almond, peach, pear, potato, straw-berry, tomato; bacterial bl-otch disease on mushrooms | In USA. Blight Ban, A506™ |
| *Pseudomonas syringae* | Postharvest pathogens on apples, pear, citrus | In USA. Bio-Save 10, 11, , 100, 110, 1000™ |
| *Serratia entomophila* | Grass grub (White); root knot nematode on tomato (*Meloidogyne i ncognita*) | In New Zealand. Invade® Mechanism: Infected insects will cease feeding clear their gut, depleting fat bodies |
| *Streptomyces griseoviridis* | Soil pathogens -- *Fusarium, Alternaria, Rhizoctonia, Phomopsis, Phythium, Phytopthora, Botrytis* -- that cause wilt, seed, root & stem rots | Mycostop™ USA and Finland |
| *Xanthomonas campestris* | Turf weeds | X-PO™ In USA |

From the database of microbial biopesticides, 2002

There are more than 1600 different viruses that infect 1100 species of insects and mites. A special group of viruses, called baculovirus (Nuclear Polyhedrosis Virus, NPV), to which about 100 insect species are susceptible, account for more than 10 percent of all insect pathogenic viruses (11). Most NPVs have been isolated from Lepidoptera (88%), Hymenoptera (6%), and Diptera (5%). These viruses develop in the nuclei of fat body, haemolymph, hypodermis, and tracheal matrix. The oral ingestion of polyhedra transmits the disease. The infections may also be transmitted from an infected female via the egg to her progeny. Few commercial preparations available are: VFN80™ to control Alfalfa looper (*Autographica californica*); Gemstar LC, Biotrol, Elcar to control *Helicoverpa zea* in USA; in Europe for the control of *Trichoplusia, Helicoverpa, Diparopsis, Phthorimaea operculella* (potato tuber-moth) a viral preparation is available under the trade name Mamestrin™ (from Database of Microbial Biopesticides (DMB) 2002). Although viruses are being used in fields for the pest control, their disadvantages are: rapid deactivation by environment, requirement of living host for the production, their effectiveness only when ingested by the host and longer time from ingestion of a lethal dose until death under field conditions (3-15 days). Similarly, bacterial pesticides are effective after ingestion and they have wide host range, therefore must be used with some care (12).

The presence of one or more specific segments of virus-like double-stranded RNA (dsRNA) were found to cause diseases of plant pathogenic fungi characterized by reductions in pathogenicity, ability to form sexual and asexual spores, spore viability and growth (6). It was also reported that double-stranded RNA viruses are ubiquitous in fungi transmitted to compatible strains via cell fusion (33). The use of fungal viruses to control fungal plant pathogens is a relatively new strategy for biological control of plant diseases that was discovered in Europe after the European chestnut (*Castanea sativa*) was destroyed. The hypoviruses of the chestnut-blight fungus, *Cryphonectria parasitica* have been studied for their potential as biological control agents of fungi. Howitt *et al.* (23) worked on viruses of two common plant pathogens, *Botrytis cinerea* (grey mould) and *Monolinia fructicola* (brown rot) with the ultimate aim of identifying viruses that might be used as biological controls of pathogenic fungi. However, these reports though indicate the possibility of the use of viruses for biological control of fungal pathogens, notable data is not available.

## 2.2 Fungi

The information on some common fungal agents used to control plant pathogenic fungi and insects is tabulated in Table 3.

Table 3. Some common fungal agents used to control plant pathogenic fungi and insects (9)

| Fungus | Host | Crop | Comments |
|---|---|---|---|
| *Beauveria* | Spider, beetles, brown plant hoppers, *Helicoverpa* | Sugarcane, rice, cardamom, cabbage, pulses, etc. | *B. bassiana*-commercial use |
| *Cephalosporium* | Coffee green bug, brinjal leaf beetle | Coffee, Brinjal, etc. vegetables | Not very common. |

| | | | |
|---|---|---|---|
| *Conidiobolus* | Aphids and diverse insects | | *C. coronatus,* weak pathogen |
| *Entomophthora* | Aphids (alfa alfa, wheat, crucifer, etc.), chestnut beetle, coffee green bug, house fly, etc. | Alfa alfa, chestnut, cruciferous vegetables, wheat, coffee, etc. | *E. muscae, E. grylli,* Studied for commercialization |
| *Fusarium* | *R. solani, Fusarium species* | Wilt of Radish, and banana, seedling blights of wheat and maize. | Within same genus mycoparasitism is observed. *F. oxysporum, F. semitectum, F. udum* |
| *Gliocladium* | *Pythium ultimum, R. solani* | Damping-off disease of cotton | Organism that has commercial potential, *G. virens* |
| *Hirsutella* | *Pyrilla* (sugarcane leaf hopper), mango leaf hopper, coconut mite | Sugarcane, coffee, mango, coconut, etc. | Major problem of sugar cane industry; important alternative for the control of coconut mite. |
| *Metarhizium* | *Pyrilla*, coconut beetle, mealy bug, aphids, *Helicoverpa* | Sugarcane, coconut, vegetables, pulses, etc. | *M. anisopliae,* and *M. flavoviride* commercial use |
| *Nomuraea* | *Helicoverpa, Spodoptera* | Pulses | Important for commercial exploitation |
| *Pythium* | *F.solani, Pythium myriotylum, P. ultimum, Sclerotinia sclerotiorum* | Browning root rot of cereals, rhizome rot of ginger, damping off of cucumber, sugar beet & cress | *P. nunn, P. oligandrum, P. acanthophoron* can mycoparasite a range of fungi |
| *Tremella* | *Peniophora laeta Peniophora incarnata Phlebia radiata* | Occurs on hardwoods with basidiomycetes | *T. mesenterica T. encephala T. mycophaga* |
| *Trichoderma* | *R. solani, Sclerotium rolfsii, F. solani, F. culmorum, P. ultimum, S. sclerotiorum* | Potato wilt, damping off of cucumber | *T. harzianum, T. viride,* one of the well studied system for fungus- fungus interaction, dual pathogenecity |
| *Verticillium* | Leaf hopper, banana aphid, sugarcane shoot borer | Banana, sugarcane, etc. | *V. lecanii* commercially important, exhibits dual pathogenecity. |

### 2.2.1 *Insect pathogenic fungi*

Fungi infect a broader range of insects than do bacteria and viruses. For instance, they can infect lepidopterans (moths and butterflies), homopterans (aphids and scale insects), hymenopterans (bees and wasps), coleopterans (beetles), and dipterans (flies and mosquitoes). In fact, some fungi such as *Beauveria bassiana* (Balsamo) Vuillemin, *Metarhizium anisopliae* (Metschnikoff) Sorokin, *Verticillium lecanii* (Zimmerman) Viegas, and *Paecilomyces* sp., have broad host ranges that encompass most of those insect groups (8,9). *B. bassiana* has been identified in many insect species in temperate and tropical regions and is used for pest control on a moderate scale in eastern Europe and China. A *B. bassiana* - based mycoinsecticide has been shown to be highly effective against coleopterans. While *M. anisopliae* has been most extensively used in Brazil for the control of spittlebugs on sugar cane. *V. lecanii* is a pathogen that has demonstrated good control of green house pests, such as *Myzus persicae* (Sulcer) aphids, on chrysanthemums. A distinct isolate of *V. lecanii* was obtained from whitefly and provided excellent control of greenhouse whitefly, *Trialeurodes vaporariorum* (Westwood), and of *Thrips tabaci* Lindeman on cucumber. During 1990s *V. lecanii* was produced commercially as Vertalec for aphid control and Mycotal for control of whitefly. Now there is a resurgence of commercial interest in its use for control of aphids, whiteflies, and thrips because these greenhouse pests have developed resistance to chemical pesticides typically used for their control. Similarly for the last more than 15 years the fungal pathogen *Entomophaga maimaiga* has been recognized and used in USA as a control for gypsy moth (14).

The viral and the bacterial control agents infect insects via their digestive tract while fungi make entry into the host through the cuticle, which is the outermost covering in insects and fungal cell wall. Moreover, the fungal biomass can be obtained easily and can also be applied easily. The other features are: they do not leave toxic residues, environmentally safe, and no known resistance in insect community against the mycopesticides. This can be attributed to the involvement of multiple factors due to the nature of their mechanism of action. However, research inputs are necessary to make them effective on a wide range of temperature and humidity and to have longer shelf-life.

Three groups of insect pathogenic fungi are observed in the field. Commonly encountered fungi with dry conidia are *Metarhizium, Beauveria, Nomuraea* and *Paecilomyces*. They possess lipophilic cell walls. Others though frequently encountered but have less potential are *Verticillium* and *Entomophaga*. Saprophytic *Aspergillus, Fusarium* and *Penicillium* are also sometimes mistaken as entomopathogens. The fungi such as *Metarhizium, Beauveria, Nomuraea, Verticillium,* and *Paecilomyces* are the favoured candidates in the area of pest control (7). Epizootics caused by *Nomuraea rileyi* and *B. bassiana* have been observed on *Helicoverpa armigera* and *Spodoptera litura* in Indian fields since last 15 years as a result the insect population is highly susceptible to these fungi (55). Different strains of *B. bassiana* were reported in the control of *Locusta migratoria* (42). The variation in their effectiveness was mainly due to the variation in their virulence and their ability to produce *in vitro* toxic metabolites. The virulence towards insect pest was also found to be affected by the repeated conidial sub-culturing of *Metarhizium anisopliae* (37)(38). Significant variation in appressorium formation and cuticle-degrading enzyme production such as chitinase, chitin deacetylase, chitosanase and protease during subsequent sub-culturing and passage through *H. armigera* was reported. As a stable virulence towards the insect pest is desirable for commercialisation of a mycoinsecticide, such changes in virulence due to repeated in vitro transfer need to be monitored and minimised.

### *2.2.2 Mycoparasitic fungi*

A wide spectrum of fungi is known to show antogonism among themselves. The potential of *Trichoderma* and *Gliocladium* in controlling plant-pathogenic fungi was reported in the beginning of 20$^{th}$ century (12). It was due to the toxic metabolites secreted by these two fungi. Initially parasitic fungus has to locate the host mycelium or spores, using chemical signals originating from the host fungus. There are two types of mycoparasitism. *Trichoderma* exhibits necrotrophic parasitism. It attaches to the host hyphae by coiling, hooks or appressorium-like bodies and penetrates the host by secreting cell wall lytic enzymes. On the other hand, fungi like *Sporidesmium sclerotivorum* show biotrophic mycoparasitism in response to the persistent contact with the host. The haustoria produced in the hyphae of a host *Sclerotinia minor*, trigger the production of cell wall lytic activities by the host itself.

One of the important aspects in the fungus-insect and fungus-fungus interactions is the possibility of identifying a strain, which has capability to show dual pathogenesis. For instance, Jassim *et al.* (25) have shown that *Trichoderma harzianum,* a mycoparasitic species, can parasitize the elm bark beetle, *Scolytus*. This is mediated through the production of specific hydrolytic enzymes. *V. lecanii*, pathogen of aphids, can parasitize rust fungi. It was suggested that entomopathogenic and mycoparasitic strains within the species can be separated on the basis of morphological and biochemical characteristics (26).

### 2.2.3 Living propagules of fungal agents

The fungal kingdom exhibits a great variety of structural types in their members (12). The hypha and unicellular yeast are vegetative growth forms, while conidia and sporangiospores are produced asexually. These are common structures as an infective propagule in mycoinsecticide preparation. The sexually produced resistant zygospores are also used in the biocontrol formulations. Blastospores and chlamydospores are also candidates in the mycopesticide formulations. For instance, *Paecilomyces fumosoroseus,* a biocontrol agent against sweet potato whitefly can be grown in form of blastospores, which are resistant to simple drying (24). Different mycoparasitic *T. harzianum* strains develop chlamydospores in submerged fermentation while conidia on solid substrates. Chlamydospores of *G. virens* were produced in stirred tank fermentor and the different fermentation stages were used to make the effective biocontrol formulation (12). While sexually produced spores of *Pythium oligandrum,* oospores were used for the seed treatment of sugar beet, cress and carrot to control damping-off diseases. *Entomophthora virulenta* which can produce zygospores in surface, and submerged fermentation also has potential to control infestation of aphids (12).

### 2.2.4 Fungus-insect and -fungus interaction

As fungi are effective by contact the considerations such as adhesion of an infective propagule to the insect body, entry of the organism in to host and the killing are important in developing effective mycoinsecticide. As mentioned earlier, in case of *Metarhizium*, *Beauveria*, *Nomuraea* and *Verticillium* spores are the means for attacking an insect population. Pre-infection stages are, adhesion and germination of spores to form germ tubes or appressoria. Entomopathogenic fungi display different strategies in their attachment to insects. Infective propagules are either dry hydrophobic conidia as in case of *B. bassiana, M. anisopliae,* and *N. rileyi,* or sticky hydrophilic conidia seen in *V. lecanii* and *H. thompsonii* (50). The adhesion is either passive (adsorption) or active host-specific (germination and penetration) phenomenon. In many cases, various fungistatic activities have been associated with the cuticle of insects. For example, the presence of antagonistic microbes, chitinases produced during moulting, toxic lipids, and cuticle phenolic compounds have been suggested to inhibit fungal growth. For instance, cuticular fatty amides may contribute to book lice or psocid *Liposcelis bostrychophila* (Psocoptera: Liposcelidae) in hampering conidial adhesion of entomopathogenic fungi by decreasing hydrophobicity and static charge (31). Once the fungus breaks through the cuticle and underlying epidermis, it may grow profusely in the haemolymph, by forming the hyphal bodies, in which case death is probably the result of starvation or physiological/biochemical disruption brought about by the fungus. The secondary metabolites of the attacking fungus may contribute to the demise of the insect. As a result, one can notice the sporulated mycelial growth of the fungus on the cadaver of the host. The condition which favours sporulation is a high relative humidity. This enhances the possibility of horizontal or vertical transmission of the disease within the insect population.

On the other hand fungus- fungus interaction is apparently less complicated. For instance, recent studies of Gao *et al.* (15) showed the interaction between *Chaetomium spirale* and *Rhizoctonia solani* in dual cultures. They reported that pathogen growth inhibition occurred soon after contact with the antagonist, followed by the overgrowth of *C. spirale* on the colony of *R. solani*. The coiling of *C. spirale* around *R. solani* and intracellular growth of the antagonist in its host occurred frequently. In case of *Gliocladium roseum* interaction with conidia and germ tubes of *B. cinerea*, the electron microscopy studies revealed that infection occurred through direct penetration by hyphal tips of *G. roseum* without the formation of appressoria (28).

### 2.2.5 Killing components

The dual pathogenecity of some fungi towards insects and fungi has been attributed to the hydrolytic enzymes, which facilitate the entry of the pathogen in the respective hosts. The insect cuticle is composed of wax, lipids, protein and chitin. The prime component is being chitin, a β-1,4-linked *N*-acetylglucos-amine polymer, its share is 25-50% of the dry weight of the cuticle. While fungal cell wall components are divided into two groups, such as skeletal and matrix components. Chitin and R-glucans are the main structural components while S-glucans and manno-proteins are the matrix components. Furthermore, chitin is absent in plants and mammals and thus its metabolism too presents an attractive target for the control of plant pathogenic fungi and insects.

The cuticle degrading enzymes (CDEs) mainly include proteases, lipases and chitinases, and enzymes such as chitinases, proteases and glucanases are the main components of the mycolytic enzyme (ME) complex (12). The composition of CDE and ME is not drastically different. It has been suggested that among these, chitinases contribute significantly in the degradation of protective covers. Therefore, most of the studies for the production of CDEs and MEs and their use in the field are centered around chitinases.

*In vitro* experiments showed that insect pathogens in submerged fermentation produce extracellularly cuticle degrading enzymes mainly chitinases, proteases and lipases when locust cuticle was used as a sole carbon source (52). In case of sequence of appearance of extracellular enzymes in *Metarhizium* and *Beauveria* cultures lot studies have been reported. Proteolytic enzymes such as esterase, endopeptidase, aminopeptidase and carboxypeptidase were produced within first 24 h of growth. *N*-Acetylglucosaminidase appeared next and endo-chitinase, which attacks chitin polymer randomly, was produced in significant quantities after 4 days. Lipases were detectable after 5 days (8). Though chitin forms the main structural mesh, In view of this sequence of extracellular appearance of components of CDE in accordance with the cuticular structure, St. Leger *et al.* (52)

assigned major role to protease in cuticle degradation by *M. anisopliae*. Recently, Nahar *et al.* (37) however, suggested an alternate mechanism of cuticle degradation. With the involvement of constitutively produced chitin deacetylase and chitosanase the process of cuticle degradation in the fungi that show low or delayed chitinase production was initiated.

Similarly, in case of mycoparasitic organisms the appearance of the extracellular enzymes was found to be corresponding to the main chemical constituents of the fungal cell wall, i.e., chitin, glucans, and proteins (18). The enzymes appeared sequentially when *Trichoderma harzianum* was grown on a mycelium of *Rhizoctonia solani*; an alkaline proteinase was produced first, followed by glucanases and chitinases. Chitinolytic enzymes of *T. harzianum* were found to be inhibitory to a wider range of deleterious fungi than similar enzymes from other sources. No other chitinases have been reported with a broad range of activity, synergy between different enzymes, or synergy with a number of other biological and chemical control agents(41). A *Fusarium chlamydosporum* strain, a mycoparasite of groundnut rust, *Puccinia arachidis* produces an endo-chitinase, which inhibits germination of uredospores of the rust fungus. This indicates the significant contribution of chitinase in the biocontrol of groundnut rust. *Myrothecium verrucaria* produced high levels of extracellular mycolytic enzymes, viz. chitinase, β-1,3-glucanase and proteinase. The crude mycolytic preparation was found to be effective against *Sclerotium rolfsii*, a root-infecting fungus of groundnut (40).

In addition to the hydrolytic enzymes, fungi secrete number of compounds that exhibit biological activity against pests (57). The entomopathogenic fungi, *B. bassiana, B. brongniartii, H. thompsonii, M. anisopliae, , P. fumosoroseus, Tolypocladium* and *V. lecanii* produce different toxins such as, bassianin, beuvericin, cytochalasin C, destruxins, dipcolonic acid, enfrapeptins, hirsutellin A, B, oosporein, swainsinone, etc. (2, 57). These toxins are not effective to the same extent against different insects. For example, lepidopterean insects are highly susceptible to destruxins produced by *M. anisopliae*. Efrapeptins of *Tolypocladium* show insecticidal and miticidal effects against potato beetle, mites, budworm and diamondback moth and others. Oosporein produced by *B. brongniartii* affects enzyme malfunctioning in cockchafer larvae by redox reactions. The hyphomycete, *H.*

*thompsonii* produced an extracellular insecticidal protein, hirsutellin A which was reported to be effective against citrus rust mite, mosquito larvae and others. (57).

*Trichoderma* and *Gliocladium* sp. produce diverse secondary metabolites with antibiotic properties including polyketides, terpenoids, and polypeptides (57). *T. harzianum* produces a novel trichothecene, harzianum A, which is one of the killing components for the biocontrol of plant pathogenic fungi. *T. viride*, species from 'harzianum' group produces tricholin, a ribosome-inactivating protein, which is active against *R.solani*. *Gliocladium* sp. also produces different enzyme inhibitors such as argifin, a potent chitinase inhibitor. As chitinase is one of the enzymes involved in the cell wall synthesis such type of inhibitor must be enhancing mycoparasitic potential of the biocontrol organism (40). Interestingly argifin also retards the moulting process in cockroach larvae that again can be attributed to the inhibition of insect chitinase activity. Nevertheless, antifungal antibiotic production does not sufficiently explain the process of mycoparasitism. Graeme-Cook and Faull (20) using UV-induced mutants of *T. harzianum* with enhanced antibiotic potential suggested that the activity can not be directly correlated with its potential as a mycoparasitic organism. In other words, the biocontrol efficacy of the mycoparasitic strains is the combination of antibiosis, lysis, competition, mycoparasitism and promotion of plant growth.

## 3. PRODUCTION OF FUNGAL BIOCONTROL AGENTS

### 3.1 Isolation of entomopathogenic and mycoparasitic fungi

For the isolation of both entomopathogenic and mycoparasitic fungi soil dilution method is common (17). For the isolation of insect pathogens *Galleria* bait method is effective (61). In *Galleria* bait method, the larvae are allowed to move through the soil in a small vial, which is moved upside down everyday. After 14 days, one can find dead or even mycosed *Galleria* larvae from which fungus can be isolated and further identified (30). The method was found to be more useful to isolate maximum number of *Beauveria* strains as compared to other entomopathogenic fungi such as *Metarhizium* and *Nomuraea* (32). Isolation of mycoparasitic strains is usually by dual culture method. It will be interesting to develop a method, which can use immobilized mycelium of a host fungus.

### 3.2 Habitat association

The isolation of the entomopathogenic fungi is affected by the locality from where the soil is taken and also by the method of isolation selected (5). It has been reported that *N. rileyi* isolates from different geographical locations showed different levels of virulence against two insect hosts, *H. armigera* and *S. litura* (58). In general, the soil samples collected from the rhizosphere of crops where chemical insecticides are not routinely sprayed one can expect more number of isolates that had high infectivity against insect pest. For instance, *Metarhizium* isolates from fields of tomato, okra and other vegetables heavily sprayed with chemicals were less virulent to *H. armigera* than the isolates from the custard apple field rarely sprayed with chemicals (39). This can also be correlated with the natural insect flora of the host plant, not necessarily pest which affects the virulence of the entomopathogenic fungi.

Many past efforts to select biocontrol organisms were empirical. The random selection of organisms from soil and plant surfaces to chance encounters on culture media showing zones of inhibition was followed initially. Second approach was to examine healthy plants and their rhizosphere in otherwise heavily infected agricultural sites. Some agricultural fields were found to be disease suppressive, these were also used to isolate biocontrol organisms, such as mycoparasitc species. In the Rhone valley in France, *Fusarium* suppressive soils specific against wilt causing *Fusarium* were found. The saprophytic isolates of *Fusarium oxysporum* and *F. solani* repressed activity of pathogenic strains of *F. oxysporum* on melons. Another example is a soil suppressive to *Pythium ultimum* in California, where prior colonization of crop debris by *Pythium oligandrum* prevented damping off disease of sugar beet caused by *P. ultimum* (12).

### 3.3 Criteria for strain selection

The performance of the fungal biocontrol agents can be attributed to their ability to kill the target pest. This involves the comparative study under laboratory and field conditions regarding the killing time, comparison with other potential strains and importantly effect of laboratory maintenance (serial transfer) on the virulence of the fungus. Attenuation of especially hyphomycetous entomopathogens after serial transfer *in vitro* was reported. However, passage through a host may improve virulence and expand host range (56). Similar type of reports on the mycoparasitic organisms are sparse.

The ability of the strains to produce differential levels of various cuticle degrading and mycolytic enzymes plays an important role in the strain selection (12, 52, 37, 18, 41, 39, 27). Interestingly, the spore related characteristics such as the size, viability, production, speed of germination, relative hyphal growth, and their response to the environmental perturbations are important for strain selection too (29). It was reported that in case of *P. fumosoroseus* the spore size and virulence towards diamond back moth, *Plutella xylostella* were correlated with each other (1). While Sugimoto *et al.* (53) reported relatedness among genetic properties, conidial morphology and virulence of *V. lecanii* towards different insect hosts. Thus, based upon the insect bioassay and level of cuticle degrading enzyme activities, selection of the best possible fungal strain along with the backup strain would be easier.

### 3.4 Preservation of fungi

The fungal spores are normally stored at -20°C or -80°C either using skimmed milk, water or mineral oil. *P. fumosoroseus* spores were reported to remain viable when stored with water or mineral oil and preserved at -80°C. Alternatively, fungal isolates can be stored for long period as dry alginate pellets (biomass plus sterile wheat bran and aqueous sodium alginate) at -23 ± 1.0°C. The fungal conidia can be stored in carriers such as mineral oils, wheat bran, talc powder, silica etc. to retain the viability and virulence of the conidia, which is an important requisite for the performance of the biocontrol preparation. Also, to retain the virulence, regular serial passage of the fungus through the insect host for entomopathogens and for mycoparasitic ones growth in dual culture are essential.

### 3.5 Large-scale production of the conidia of fungal strains

Different substrates like sorghum, wheat, barley, bajra, beaten rice, and rice can be used to grow the fungal strains for conidia production (59). One of the major constraints in the solid state fermentation is contamination. The attempts are being made for the production of conidia in submerged fermentation by manipulating different nutritional parameters (12). However, in case of mycoparasitic organisms, biomass produced in the submerged fermentation as such can be used for soil treatment. The mass production of *M. anisopliae*, *B. bassiana* and *N. rileyi* conidia can be carried out in bags with a membrane for gaseous exchange (22). Different methods are used to separate spores from the mycelium. The use of water with surfactant like Tween 80 to suspend the sporulated mass is a common practice. Hassani *et al.* (22) harvested conidia from the dried sporulated mass (moisture content <20%) with a myco-harvester, a unit specially designed to remove conidia from the biomass under vacuum. Milner (36) reported separation of *M. anisopliae* conidia using industrial electric sieve separator.

### 3.6 Formulation

The performance of any biocontrol agent under field condit-ions is dependent on the formulation used. Fungal preparations are usually supplied either as concentrates or powders. Though the aqueous formulation for foliage spray is common, vegetable oil: diesel formulation is considered as a better option, because the oil protects conidia from desiccation without affecting germination rate and diesel reduces the viscosity of oil there by facilitates easy spraying (30,10). However, phytotoxicity of diesel, if any, is a major concern (39, 13). For oil formulation the ultra low volume (ULV) sprayer is needed which can control the droplet size. The controlled droplet application (CDA) is a specialized mycopesticide delivery system, presently used for oil-based formulations only. The development of water-miscible formulations for CDA will be useful in coming years.

For the control of pathogenic fungi, application of laboratory – grown cultures is through the drip irrigation system with a liquid formulation. A granular material would be more appropriate for combining with potting mix, while a wettable powder would be more appropriate for root dips or sprays. To control root-infecting fungi seed-dressing with a mycoparasitic organism is the acceptable method. Usually 10 kg of seed are mixed with 10 g *Trichoderma* conidia in jaggery or rice kanji. The application with the seedlings is done by mixing spores with molasses. Mostly soil application along with fertilizers is the acceptable practice (51).

### 3.7 Storage and shelf- life

This is one of the major concerns of biopesticide industry due to its inevitable comparison with the chemical industry. To prevent the conidia from desiccation, they are stored in different carriers such as mineral oils, talc, bentonite, silica etc. at different temperatures. The carrier that retains the maximum viability and virulence of conidia for a minimum of 6 months at 28°C, ideally 18 months at 20°C, is suitable for the further commercialization. In fact it is now necessary to develop a food industry (perishable product) model for bio-products which are used in agriculture too.

## 4. MICROBIAL ENZYMES AND ENZYME INHIBITORS

Number of soil bacteria also produces chitinolytic enzymes. Roberts and Selitrennikoff (43) studied plant and bacterial chitinases for anti-fungal activity and enzyme specificity. The plant chitinases exhibited higher proportion of randomly acting enzymes, as a result, the plant chitinases retarded hyphal tip growth drastically. While bacterial chitinases from *Serratia marcescens, Streptomyces griseus* and *Psuedomonas stutzeri* act as exo-enzymes and had no effect on hyphal extension of test fungi like *Trichoderma reesei* and *Phycomyces blackesleeanus*. However, Shapira *et al.* (45) cloned *S. marcescens* chitinase gene in *Escherichia coli* and the chitinase preparation obtained was found to be effective in reducing disease incidence caused by *S. rolfsii* in beans and *R. solani* in cotton under greenhouse conditions. This contradiction in the observations may be attributed to the difference in the test fungi used and other factors, which contribute in the killing of pathogenic fungus (41).

To control pests such as longhorn beetles and aphids, the enzyme treatment before or simultaneously along with the entomopathogenic fungus, itself was successfully tried. The insect-pathogenic fungi, *B. bassiana, B. brongniartii* and *V. lecanii* produced cuticle degrading enzymes when grown in chitin containing medium. The pretreatment of insects with the enzyme solution was reported to be useful. Nahar *et al.* (37) have suggested the combination of chitin deacetylase and chitosanase along with *M. anisopliae* spores to enhance the *H.armigera* control in the field.

Several microbial chitinases, with β-1,3-glucanases, propane – 2-ol and polyoxyethelene lauryl ether have been sprayed on rice field to control rice blight caused by *Pyricularia oryzae* (54). Similarly, a chitinase producing *Arthobactor sp.* inhibited the growth of *Fusarium monilliforme* var. *subglutinans,* the causal agent of pine pitch canker (3). Furthermore chitinase alone and in combination with *B. thuringiensis* was reported to control larvae of spruce budworm (48, 49). Recently chitinase (1%) was used under field conditions for control of raspberry midge blight in Siberia (46). Govindsamy *et al.* (19) also reported use of purified preparation of *M. verrucaria* chitinase to control groundnut rust, *Puccinia arachidis. M. verrucaria* produced high levels of extracellular mycolytic enzymes, viz. chitinase, β-1,3-glucanase and proteinase. The crude mycolytic preparation was found to be effective against *S. rolfsii*, a root-infecting fungus of groundnut (40).

The interactions between fungi and bacteria involved in the biocontrol may play a key role in the natural process of biological control, although the molecular mechanisms involved are still largely unknown (60). For instance, cell wall degrading enzymes produced by *Trichoderma* sp. increased the efficacy of bacterium, *Pseudomonas* sp., which produced syringomycins against *B. cinerea* and other phytopathogenic fungi. A combined effect of mycolytic enzyme mixture of *M. verrucaria* and the cell wall chitin metabolism inhibitors to control the growth of a root infecting fungus, *S. rolfsii* on peanut was reported by Patil *et al.* (40). Different microbial cultures and organic molecules were screened for their potential to produce extracellularly cell wall synthesis inhibitors, using hyphal tip bursting test with different fungi, and inhibition of germ tube formation during yeast-hypha transition in a dimorphic saprophytic fungus, *Benjaminiella poitrasii* (16,40). Most of them exhibited significant *in vitro* inhibition of enzymes *viz.* chitin synthase, endo-chitinase or *N*-acetylglucosaminidase activities. Shirazi et al (47) recently reported a method for the rapid and sensitive screening of chitinase inhibitors.

## 5. SOME PRACTICAL CONSIDERATIONS TO MAKE THE COST EFFECTIVE USE OF BIOPESTICIDES

### 5.1 Strain

Though number of different fungal structures are being used, the major infective propagule used for foliage sprays as well as for seed treatment is conidia. The cost of conidia production is solely dependent on the yield of viable aerial conidia per kg of solid substrate used. Most of the fungal strains kill their hosts, either fungus or insect in 72-96 h. The strains which exhibit rapid germination of conidia on the insect host, and highly virulent as evidenced by the reduced killing time have greater potential as an effective biocontrol agents. The additional selection criteria can be, ability for germination at low water activity (aW), high levels of hydrolytic enzyme activities, and $CO_2$ production rate. In other words, apart from the killing potential, stability of the strain under natural environmental perturbations in temperature and relative humidity, broad host range are some of the characteristics which need some considerations too. Finally, as mentioned earlier the dual pathogenicity of the strains towards plant pathogenic fungi and insects can have added advantage too.

### 5.2 Registration and patent

Before the commercial exploitation of a microbial pesticide, its legal protection as a biotechnological invention can be assured by means of a patent. A patent is a temporary privilege for industrial or commercial exploitation given by the administration to the owner for 20 years after the application date, according to a series of claims. Biopesticide patents are considered biotechnological inventions because they include microbial products and processes. However, a patent is not an authorization for commercial use. Therefore, every microbial pesticide, which is used, distributed, sold, or offered for sale needs official registration of the product with Central Insecticide Board.

### 5.3 Major issues in biopesticide use

The market development of biopesticides faces several technical hurdles. Two of the most significant are relative safety and the future of genetic modification. One of the most important issues relating to biopesticides is their potential hazard to humans. A general consensus seems to be that the efforts are being made to develop biopesticides will be less hazardous than the currently used products. Regarding the use of genetically modified microbes and transgenics, there are many supporters as well as many detractors. This needs a separate exercise and discussion. The fact that one of the first applications of genetic modification essentially aimed to replace a widely used natural pesticide (Bt corn) illustrates the intimate relationship between genetic modification and biopesticides. Several economic and geo-political factors also influence the market potential for biopesticides, ranging from international trade issues to regulatory initiatives. The major concern is of course the acceptability by the farming community.

## 6. FUTURE CONSIDERATIONS

A lot of work has been carried out with respect to the successful application of fungi and their products and economic feasibility of pilot scale production plants. Indeed the efforts are more to develop *Trichoderma* as a biofertilizer as well as biocontrol agent. Foremost need of India, which is transforming from developing to developed country will be more and more tie-ups of research and development institutions with the small scale industries (13, 21). Market for Fungal agents is considered to be 10-15% of the total biopesticide market in India. The data available from personal communications, internet and published reports of different organized industries show that around 125-150 tons of different fungal formulations per year are being marketed which are approximately Rs. 2-2.5 crores. Indeed the figures may not be realistic possibly due to the involvement of big unorganized sector in this field. One of the state governments report indicates a serious concern about this so called "garage technology".

If biopesticide industry continues to be a chemical industry model, then the efforts are necessary to increase shelf-life of the preparation. The future studies regarding identification of the fast killing strain, use of consortium (one or two strains in combination), suitable formulation and application strategies for consumption commodities that mainly include the vegetable crops are essential to broaden the scope of the fungi-based technology. In case of vegetable crops, the direct application of the formulation on the vegetable is not desirable, therefore the soil application could be a better alternative even for insect control. Most of the insects undergo the pupation stage, an important stage of their life cycle in soil. The strains, which can infect pupa will be more useful in certain cases. The question now to be addressed is "Would extensive inundation by the fungal strain affect the local ecosystem?" The possible solution, if the answer is 'Yes', is to select highly virulent strain and decrease the dosages. Moreover persistence of the microbial agents in the soil will also be useful to decrease the successive dosages (13). Secondly, the feasibility of the products in an integrated pest management strategies is important too.

Lastly, in view of the increasing knowledge of organic farming, it would certainly be beneficial if the technology is translated to "Do it yourself (DIY)" mode to enable the Indian farmers to have the cake and eat it too. In this regard, a Food industry model technology can ease the way to translate biopesticide technology to DIY. Availability of cheaper raw material, simple production process, convenient time for the production and less biohazard will be the essential features.

## *Literature Cited*

1. Altre JA, Vandenberg JD, Cantone F.A. 1999. Pathogenicity of *Paecilomyces fumosoroseus* isolates to diamondback moth, *Plutella xylostella:* Correlation with spore size, germination speed, and attachment to cuticle. *J. Invertebr. Pathol.* 73: 332-338

2. Amiri-Besheli B, Khambay B, Cameron S, Deadman MI, Butt TM. 2000. Inter- and intraspecific variation in destruxin production by insect pathogenic *Metarhizium* sp., and its significance to pathogenesis. *Mycol. Res.* 104: 447-452

3. Barrows B, Kerr T. 1981. Inhibition of *Fusarium moniliforme* var. *subglutinans*, the causal agent of pitch canker by the soil bacterium *Arthiobacter* sp. *Can. J. Microbiol.* 27: 20-27

4. Baum JA, Johnson TB, Carlton BC. 1999. *Bacillus thuringiensis*: Natural and recombinant bioinsecticide products. In *Biopesticides, use and delivery*, eds. FR Hall, JJ Menn, pp. 189-210. NJ, USA: Humana Press

5. Bidochka MJ, *et al.* 2001. Habitat association in two genetic groups of the insect-pathogenic fungus *Metarhizium anisopliae*: uncovering cryptic species? *Appl. Environ. Microbiol.* 67: 1335-1342

6. Buck KW. 1988. Control of plant pathogens with viruses and related agents. *Phil. Trans. R. Soc. London B* 318: 1189

7. Butt TM, Jackson CW, Magan N. 2000. Fungal biological control agents: progress, problems and potential. In *Fungi as biocontrol agents: Progress, Problems and Potential*, eds. TM Butt, CW Jackson, N Magan, pp. 1-8. UK: CABI Publishing Co.

8. Charnley AK. 1989. Mechanisms of fungal pathogenesis in insects. In *Biotechnology of fungi for improving plant growth,* eds. JA Whipps, RD Lumsden, pp. 85-117.UK: Cambridge University Press

9. Chavan S, Ghormade V, Nahar P, Deshpande MV. 2006. Entomopathogenic fungi: A valuable tool to fight against insect pests. In: *Plant Protection for the New Millennium.Vol.II* eds. AV Gadewar, and BP Singh, pp. 227-243 Delhi: Satish Serial Publishing House

10. Deshpande MV, *et al.* 2004. Comparative evaluation of indigenous fungal isolates, *Metarhizium anisopliae* M34412, *Beauveria bassiana* B3301 and *Nomuraea rileyi* N812 for the control of *Helicoverpa armigera* (Hüb.) on pulses. In: *Proc. International Workshop on Entomopathogenic fungi: a valuable alternative to fight against insect pests,* ed. MV Deshpande, pp.51-59 Pune, India: NCL

11. Deshpande MV. 1998. In: *Novel Biopesticides*. CSIR Golden Jubilee Series, CSIR, New Delhi

12. Deshpande MV. 1999. Mycopesticide production by fermenta tion: Potential and challenges. *Crit. Rev. Microbiol.* 25:229-243

13. Deshpande MV. 2005. Formulations and Applications of Mycopathogens. In *Proc. ICAR-CABI Workshop on biopesticide formulations and application,* eds. RJ Rabindra, SS Hussaini, B Ramanujam, pp. 150-158. Bangalore, India: PDBC

14. Elkinton JS, Hajek AE, Boettner GH, Simons EE. 1991. Distribution and apparent spread of *Entomophaga maimaiga* (Zygomycetes: Entomoph-

thorales) in gypsy moth (Lepidoptera: Lymantriidae) populations in North America. *Environ. Entomol.* 20:1601-1605

15. Gao K, X. Liu, Z Kang, K. Mendgen. 2005. Mycoparasitism of *Rhizoctonia solani* by endophytic *Chaetomium spirale* ND35: Ultrastructure and cytochemistry of the interaction. *J. Phytopathol.* 153: 280 -290

16. Gholap AR, *et al.* 2007. Synthesis and evaluation of antifungal properties of a series of the novel 2-amino-5-oxo-4-phenyl-5, 6, 7, 8-tetrahyd roquinoline-3-carbonitrile and its analogues. *Bioorg. Medici-nal Chem.* 15: 6705-6715

17. Goettel MS, Inglis GD. 1996. Fungi: Hyphomycetes. In *Manual of techniques in insect pathology,* pp. 213-249. Sandiego, USA: Academic Press, USA

18. Goldman MHS, Goldman GH. 1998. *Trichoderma harzianum* transformant has high extracellular alkaline proteinase expression during specific mycoparasitic interactions. *Genet. Mol. Biol.* 21:ISSN 1415-4757

19. Govindsamy V, Gunratna KR, Balsubramanian R. 1998. Properties of extracellular chitinase from *Myrothecium verrucaria,* an antagonist to the groundnut rust *Puccinia arachidis. Can. J. Plant. Pathol.* 20: 62-8

20. Graeme-Cook KA, Faull JL. 1991. Effect of ultraviolet-induced mutants of *Trichoderma harzianum* with altered antibiotic production on selected pathogens *in vitro. Can J. Microbiol.* 37: 659-664

21. Grimm C. 2001. Economic feasibility of a small-scale production plant for entomopathogenic fungi in Nicaragua. *Crop Protection* 20:623-630.

22. Hassani M, *et al.* 2004. Mass production of entomogenous fungus *Metarhizium anisopliae* (Deuteromycotina: Hyphomycetes) for biological control of *Helicoverpa armigera* (Lepidoptera: Noctuidae) on pulses. In *Proc. International Workshop on Entomopathogenic fungi: a valuable alternative to fight against insect pests,* ed MV Deshpande, pp. 37-49. Pune, India: NCL

23. Howitt RLJ, Beever RE, Pearson MN, Forster RLS. 1995. Presence of double stranded RNA and virus-like particles in *Botrytis cinerea. Mycol. Res.* 99: 1472-1478

24. Jackson MA, Mcguire MR, Lacey LA, Wraight SP. 1997. Liquid culture production of desiccation tolerant blastospores of the bioinsectcidal

fungus *Paecilomyces fumosoroseus*. *Mycol. Res.*101: 35-41

25. Jassim HK, Foster HA, Fairhurst CP. 1990. Biological control of dutch elm disease larvicidal activity of *Trichoderma-harzianum Trichoderma-polysporum* and *Scytalidium-lignicola* in *Scolytus-scolytus* and *Scolytus-multistriatus* reared in artificial culture. *Ann. Appl. Biol.* 117: 187-196
26. Jun Y, Bridge PD, Evans HC. 1991. An integrated approach to the taxonomy of the genus *Verticillium*. *J. Gen. Microbiol.* 137: 1437-
27. Krieger de MC, Schrank A, Vainstein MH. 2003. Regulation of extracellular chitinases and proteases in the entomopathogen and acaricide *Metarhizium anisopliae*. *Curr. Microbiol.* 46: 205- 210
28. Li GQ, Huang HC, Kokko EG, Acharya SN. 2002. Ultrastructural study of mycoparasitism of *Gliocladium roseum* on *Botrytis cinerea*. *Bot. Bull. Acad. Sin.* 43: 211-218
29. Liu H, Skinner M, Brownbridge M, Parker BL. 2003. Characterization of *Beauveria bassiana* and *Metarhizium anisopliae* isolates for management of tarnished plant bug, *Lygus lineolaris* (Hemiptera: Miridae). *J. Invertebr. Pathol.* 82: 139-147
30. Lomer CH, Lomer CJ. 2001. Formulations and applications of mycopathogens. In *LUBILOSA Insect Pathology Manual*. UK: CABI Bioscience Publication
31. Lord JC, Howard RW. 2004. A proposed role for the cuticular fatty amides of *Liposcelis bostrychophila* (Psocoptera: Liposcelidae) in preventing adhesion of dry-conidia entomopathogenic fungi. *Mycopathologia* 158: 211-217
32. Maranhao EAA, Santiago-Alvarez C. 2003. Occurrence of entomopathogenic fungi in soils from different parts of Spain. *IOBC WPRS Bull* 26: 59- 62
33. McCabe PM, Pfeiffer P, Van Alfen NK. 1999. The influence of dsRNA viruses on the biology of plant pathogenic fungi. *Trends Microbiol.* 7: 377-381
34. McCoy CW. 1990. Entomogenous fungi as microbial pesticides. In *New Directions in Biological Control,* eds. RR Baker, PE Dunn, pp. 139-159. New York: Alan R. Liss
35. Milner RJ. 1994. History of *Bacillus thuringiensis*. *Agric. Ecosys. Environ.* 49: 9-13
36. Milner RJ. 2004. From lab to field: the critical factors in the development and application of mycoinsecticides. In *Proc. International Workshop on Entomopathogenic fungi: a*

*valuable alternative to fight against insect pests,* ed. MV Deshpande, pp. 3-15, Pune, India: NCL

37. Nahar P, Ghormade V, Deshpande MV. 2004. The extracellular constitutive production of chitin deacetylase in *Metarhizium anisopliae*: a possible edge to entomopathogenic fungi in the biological control of insect pests. *J. Invertebr. Pathol.* 85: 80-88

38. Nahar PB, Kulkarni SA, Kulye MS, Chavan SB, Kulkarni G, Rajendran A, Yadav PD, Shouche Y, Deshpande MV. 2008. Effect of repeated *in vitro* sub-culturing on the virulence of *Metarhizium anisopliae* against *Helicoverpa armigera* (Lepidoptera: Noctuidae). *Biocontrol Sci. Technol.* 18: 337-355

39. Nahar P, *et al.* 2003. Comparative evaluation of indigenous fungal isolates *Metarhizium anisopliae* M 34412, *Beauveria bassiana* B 3301 and *Nomuraea rileyi* N 812 for the control of *Helicoverpa armigera* (Hub) on chickpea. *J. Mycol. Plant Pathol.* 33: 372-377

40. Patil RS, *et al.* 2001. Biocontrol of root infecting plant pathogenic fungus, *Sclerotium rolfsii* using mycolytic enzymes and chitin metabolism inhibitors singly and in combination. *J. Biol. Control* 15: 157-164

41. Patil RS, Ghormade V, Deshpande MV. 2000. Chitinolytic enzymes: An exploration. *Enz. Microb. Technol.* 26: 473-483

42. Quesada-Moraga E, Vey A. 2003. Intra-specific variation in virulence and *in vitro* production of macromolecular toxins active against locust among *Beauveria bassiana* strains, and effects of *in vivo* and *in vitro* passages on these factors. *Biocontrol Sci. Technol.* 13: 323–340

43. Roberts WA, Selitrennikoff CP. 1988. Plant and bacterial chitinases differ in antifungal activity. *J. Gen. Microbiol.* 134: 169-176

44. Schmidt-Tiedemann A, *et al.*, 1999. Mating disruption tests to control *Sparganothis pilleriana* (Schiff.) (Lepidoptera: Tortricidae) *IOBC wprs Bulletin* 22(9)

45. Shapira R, Ordentlich A, Chet I, Oppenheim AB. 1989. Control of plant disease by chitinase expressed from cloned DNA in *Escherichia coli. Phytopathol.* 79: 1246-1249

46. Shternshis MV, Beljaev AA, Shpatova TV, Bokova JV, Duzhak AB. 2002. Field testing of bacticide, phytoverm and chitinase for the control of

raspberry midge blight in Siberia. *Biocontrol*. 47: 696-706.

47. Shirazi F, Kulkarni M, Deshpande MV. 2007. A rapid and sensitive method for screening of chitinase inhibitors using Ostazin Brilliant Red labeled chitin as a substrate for chitinase assay. *Lett. Appl. Microbiol*. 44: 660-665.
48. Smirnnoff WA. 1973. Results of tests with *Bacillus thuringiensis* and chitinase on larvae of the spruce budworm. *J. Invertebr. Pathol.* 21: 116-118
49. Smirnnoff, WA. 1974. Three years of aerial field experiments with *Bacillus thuringiensis* plus chitinase formulation against the spruce budworm. *J. Invertebr. Pathol*, 24: 344-348
50. Sosa-Gomez DR, Boucias DG, Nation JL. 1997. Attachment of *Metarhizium anisopliae* to the southern green stink bug *Nezara viridula* cuticle and fungistatic effect of cuticular lipids and aldehydes. *J. Invertebr. Pathol.* 69: 31-39
51. Spadaro D, Lodovica MG. 2005. Improving the efficacy of biocontrol agents against soil-borne pathogens. *Crop Protection* 24: 601-613
52. St. Leger RJ, Cooper RM, Charnley AK. 1986. Cuticle-degrading enzymes of entomopathogenic fungi: regulation of production of chitinolytic enzymes *J. Gen. Microbiol.* 132: 1509-1517
53. Sugimoto M, Koike M, Hiyama N, Nagao H. 2003. Genetic, morphological, and virulence characterization of the entomopathogenic fungus *Verticillium lecanii J. Invertebr. Pathol.* 82: 176-187
54. Tanaka H, Ogasawara N, Nakajima T, Tamari K. 1970. Cell walls of *Pyricularia oryzae*. I Selective enzymolysis of *Pyricularia oryzae* walls by wall lytic enzymes of *Bacillus circulans* WL 12. *J. Gen. Appl. Microbiol.* 16: 39-60
55. Uma Devi K, Murali Mohan CH, Padmavathi J, Ramesh K. 2003. Susceptibility to fungi of cotton boll worms before and after a natural epizootic of the entomopathogenic fungus *Nomuraea rileyi* (Hyphomycetes). *Biocontrol Sci. Technol.* 13: 367-371
56. Vandenberg, JD, Cantone FA. 2004. Effect of serial transfer of three strains of *Paecilomyces fumosoroseus* on growth in vitro, virulence, and host specificity *J. Invertebr. Pathol.* 85: 40-45
57. Vey A, Hoagland RE, Butt TM. 2001. Toxic metabolites of fungal biocontrol agents. In *Fungi as biocontrol agents: progress, problems and potential,* eds. TM Butt, CW

Jackson, N Magan, pp. 311-346. UK: CAB Publishing Co.

58. Vimala Devi PS, *et al.* 2003. Identification of virulent isolates of the entomopathogenic fungus *Nomuraea rileyi* (F) Samson for the management of *Helicoverpa armigera* and *Spodoptera litura*. *Mycopathologia* 156: 365-373

59. Vimala Devi PS. 1994. Conidia production of the entomopathogenic fungus *Nomuraea rileyi* and its evaluation for control of *Spodoptera litura* (Fab) on *Ricinus communis*. *J. Invertebr. Pathol.* 63: 145-150

60. Woo S, Fogliano V, Scala F, Lorito M. 2002. Synergism between fungal enzymes and bacterial antibiotics may enhance biocontrol. *Ant. Van Leeuwenhoek* 81: 353-356

61. Zimmermann G. 1986. The *Galleria* bait method for detection of entomopathogenic fungi in soil. *J. Appl. Entomol.* 102: 213-215.

*Rev. Plant Pathol. Vol. 4, 2008*
*Indian Society of Mycology and Plant Pathology*
*Scientific Publishers (India), Jodhpur pp. 421- 466*

# SUPPRESSION OF PLANT PARASITES INCLUDING NEMATODES BY A M FUNGI INDUCED RESISTANCE IN PLANTS

***Subhash Bhargava[1], Mahaveer P. Sharma[2], Rakesh Pandey[3] and H.N. Gour[4]***

[1]Department of Nematology, Rajasthan College of Agriculture, MPUAT, Udaipur-313001, India (e-mail subhash_bhargav@yahoo.co.in)
[2]Microbiology Section, National Research Centre for Soybean (NRCS-ICAR), Khandwa Road, Indore-452 001, India (e-mail mahaveer620@yahoo.com)
[3]Central Institute of Medicinal & Aromatic Plants (CIMAP-CSIR) PO CMAP, Lucknow- 226015 India (e-mail rakeshpandey66@hotmail.com)
[4]Department of Plant Pathology, Rajasthan College of Agriculture, MPUAT, Udaipur- 313001, India (e-mail gour_hn@sancharnet.in)

KEY WORDS: Mycorrhiza, nematode, biocontrol, resistance, plant diseases

## ABSTRACT

Most plant species form symbiotic associations with arbuscular mycorrhizal fungi. These fungi could be the most important and poorly understood resources for its biocontrol potential in agriculture. Plant parasitic nematodes and arbuscular mycorrhizal fungi (AMF) share a common plant roots as a source of food and space. The interest in AMF-nematode interactions lies in the possibility of enhanced resistance or tolerance of AMF-infected plants to nematodes and the potentials of this for control of soil-borne plant diseases. Better understanding of the interactions between arbuscular mycorrhizal fungi and plant parastic nematodes is necessary for the development of sustainable management of crop protection. This review deals with the role of AM fungi in suppression of plant parasitic nematodes, understands their interaction with various plants and production and commercialization of AMF in order to develop strategies for achieving commercial mycorrhizal biocontrol agents against soil-borne diseases. Mycorrhizal associations enable a better use of sparingly soluble phosphorus pools, thereby increasing the efficiency of added phosphorus fertilizer and of the large relatively immobile phosphorus pools, which confer tolerance against plant parasitic nematodes. The crop and soil management practices are also observed to influence the efficacy of native AMF potential to protect plant against soil-borne diseases. Varieties of examples dealing with use of AMF to control plant parasitic nematodes associated with variety of plants under different situations have been provided. Strategies that are needed to increase the efficacy of mycorrhizal fungi to suppress nematodes infections are narrated. Practical commercial application and utilization of mycorrhizal fungi as biocontrol agents are tremendous and would still seem to have a long way to go.

## 1. INTRODUCTION

Albert Bernard Frank in 1885 introduced the Greek word "mycorrhiza" which literally means "fungus roots". Mycorrhizal fungi form a symbiotic relationship with plant roots in a similar fashion to root-nodule bacteria in legumes. The fungus takes carbohydrates from the plant and in return supplies the plant with nutrients, hormones, etc. Arbuscular mycorrhizal fungi (AMF) are found to occur on roots of most of the food, medicinal and aromatic, horticultural crops and tropical trees. It has also been estimated that about 95 per cent of the world's present species of vascular plants belong to families that are characteristically mycorrhizal, which demonstrate the usefulness of such organisms as biological control agents. The presence of arbuscules and in some cases vesicles is a diagnostic criterion for identifying arbuscular mycorrhizal fungi in roots. AMF belong to the fungal phylum Glomeromycota (115), which currently comprises approximately of 150 described species distributed among ten genera, most of which have been defined mainly by spore morphology. A large proportion of the AM fungi were previously placed in the phylum Zygomycota, but several facts indicate that they form a monophyletic group distinct from other Zygomycotan lineages, among them their symbiotic habit, apparently lack zygospores and rDNA gene phylogeny. Evidently the scientists (115) erected the phylum Glomeromycota. These authors also corrected the previously used, grammatically incorrect ordinal name "Glomales" to "Glomerales". In phylogenetic trees based on rDNA, the Glomeromycota are the sister group to Asco- and Basidiomycota. AM fungi are thought to be the oldest group of asexual multicellular organisms. Fossil records have shown that they are at least 460 million years old (116). The AM fungi are considered asexual organisms, since no evidence exists that they are able to reproduce sexually. Therefore mutation and possibly heterokaryosis probably provide the main basis for the variation necessary to permit adaptation to environmental changes and continuing evolution (134)

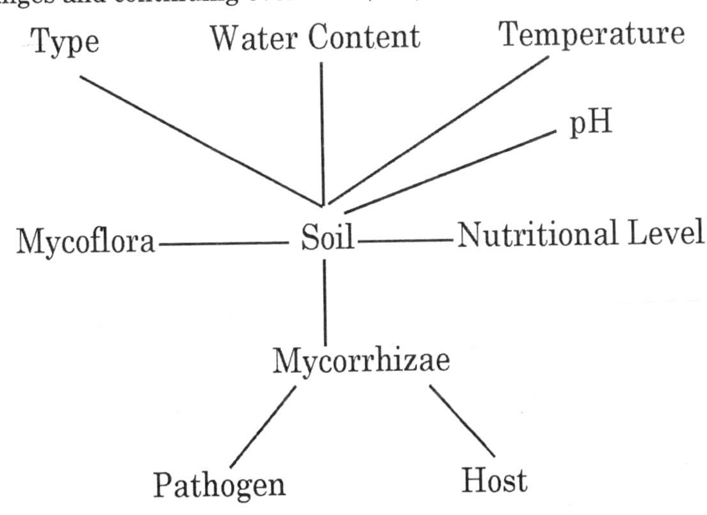


**Fig. 1.** Interaction between the soil, the pathogen, the host and the mycorrhizae as influenced by various factors

The name "nematode" comes from the Greek words: *nema*, which means "thread", and *toid*, means "form". Nematodes are multicellular, invertebrate, unsegmented, vermiform, long and slender, but some species are swollen. Most people know them as roundworms because their cross-section is round. These nematodes occupy different ecological niches like soil, marine, fresh water form as well as predator and parasite of vertebrate, invertebrate and plants. Several of parasitic forms of nematode causes disease in plants, humans, and animals. These nematodes also serve as food for different organisms like bacteria, fungi, insects and others. Some nematodes are parasites to insects and used as biocontrol agents to manage insect problem in plants, while other nematodes play some beneficial role like cycling of mineral nutrients, decomposition of organic wastes including biodegradation of toxic compounds, acting as indicator of soil and water quality.

Plant parasitic nematodes generally fall in two major groups, nematodes that spend their most of life stages in roots and soil and other which spend at least part of their life stages in foliage of plants. The soil form are: root-knot nematodes, *Meloidogyne* spp., which have wide host range and other commonly root feeding nematodes are stunt nematodes (*Tylenchorhynchus* spp.), lance nematodes (*Hoplolaimus* spp.), spiral nematodes (*Helicotylenchus* spp.), lesion nematodes (*Pratylenchus* spp.), cyst nematodes (*Heterodera* spp.), and ring nematodes (*Criconema* spp.). The aerial parasitic nematodes generally belong to the genera *Aphelenchus* or *Aphelenchoides or Ditylenchus,* etc.

The average losses caused by phytonematodes to different agricultural crops including vegetables, sugar, cotton, cereals, pulses, oil seeds, tubers, medicinal and aromatic plants are more than 100 billion dollars every year. Due to hazardous effects of chemical nematicides on soil, water, air and human, most of effective nematicides have been banned since 2005. Recently nematological researchers are engaged to develop some new nematicides through use of different phytoextracts (92), and micro-organisms (127) having achieved significant success in this direction, consequiently Pasteuria penetrans, *a* parasitic soil bacterium that infects and kills the nematodes, has gained popularity in USA. This bacterium attacks targeted phytonematodes and is highly durable. The active component of Pasteuria is its endospore, the most robust life form known, resistant to heat, drying, shearing and freezing. As such in USA, a Pasteuria Bioscience Company (PBC) has developed a formulation which will be soon available in Asian markets. The safest way of phytonematode management is by using different other biological strategies like use of nematode parasitic bacteria, PGPR, nematophagus fungi, egg parasitic fungi and arbuscular mycorrhizal fungi.

The mycorrhiza and plant parasitic nematodes both infest root systems, but mycorrhizae are useful to plants, whereas phytonematodes are detrimental to plant health. Mycorrhizal fungi become a useful tool to manage nematode population in one way while in other way it provides several useful effects on host plant to increase crop yield at significant level. Phytonematodes and arbuscular mycorrhizal fungi both remain associated with plant roots for their nutrition and space. The major interest in such association is to provide increased plant resistance against phytonematodes. Symbiotic association of mycorrhizal fungi with plants provides a range of beneficial effects like (a) enhanced micro and macro nutrient availability, which provides resistance to plant against different biotic and abiotic stresses making soil rich in micronutrients; and (b) make plants more tolerant to heavy metal, and drought resistance. However, several agricultural practices like tillage, use of chemical fertilizers, monocultures, herbicides, nematicides, fungicides as well as growing non host crop of mycorrhizal fungi are highly detrimental to such useful AM fungi. Even in organic cultivation, these AMF provide sufficient nutrients to plants, well colonize the root system thus reducing several inputs in crop cultivation. There is full evidence that these AM fungi reduce nematode problem at bay, if used with other management practices carefully(64).

Biological control of plant pathogens is currently accepted as a key practice in sustainable agriculture because it is based on the management of natural resources, i.e. certain rhizosphere organisms, common components of ecosystems, known to develop antagonistic activities against harmful organisms (nematode, bacteria, fungi, etc.). Arbuscular Mycorrhizal (AM) association has been shown to reduce damage caused by soil borne plant pathogens. Although few AM isolates have been tested in this regard, few appear to be more effective than others. Furthermore, the degree of protection varies with the pathogens involved and can be modified by soil and other environmental conditions. This prophylactic ability of AM fungi could be exploited in cooperation with other rhizospheric microbial antagonists to improve plant growth and health.

It has been recognised that mycorrhizal symbiosis plays a vital role in nutrient cycling in the ecosystem and also protects plants against environmental and cultural stresses (8). Generally, AMF cause few changes in root morphology, but the physiology of the host plant changes significantly. The improved potential for mineral uptake from the soil accounts for changes in the nutritional status of the host tissues, which in turn, changes the structural and biochemical aspects of root cells. This can alter membrane permeability and thus the quality and quantity of root exudation induces changes in the composition of microorganisms in the rhizospheric soil, which could be known as 'mycorrhizosphere'. The net effect of these alterations is a tolerant plant better able to withstand environmental stresses and plant diseases. To evaluate the influence of AM fungi on diseases incidence and development, the variable factors *viz.* plant pathogens, the symbiotic fungus and environmental conditions, have been considered. Mostly, the host plant mediates the interactions between pathogen and the symbiont. The characterization of these interactions should, therefore, include information on the mechanisms involved because interactions vary with specific host-symbiont- pathogen combination; it is difficult to make generalizations on the impact of AMF on plant diseases.

The purpose of this review is to attempt to (a) discuss the potentiality of AMF to suppress the soil-borne plant parasitic nematodes and understand their interactions and behaviour and (b) exploit the possibilities of mass production and commercialization of AMF.

## 2. ARBUSCULAR MYCORRHIZAE IN THE SUPPRESSION OF SOIL-BORNE NEMATODES

AM fungi have the potential to reduce damage caused by soil-borne pathogenic fungi, nematodes and bacteria. Meta-analysis showed that AM fungi generally decreased the effects of fungal pathogens (18,120). Effects on nematodes were more variable and thus a general pattern is not evident and pattern therefore needs to be tested over a wider range of plants and nematodes of different feeding strategies. The beneficial effect of mycorrhizal symbiosis on plant growth and interaction with nematodes has led to investigations into the potential of AM fungi to limit yield losses due to nematodes (Tables 1 & 3). The effect of AM fungi on the development and nutrition of the peach almond hybrid GF-677 in a replant soil heavily infested with *M. javanica* was studied (17) and significant beneficial effects of mycorrhizal inoculation on plant growth and nutrition in soil and also suppression of root-knot nematode reproduction were found. Symptoms of nematode infection are generally reduced and often, but not always, nematode populations are also reduced (as indicated by number of galls, juveniles or eggs per unit root length) (64). Inoculations of tomato transplants with *G.fasciculatum* significantly reduced root penetration by juveniles and the development of *R.reniformis* compared to controls. Development of the gelatinous matrix was delayed and fewer eggs per eggmass were produced on inoculated plants. The number of galls formed by *M.incognita* on tomato was significantly lower in AM-inoculated tomato plants. However, the AM colonization did not prevent the penetration by the larvae (139). Inoculation of tomato roots with increasing levels of *G. fasciculatum* progressively enhanced the plant growth along with its own reproduction in term of root colonization and chlamydospore production. Highest percentage of AM colonization, maximum number of chamydospores and maximum reduction in number and size of nematode galls were observed in the plants inoculated simultaneously with 500 second-stage larvae and 200 chlamydospores per plant/ pot. Similarly, uptake of nutrients viz., nitrogen, phosphorus and potassium was also enhanced considerably with the simultaneous inoculation than with the plants inoculated with the nematode alone (84). Talavera *et. al.* (141) found that when tomato seedlings were inoculated with the nematode three weeks after mycorrhization, colonization of tomato roots by *G.mosseae* compensated for the reduction of plant growth caused by *M. incognita* infection and final soil densities of nematode were reduced drastically by 85%. AM fungi also conferred greater beneficial effects when applied in combination with bacterium. Some workers (142) found that tomato shoot and fruit weights in *M. incognita* – infected plants receiving a combined application of *Glomus* sp. and *P. penetrans* were greater than untreated nematode plants. Siddiqui and Mahmood (124) observed simultaneous use of biocontrol agents against pathogens to give better control than individual application. The highest reduction in nematode multiplication was observed when all three biocontrol agents namely *G.mosseae,Trichoderma harzianum* and *Verticillium chlamydosporum* were applied in combination. The inoculation of *G. mosseae* stimulated the growth of citrus jambhiri seedlings. Simultaneous inoculations of AM fungus and the citrus nematode, *Tylenchulus semipenetrans* partly neutralized the adverse effect of the nematode (5). Pre-inoculations of *Piper nigrum* cv. *panniyar* colonizing with *G. fasciculatum* or *G. etunicatum* significantly reduced root-knot index (*M. incognita*) by 32.4 and 36.0% respectively and also improved the growth of piper plants (123). Krishna Prasad (72) reported that the percentage of root-knot nematode (*M. incognita*) infestations in tobacco (*Nicotiana tabaccum*L.) seedlings was 67.5% at 50 days and 95% at 75 days after sowing in non-mycorrhizal plants and 50% at 50 days and 73% at 75 days after sowing in mycorrhiza (*G. fasciculatum*) inoculated plots. The number of galls, egg masses per plant, and eggs per egg mass of infested plants were reduced by 61-89% as a result of AM inoculation. Transplanting of mycorrhizal tobacco seedlings into root-knot nematode infested soil showed improved growth and yield compared to non-mycorrhizal plants. Al-Raddad and Ahmad (1) showed that pre-inoculations of *G. mosseae* significantly reduced root-knot infection and reproduction of root-knot nematode on tomato. The ability of mycorrhizal plants to grow well despite infection by nematodes is generally considered to be the principal affect of mycorrhizal fungi or the interaction of host plants with parasitic nematodes (65). Furthermore, a comparison of mycorrhizal and P-fertilized non-mycorrhizal plants showed that the latter are more susceptible to nematode attack indicating the likely involvement of factors other than P nutrition in the interactions. On the other hand, mycorrhizal inoculations of tamarillo (*Cyphomandra betacea*) against root-knot nematodes *M. incognita* could not be duplicated by adding P fertilizer and was not therefore due to merely improved P nutrition of the host (26). Heald *et al.* (57) showed that *M. incognita* suppressed the growth of non-mycorrhizal *Cucumis melo* plants by 84% as compared to 21% in AM inoculated plants at 50 mg/g P. A similar trend was observed in soil with 100 mg/g phosphorus. Pre-inoculations of *G. fasciculatum* in tomato var. Pusa Ruby significantly reduced the number of galls per plant, egg masses per plant and eggs per egg mass due to *M. incognita*. It was also observed that the effect of AM fungi on reducing root-knot nematode was not the same in pots and in the field. Nematode population of AM-mediated tomato grown in the plots was reduced up to 36% where as in the pots, the nematode reduction was 58% (118).

**Table 1.** Influence of arbuscular mycorrhizal fungi (AMF) on nematode diseases of Fruit, vegetable and other crops

| Nematode | Host Plant | Incidence | Reference |
|---|---|---|---|
| *Meloidogyne arenaria* | Clover | - | 71 |
| *M. hapla* | Tomato | - | 25 |
| *M. hapla* | Alfalfa | - | 52 |
| *M. incognita* | Cucumber | - | 104 |
| *M. incognita* | Tomato | - | 6 |
| *M. incognita* | | - | 138 |
| *M. incognita* | | - | 19 |
| *M. incognita* | Eggplant | - | 107 |
| *M. javanica* | Tomato | - | 6 |
| *M. javanica* | Tomato | - | 130 |
| *M. javanica* | Tomato | - | 1 |
| *M. incognita* | Tomato | - | 142 |
| *Pratylenchus penetrans* | Cucumber | .... | 104 |
| *R. reniformis* | Tomato | - | 131 |
| *Meloidogyne* sp. | Egg plant | - | 56 |
| | Lady finger | - | 56 |
| | Tomato | - | 56 |
| *M. incognita* | | - | 67 |
| *Pratylenchus coffeae* & *Radopholus similis* | Banana | - | 41,42,43 |

- decrease; ..... increase

The sequence in which plants become mycorrhized and infected by nematodes may affect the interaction between these organisms on certain plants. Pre-inoculation of tomato and other transplanted crops with AM fungi to allow this slow growing symbiont to become established in the roots before the attack of *M.*

*incognita* resulted in fewer juveniles penetrating and developing to maturity in the roots of mycorrhized plants than in the roots of control plants (119). Studies conducted by several workers demonstrated that pre-inoculations of AM fungus suppressed nematode reproduction and development in roots to a greater degree as compared to plants inoculated simultaneously with both the organisms (66,138). Apart from reduced penetration of roots by nematodes, mycorrhization also resulted in reduced development of the pathogen (119). Priestel (104) reported specific decrease in reproduction rate of larvae of *Meloidogyne*, which resulted in strong reduction in egg production. This observation assume great significance in the field and helped to explain their changed population dynamics in the mycorrhizal root system (131). While evaluating three species of arbuscular mycorrhizal fungi, *G. aggregatum, G. intraradices* and *G. mosseae* for their effectiveness to suppress plant parasitic nematode *Meloidogyne incognita* in white clover found that growth of plant was significantly stimulated by mycorrhizal colonization (53). The degree to which mycorrhizal fungi reduced nematode damage varied with the species of mycorrhizal fungus; the extent of damage reduction ranged from 19 to 49.8% based on shoot mass. Correlation between *G. fasiculatum* spore density, root colonization and *Meloidogyne incognita* infection on tomato was studied and found that at least two spores/g soil were required for 50% root colonization under pot conditions. However under nursery conditions, four-spores/g soil were required to obtain 50% root colonization which reduced nematode populations in root (in terms of egg masses/plant) by 59 and 61%, respectively (89).

Azcón-Aguilar and Barea (3) concluded that AM-induced increase in resistance or a decrease in susceptibility required the pre-establishment of AM and extensive development of the symbiosis before the pathogen attack. The incorporation of mycorrhiza and oil cakes significantly increased plant growth characters of *Crossandra* compared to that of nematode inoculated checks. However, the magnitude of increase in each character under report varied with the combinations of *G. mosseae* and oil cakes. *G. mosseae* promoted better plant growth over their individual applications.The root galls were least on the plants treated with *G. mosseae* + neem cake followed by *G. mosseae* + karanj cake and *G. mosseae* + castor cake treated plants (88). The inoculation of mulberry with appropriate AM fungus yielded a considerable improvement in plant growth characters. The AM colonization induced a sort of resistance against root-knot nematode infestation and multiplication (23). Recently, De la Pena *et al.* (33) reported that infection and multiplication of *P. penetrans* were significantly reduced by the native inoculum of AMF in pioneer dune grass *Ammophila arenaria*. Plant preinoculation with AMF further decreased nematode colonization and reproduction. Nematode suppression by AMF did not occur through a systemic plant response but through local mechanisms.

## 3. INTERACTION OF ARBUSCULAR MYCORRHIZAL FUNGI WITH SOIL BORNE NEMATODES ASSOCIATED WITH PLANTS

Protective ability of mycorrhizae is generally observed against soil borne diseases and is often related to the nature of the host plant, mycorrhizal symbionts, plant pathogen and condition of the soil environment. Mycorrhizal fungi and plant parasitic nematodes are commonly found inhabiting the rhizosphere and colonising the roots of their host plants. These two groups of microorganisms exert opposite effects on the plant growth (65). As a result of interactions, in general the severity of nematode diseases is reduced in mycorrhizal plants. However in some cases, root colonisation by mycorrhizal fungi had no effect on nematode reproduction and in some found it to be stimulated (124) (Table 2). Parasitism of nematode eggs with AM fungus has been demonstrated but the level of parasitism was not considered sufficient enough to affect nematode activities negatively (46). In most of cases, AM fungi colonize only stressed or weakened nematode eggs (124). The nematode parasitism by AM fungi is an opportunism and depends on carbon nutrition from autotrophic symbionts, rather than being representative of a true host-parasitic relationship (124). The negative effect of AM fungi on nematode development and reproduction is mainly due to histopathological changes occurred in cells like smaller syncytia and fewer giant cells which confer resistance against *Meloidogyne* sp. on the host plants. The alteration of host physiology and biochemical changes brought by mycorrhizal plants has been found to be one of the factors conferring the resistance against nematodes (34,17).

**Table 2.** Interactions between plant-parasitic nematodes and AM Fungi.

| Type of interaction | Component | Effect on component |
|---|---|---|
| Positive | Fungus | Root infection or sporulation increased. |
| | Host | Nematode suppression of vegetable growth and yield offset. |
| | Nematode | Attraction to roots, penetration or subsequent reproduction and development suppressed. |
| Neutral | Fungus | Root infection or sporulation not altered. |
| | Host | Mycorrhizal stimulation of vegetative growth or yield not altered; nematode suppression of vegetable growth or yield not offset. |
| | Nematode | Attraction to roots, penetration, or subsequent development and reproduction not altered. |
| Negative | Fungus | Root infection or sporulation suppressed. |
| | Host | Yield response to mycorrhiza suppressed. |
| | Nematode | Attraction to roots, penetration or subsequent reproduction and development increased. |

Source: Hussey and Roncadori (1982)

Apart from AM fungi alone, the role of fertilisation has been also worked out on the efficacy of AM fungi to reduce nematode infestations (22). Recently, Waceke *et al.* (150) worked on inorganic phosphatic fertilizers on the efficacy of an arbuscular mycorrhiza fungus against root-knot nematodes on pyrethrum. The fertilizers doses were applied @ 150 and 300 kg $ha^{-1}$ alongwith fungus inoculation after two months of which plants inoculated with the nematodes. The fertilizers at both levels found to improve plant growth in all treatments. In general, the fungus improved plant growth on its own or in the presence of nematodes but not in the presence of fertilizers. Both fertilizers at both the levels were more effective in improving plant growth compared to fungus. The fungus showed sensitivity to inorganic P fertilizers in that the fertilizers significantly reduced fungal root colonization and its pyrethrum growth stimulative effects. The fungus suppressed nematode disease severity unlike the fertilizers. The suppressive effects of the fungus on the nematodes were in most cases reduced by the fertilizers. The nematodes, unlike the fertilizers, did not have any significant effects on root colonization by the fungus or on its ability to improve pyrethrum growth. The presence of nematodes in fertilizer or fertilizer-fungus-treated plants, however, significantly reduced pyrethrum growth. A number of recent reports are available on fruit and vegetable crops where AM fungi found to reduce nematode infestations however, the degree of reduction found to be varied due to AM application combinations (17,41,42,43, 53,105,141,142).

Interactions between AMF and nematodes have been studied in banana by Jaizme-Vega *et al.* (67) who stated that inoculation with the fungus *G.mosseae* increased banana performance and reduced *M. incognita* reproduction. There was however no effect of the nematode on mycorrhizal development. The interaction between mycorrhizal banana and the *Meloidogyne javanica* (Treub) Chitwood although did not reduce the reproduction of the nematode, a positive growth response of banana was observed (96). Timings of the interactions between AMF and nematodes can be of prime importance. Vaast *et al.* (147) noted that simultaneous inoculation of AMF with root lesion nematode *Pratylenchus coffeae* Goodey on coffee did not enhance tolerance in coffee, but early inoculation (4 months before coffee plants were challenged with the nematodes) significantly improved the tolerance in coffee. In early, inoculation the nematode did not affect AM colonization, while in simultaneous inoculation, the nematode suppressed mycorrhizal colonization. An experiment involving banana and two nematode species (*P. coffeae* and *R. similis*) showed that if the AM fungus *G. mosseae* added 2 months before the nematodes it affects the nematodes negatively. The effects were to a large extent due to the improvement of the nutritional status of banana (43). Finally, the AM symbiosis with *G. intraradices* annulled differences in nematode species composition as found in the rhizosphere of different Sahelian legumes (148). A brief account of work published on interaction between AMF colonized plants and nematode has been presented in table 3.

**Table 3.** Interactions between arbuscular mycorrhizal fungi (AMF) and plant parasitic nematodes associated with important crop plants*.

| Plant | Nematode species | AMF | AMF effect on nematode damage root (%) | Nematode response (%) | AMF response (%) | Reference |
|---|---|---|---|---|---|---|
| Pigeon pea | *H. cajani* | *G. mosseae* | ND | 18-42 | 5-11 | 26 |
| | *H. glycine* | *Glomus* spp. + *Gigaspora* spp. | -27-11 | -142 -38 | ND | 46 |
| Soybean | *H. glycine* | *G. intraradices* | -8 | ND | 10 | 103 |
| | *H. glycine* | *G. mosseae* | -17-10 | ND | 27-28 | 144 |
| Potato | *G. pallida, G. rostochiensis* | Vaminoc (commercial AMF) | ND | ND | 2-3 | 111 |
| | *G. pallida* | | 18 | 6 | ND | 110 |
| Green gram | *M. incognita* | *G. fasiculatum* | ND | 70 | 22 | 113 |
| Lady finger | | *G. mosseae* | 47 | 2 | 7 | 122 |
| | *M. javanica* | *G. mosseae* | -6-39 | 8-37 | 3-6 | 25 |
| Tomato | *M. incognita* | *G. mosseae* | 15-19 | 6-12 | -13-10 | 12 |
| | | *G. etunicatum* | -5 | 60 | 13 | 11 |
| | | *G. fasiculatum* | ND | 5-77 | ND | 89 |
| | | *G. coronatum* | -10-19 | -20-72 | -6-22 | 36 |
| *Mentha* | *M. incognita* | *G. aggregatum, G. fasiculatum, G. mosseae* | ND | ND | ND | 95 |
| | | *G. mosseae* | -6-2 | 5-27 | 10-19 | 108 |
| Chamomile *(Matricaria chamomilla)* | *M. incognita* | *G. mosseae, G. fasiculatum* | ND | ND | ND | 94 |
| Henbane *(Hyoscyamus niger)* | *M. incognita* | *G. aggregatum* | ND | ND | ND | 93 |

| | | | | | | |
|---|---|---|---|---|---|---|
| Pyrethrum (*Chrysanthemum cinerariefolium*) | *M. hapla* | *G. etunicatum, Glomus* sp. K14 | -3-29<br>9-20 | 54-57<br>63-71 | 10-11<br>4-7 | 149 |
| *Capsicum* | *M. incognita* | *G. mosseae* | ND | 17-53 | ND | 137 |
| Zinger | *M. incognita* | *G. mosseae,*<br>*G. fasiculatum* | ND | 18-23<br>64<br>68 | ND | 90 |
| Muskmelon | | *G. fasiculatum* | ND | 18-29 | ND | 106 |
| Banana | *R. similis*<br>*P. coffeae* | *G. mosseae* | -75-67<br>-10-0 | -340-48<br>58-63 | -5-22<br>17-24 | 43 |

*Modified from (62); ND- not presented; Gi, *Glomus intraradices*; Gm, *Glomus mosseae*; Gf, *Glomus fasciculatum*

## 4. MECHANISMS INVOLVED IN PATHOGEN SUPPRESSION BY AM FUNGI

It is evident that an increased capacity for nutrient acquisition resulting from mycorrhiza association could help the plant to become stronger to resist stress. However, AM symbiosis may also improve plant health through a more specific increase in protection (8).

In fact, an AM-induced increase in resistance or a decrease in susceptibility requires the pre-establishment of AM and extensive development of the symbiosis before pathogen attack. The hypotheses proposed to explain AMF effects on soil-borne plant pathogens generally considered to have either a physical or physiological basis (65). In reviews of biological control of plant diseases, mycorrhizae are thought to be biological controlling agent against plant diseases primarily by means of stress reduction. The biological control of plant diseases may be strongly affected by AM fungi by one or more mechanisms involving different mechanisms (4). One mechanism is via the changes in microbial communities that are brought about the development of the new mycorrhizosphere. There is strong evidence that shifts in microbial community structure and the resulting microbial changes can strongly influence the growth and health of plants (78). A major mechanism is nutritional, because plants with a high phosphorus status are less sensitive to pathogen damage. Non-nutritional mechanisms are also important, because mycorrhizal and non-mycorrhiza plants with the same internal phosphorus concentration may still be differentially affected by pathogens (18). Such non-nutritional mechanisms include activation of plant defense systems, changes in exudation patterns and concomitant changes in mycorrhizosphere populations, increased lignification of cell walls, and competition for space for colonization and infection sites. Activation of plant defense mechanism including the development of systemic resistance has also been proposed by Pozo *et al.* (100), but the occurrence of this mechanism and its effect in biological control, needs further evaluation.

## 4.1. Physical mechanisms

### *4.1.1 Morphological changes in roots and root tissues*

It has been demonstrated that AM colonization induces remarkable changes in root system morphology as well as in the meristematic and nuclear activities of root cell (2) which might affect rhizosphere interaction and particularly pathogen-infection development. The most remarkable consequence of AM colonization is an increase in branching, resulting in a relatively larger proportion of higher order roots in the root system (63). Lignification is reported to prevent penetration of mycorrhizal plants by pathogens. Dehne and Schonbeck (35) showed increased lignification of cells in the endodermis of mycorrhizal tomato and cucumber plants and speculated that such responses accounted for reduced incidence of Fusarium wilt (*F. oxysporum f. sp. lycopersici*). Baker (7) reported a similar effect on pink root of onion (*Pyrenochaeta terrestris*). Increased wound-barrier formation inhibited thielaviopsis black root rot (*T. basicola*) of mycorrhizal holly (*Ilex crenata*) plants.

A stronger vascular system in mycorrhizal plants increases flow of nutrients that imparts greater mechanical strength and reduces the effect of vascular pathogens (114). Smaller syncytia with fewer cells increased the resistance of host to nematodes (44). In most studies on AM fungi and biocontrol agents, the roots have not been examined for anatomical changes. Thus, more attention needs to be given to root system morphology in future because it could modify the infection dynamics of the pathogen as well as the pattern of resistance of AM roots to pathogen attack.

## 4.2. Physiological mechanisms

### *4.2.1. Nutritional Changes*

Since increased nutrient uptake made possible by the AM symbiosis results in more vigorous plants, the plant itself may thus be more resistant or tolerant to pathogen attack. AM fungi have been shown to affect root growth, root exudations, nutrient absorptions and physiological responses of the host to environmental stresses. Many of these factors are related to P physiology and nutrition (51). An enhanced P nutrition could account for the higher tolerance of mycorrhizal plants to pathogen (nematode). AM fungi enhanced root growth, expanded absorptive capacity and affect cellular processes in roots (133). Davis and Menge (30) showed that the effect of AM fungi on thielaviopsis root rot of citrus was more than non-inoculated plants unless the latter was supplied with additional P. These mycorrhizal-induced compensatory processes may explain the increased tolerance of mycorrhizal and P- fertilized plants because many plants can compensate for loss of root mass or function caused by pathogens (151). Greater tolerance is also attributable to increased root growth and phosphate status of the plant. Host susceptibility to pathogens and tolerance of the disease process can be influenced by the nutritional status of the host and fertility status of the soil. In the absence of AM, phosphate can combine with minor elements to create deficiencies, which would predispose plants to root knot nematodes (132). Thus, AM fungi may increase tolerance to pathogens by increasing uptake of essential nutrients other than P that would be deficient in a non-mycorrhizal plant. These reports indicated improved nutrition as a mechanism for disease control and could account for the higher tolerance of mycorrhizal plants to pathogens. However the improved nutrition to confer tolerance to nematodes may vary with the types of nematode. Tylka *et al.* (146) reported that the differences in carrying capacities between mycorrhizal and nonmycorrhizal plants can mask suppressive effects and others have observed that nematode suppression can vary with fungal species and P nutrition (53). Nevertheless, it seems clear that the potential for suppression of *H. glycines* populations by mycorrhizal fungi is much less than that observed for *M. incognita* on the same host plant. A possible explanation for this differential response is suggested by the observation that improved P nutrition could be suppressive to *M. incognita* (19) but not *H. glycines* (146).

### *4.2.2. Competition for host photosynthates and infection sites*

It has been proposed that the growth of both AM fungi and root pathogens depends on host photosynthates and that they compete for carbon compounds reaching to root (77). When AM fungi have primary access to photosynthates, the higher cabon demand may inhibit pathogen growth. AM fungi are almost totally dependent on soluble carbohydrates produced by the host for their carbon source (54).

The higher carbon demand of AMF may inhibit pathogen growth, however, there is little or no evidence that competition for carbon compounds is a generalized phenomenon for pathogen - biocontrol activity of AM symbiosis (3). Dehne (34) indicated that fungal root pathogens could occupy root cortical cells adjacent to those colonized by AM fungi, indicating a lack of competitions. It has been suggested that nematode pathogens, on the other hand, require host nutrients for development and reproduction and direct competition with AM fungi has been hypothesized as a mechanism of their inhibitors. Since AM fungi, soil-borne fungal pathogens and plant parasitic nematodes occupy similar root tissues, direct competition for space has been postulated as a mechanism of pathogen inhibition by AM fungi (65). Both localized and non-localized mechanisms could be responsible for competition for colonization sites mainly depending on the pathogen (fungus, nematode) and pathogen development is reduced due to lack of resources and food supply (146). Observations by Dehne (34) have illustrated that fungal root pathogens and AM fungi, although colonizing the same host tissues, usually develop in different root cortical cells, thus, indicating some sort of competition for space. Jalali and Jalali (68) have reported a localized effect, whilst the reviews by Dehne (34) suggested that the extent of the protection couldn't be explained by a localized mechanism alone. The changes in root architecture resulting from AMF infection may provide fewer sites for infection (47), and many of these plant growth changes would have arisen as a result of altered plant hormone levels in AM plants (145). *Phytophthora* development was reduced in AM fungal-colonized and adjacent uncolonized regions of AM root systems and that in the former the pathogen does not penetrate arbuscules-containing cells (27).

### *4.2.3 Biochemical changes or changes in chemical constituents of plant tissues*

This represents an indirect contribution to biocontrol through the conservation of root-system function, both by fungal hyphae growing out into the soil and increasing the absorbing surface of the roots and by the maintenance of root cell activity through arbuscules formation in cells (2, 27) which might perhaps affect rhizosphere interactions and particularly pathogen-infection development.

AMF colonization has also conferred structural and biochemical changes in the cell walls of the plants. Synthesis of β-1,3-glucans has been reported in mycorrhizas of pea. β-1,3-glucans has also been detected within the structural host wall material around the point of penetration of AM hyphae into plant cell (49). Triggering of specific plant defense response due to AM colonization can be a most obvious basis for the protective ability of AMF. Current research using modern molecular techniques is being used to detect substances and/or reactions elicited by AM fungi involved in plant protection. Among the compounds involved in plant defence in relation to AM formation include phytoalexins, enzymes of the phenylpropanoid pathway, chitinases, β-1,3-glucanases, peroxidases, pathogenesis-related (PR) proteins, callose, hydroxyproline- rich glycoproteins (HRGP) and phenolics (50).

There is considerable evidence for increase in phytoalexins in mycorrhizal roots infected by *Rhizoctonia solani* (154). Harrison and Dixon (55) reported an increase in the level of the phytoalexin medicarpin transiently during the early stages of AM colonization in *Medicago truncatula*, but decreased to very low levels during later stages of symbiosis development. In another study (155), the effect of colonization with the vesicular arbuscular mycorrhizal fungus *G. etunicatum* on the content of rishitin and solavetivone was determined in potato plants challenged with *Rhizoctonia solani*. Mycorrhization stimulated significantly the accumulation of both phytoalexins in roots of plantlets challenged with *R. solani* but did not influence phytoalexin levels in non-challenged plantlet roots. Furthermore, Lambias (74) showed that the elicitor derived from an extract of extraradical mycelium of *G. intraradices* was able to induce phytoalexin synthesis in soybean cotyledons. Further results have shown that *G. intraradices* induced the expression of chalcone synthase, the first enzyme in metabolism of flavonoid compound, such as phytoalexin, in *Medicago truncatula* (15). Increased levels of phytoalexins in mature mycorrhizal soybean roots could be elicited by the release of fungal cell wall molecules in senescent arbuscules.

Pathogenesis related proteins (PR proteins) are also induced in plants during resistance to pathogen infection. PR proteins are synthesised locally and in small amounts. For example, PR–$b_1$ proteins are only synthesised in cells containing living arbuscules. Measurements of specific enzymes allow selective measurements of different fungi in the same root system (59). Alterations in the isoenzymatic patterns and biochemical properties of some defence-related enzymes such as chitinases (102), chitosanases and ß-1,3-glucanases (99) have previously been shown during mycorrhizal colonization of tomato roots with the induction of new isoforms. These hydrolytic enzymes are believed to have a role in defence against invading fungal pathogens because of their potential to hydrolyse fungal cell wall polysaccharides (128).

Plant chitinases and/or β-1,3 glucanase elicitation has been reported in AM roots (14) and these are antifungal against soil/root pathogens. Mycorrhiza induced chitinases isoforms appear to be a general phenomenon in AM roots. These chitinases release oligosaccharide elicitors from chitinous AM fungal cell walls which inturn stimulates the general defence responses of plants. Dumas-Gaudot *et al.* (40) found that chitinases were little or only transiently induced by AM colonization. These chitinase isoforms differ from those elicited by root fungal pathogens (40), indicating a different pattern of plant response to pathogenic and mutualistic fungi. Furthermore, it was shown that increased levels of chitinase activity are only detected in AM roots at the beginning of colonization (75). A decrease in β-1,3-endoglucanase activity has also been reported at specific stages during mycorrhiza development (75). The decreases were accompanied by differential reductions in the levels of mRNAs encoding for different endochitinase and endoglucanase isoforms. These observations indicated a systemic suppression of the defence reaction in the presence of AM symbiotic interactions. In another study by Pozo *et al.* (100), the ability of two AMF (*G.mosseae* and *G.intraradices*) to induce local or systemic resistance to *Phytophthora parasitica* in tomato roots was compared using a split root experimental system. *G. mosseae* was effective in reducing disease symptoms produced by *P. parasitica* infection, and evidence points to a combination of local and systemic mechanisms being responsible for this bioprotector effect. The biochemical analysis of different plant defence-related enzymes showed a local induction of mycorrhiza-related new isoforms of the hydrolytic enzymes chitinase, chitosanase and ß-1,3-glucanase as well as superoxide dismutase, an enzyme which is involved in cell protection against oxidative stress.

Peroxidase activity has been associated with epidermal and hypodermal cells increased in mycorrhizal roots, a process that can contribute to higher resistance to certain root pathogens. Also, Wall-bound peroxidase activity had been detected during initial stages of AM colonization, which later decreased (135). However, peroxidase activity was not detectable in cells containing intracellular arbuscular hyphae (135).

### 4.3 Biological mechanisms

#### *4.3.1 Microbial changes in the mycorrhizosphere*

The concept of 'mycorrhizosphere' implies that mycorrhizae significantly influence the microflora of the rhizosphere by altering root physiology and exudation. In addition, extra-radical hyphae of AM fungi provide a physical or nutritional substrate for bacteria. Meyer and Linderman (83) and Secilia and Bagyraj (117) showed qualitative and quantitative changes in the mycorrhizosphere soil of AM plants compared to the rhizosphere of non-AM plants. These microbial shifts were clearly time-dependent and dynamic. Changes in soil microorganism populations induced by AM formation may lead to stimulation of certain components of the microbiota, which, in turn, may be antagonistic to root pathogens. Meyer and Linderman (83) showed that the production of sporangium and zoospere by the root pathogen *Phytophthora cinnamomi* was reduced in the presence of rhizosphere leachates from sweet corn and chrysanthemum. Similarly, populations of bacteria and actinomycetes in pot cultures of different AM fungi were quantitatively and qualitatively analysed (117). They showed that pot cultures of *G. fasciculatum* harboured more actinomycetes antagonistic to *Fusarium solani* and *Pseudomonas solanacearum* than those of non-mycorrhized plants or plants colonized by other mycorrhizal species tested in the study. Similarly, Caron *et al.* (21) showed a reduction in *Fusarium* population in the mycorrhizosphere soil of tomatoes and a corresponding reduction in root rot in AM plants when compared to non-AM plants, probably due to the increased antagonism in the AM mycorrhizosphere. They also concluded that the reduction in disease was not due to P nutrition. Thus, microbiota-mediated changes may be involved in biological control by AM associations (148).

## 5. STRATEGIES FOR THE MANAGEMENT OF ARBUSCULAR MYCORRHIZAE IN DISEASE CONTROL

### 5.1 Integrated approach

Obviously when AMF contribute to disease suppression, then agricultural practices (tillage, pesticides) that affects the population of AM fungi and associated antagonists, should be followed judiciously. Since AM fungi may ameliorate the nutrient status of their hosts and provide protection against pathogens it may possible that a low degree of tillage could be beneficial to AM colonization and crop health. Conservation tillage (i.e., no tillage or less frequent tillage) tends to concentrate plant debris and consequently microbial biomass in the topsoil, and thus promotes survival of pathogens. It is generally accepted that reducing tillage of agricultural soils can increase early-season P uptake in crops, especially in low-input farming systems. Mixing of soil may negatively affect AM colonization of plant roots, due to disruption of the extraradical mycelium, however, non-mycorrhizal fungi may also be affected. Only a few studies have investigated the long-term effects of management history on microbial communities. Mader *et al.* (80) compared AM colonization in plots with 7-year identical crop rotation and tillage schemes that differed only in the amount and type of applied fertilizer. Root length colonised by AM fungi was 30–60% higher in plants grown in soils from low input farming systems than those grown in conventionally fertilized soils. To ensure that compatible combinations of AM and antagonists occur in the same soil, growers could inoculate the seeds with appropriate cultures and then use less than the recommended levels of pesticides. Saleh and Sikora (112) showed that the fungicides benomyl and carbendazim significantly reduced the antagonistic activity of *G. fasciculatum* that affected reproduction of *M. incognita* on cotton when applied as a drench at planting time. The number of nematodes increased significantly on mycorrhizal plants when benomyl was applied at planting time but no significant change was noticed with similar applications of carbendazim, which is less toxic to *G. fasciculatum* than benomyl. However, when the fungicides were applied 25 days after planting, there were fewer nematodes in mycorrhizal plants. Similarly, *G. fasciculatum* did not affect nematode population density when applied 30 days after or simultaneous inoculations at planting time but when it was applied 50 days prior to nematode inoculation, it reduced the nematode population significantly. Combined inoculation of AM fungi with growth stimulating bacteria may favour the inoculum production (129). Several studies have demonstrated that some AM fungi exhibit biocontrol properties (91) against root pathogens. Whether AM fungi could be used as biocontrol agents practically, or possibly function as vectors for associated bacteria with biocontrol properties, remains to be explored. Hence, co-inoculation efficacy of mycorrhizae with other biocontrol agents needs to be optimized for both pot culture and field conditions.

### 5.2 Host-symbiont-soil nutrient status

Research is needed to assess how AM fungi affect the host-pathogen relationship differently. Growth responses to AMF vary depending on the mycorrhizal dependency of the host, the endophyte strain and the soil fertility level (54). Preliminary research needs to determine whether a relationship is dependent on an AM fungal species or mixture of species that are indigenous to an area and stimulate plant growth at moderately fertile soils. This is necessary assessing to a closer approximation, the response of mycorrhizal plants to pathogen infection for the soil type and fertility level occuring in the field. A better approach would be to use varying soil P levels that encompass field recommendations to produce non-AM plants of equivalent size and similar P status as mycorrhizal plants. These conditions are necessary to find out the real mechanism of the tolerance to pathogen attack shown by AMF plants.

### 5.3 Statistical models for mycorrhizae - pathogen interactions

Regression analyses that evaluate the response of qualitative treatments (mycorrhizal versus non-mycorrhizal plants receiving a particular P regime) over varying quantitative treatments (Pathogen Pi) are essential for valid comparison and statistical interpretation for evolving best combinations. Also, frequency values of nematode cycles over physiological time have been used to compare the effects of AM fungi and P nutrition on nematode parasitism (32). Another quantative approach adaptable to mycorrhiza - nematode interactions is the Seinhorst damage functions, $Y=m + (1-m)\ ZP^{-T}$, to determine if the model parameters T (tolerance limit), m (minimum yield), or Z (nematode virulence) are affected differently in mycorrhizal and non-mycorrhizal P-fertilized plants (151).

### 5.4 Initial inoculum density and inoculations sequence of pathogen and symbiont

It is difficult to draw firm conclusions about the potential biocontrol unless a range of pathogen inoculum densities is used. While varying the inoculum densities of pathogen to produce different levels of disease incidence or yield reduction, one should not solely use Pi (inoculum density) to get severely stunts or kills plants otherwise. AM fungi are not given an opportunity to colonize roots and stimulate growth before the challenge of nematode attack (79). Because root infections by fungal and nematode pathogens precede mycorrhizal root colonization, the sequence in which plants are inoculated may affect the nature of the interaction. Thus, an inoculation method has been used in many studies where plants are inoculated with the AM fungus 2 - 4 weeks before inoculating them with the pathogen (25,119). The time of AM inoculations, whether before, simultaneously or after inoculations with the pathogens or nematode, greatly affects the efficacy of AM fungi in controlling nematodes. Studies conducted by several workers demonstrated that preinoculations of AM fungus suppressed nematode reproduction and development in roots to a greater degree, when compared to plants inoculated simultaneously with both the organisms (38).

This technique allows AM fungi enough time to colonize roots before they are challenged by the pathogen; however, it is restricted to containerized or transplanted hosts (65). Preinoculations of *G. fasciculatum* in tomato var. Pusa Ruby significantly reduced the number of galls per plant, egg masses per plant and number of eggs per egg mass due to *M. incognita*. They also observed that the effect of AM fungi on reducing root-knot nematode was not the same in pots and in the field. Nematode populations were reduced up to 58% in the pots and only up to 36% in the field (118). Since AM fungi are generally considered slow colonizers, it seems logical that a minimum level of AM fungal root colonization is required for the fungus to have any effect on the parasite. AM fungal inoculum dose of 0.5 - 5.0 spores per gram of soil has been used to produce optimal growth responses and maximum root colonization levels. Although such threshold levels of colonization required to affect nematode activites have been reported (52) but high levels of AM fungal root colonization have not been found to reduce the degree of root infections by fungal pathogens (30). Carling *et al.* (19) showed that AM fungi made peanut plant more tolerant to the nematode and offset the reduction in growth caused by *M. arenaria* at the two lower P levels. However, AM fungi and added P increased the no. of galls and egg production in *M. arenaria*, thereby increasing peanut susceptibility to nematode attack and had only a minimal effect on AM colonization. Furthermore, a comparison of mycorr-hizal and P-fertilized non-mycorrhizal plants of similar size showed that the latter are more susceptible to nematode attack indicating the likely involvement of factors other than P nutrition in the interactions. While in mycorrhizal inoculations of tamarillo (*Cyphomandra betacea*) against root-knot nematode, *M. incognita* could not be duplicated by adding P and fertilizer. It was not, therefore, due to merely improved P nutritions of the host (26).

## 6. PRODUCTION AND COMMERCIALIZATION OF AMF

### 6.1. Conventional methods

There are various techniques currently used to culture AM fungi on hosts such as on-farm production (39), pot culture techniques using traps (121), nutrient film technique (86) and aeroponics (70). In pot and field multiplication, there are many factors which govern AM development besides the host plant (136), such as temperature, light, pot size, soil fertility (82) and the particle size of the growth substrate (48).

The most frequently used technique for increasing propagule number has been the propagation of AM fungi on a suitable host in disinfested soil using pot cultures. Hepper (60) reviewed procedures for disinfestations and germinating spores. Other factors for creating a conducive environment for culturing of AM fungi are a balance of light intensity, adequate moisture and moderate temperature without detrimental addition of fertilizers or pesticides (69). Any moisture conditions, which inhibit primary root growth, will reduce the development of mycorrhizal colonization. Similarly, soil temperature is important for the growth of fungi and indirectly affects the development of the fungus (16). Cultures reaching high propagule density (10 spores per gram) after a number of multiplication cycles can be stored using suitable methods after air-drying (73). Furthermore, AM fungi have been cultured with plant hosts in different substrates such as sand, peat, expanded clay, perlite, vermiculite, soilrite (81), rockwool (58) and glass beads (109). Apart from the above environmental and physical factors, biological factors also play important roles in rapid culturing of AM fungi in pots. Recently, Bhowmik and Singh (13) demonstrated the role of plant growth promoting rhizobacteria in enhancing the mycorrhizal colonization of *G. mosseae* in pots on *Chloris gayana*.

AM production can also be produced aeroponically (140). The aeroponic system was adopted for mycorrhiza production by the utilization of seedlings with roots pre-colonized by an AM fungus and the use of modified Hoagland's nutrition with a very low P level (61). In the aeroponic system, colonization and sporulation were superior to that reported in soil-based pot culture. *Entrophospora kentinensis* was successfully propagated with Bahia grass and sweet potato in an aeroponic system (153). Mohammad *et al.* (85) reported the production of *G. intraradices* in an aeroponic system where they compared the conventional atomizing disc with the ultrasonic nebulizer technology as misting sources.

Nutrient film technique (NFT) was adapted for the AM fungal inoculum production (86) and culture host plants are placed on an inclined tray over which flows a layer of nutrient solution. As in the aeroponic culture, seedlings must be pre-colonized in another medium. Recently, Lee and George (76) proposed a modified nutrient film technique for large scale production of AM fungal biomass with the help of improved aeration by intermitent nutrient supply, optimum P supply and use of glass beads as support materials.

### 6.2. *In vitro*/axenic cultivation of AM fungi (root organ culture)

*In vitro* culture of AM fungi was achieved for the first time in early 1960s. Since then various pioneering steps were aimed at axenic culturing of AM fungi. Detail evaluation of Root Organ Culture (ROC) technique and basic improvements necessary for AM fungus colonization of roots, have been presented (9,31,45,97). The ROC is a most attractive mass multiplication method for providing a pure, viable, rapid and contamination free inoculum using less space and has an advantage over pot culture multiplication/conventional system (28,45,143). The mono-specific strains available can be used directly as starting material for large-scale inoculum production, a sole Petri dish culture being enough to generate several thousand spores and meters of hyphae within four months (29).

Recently, Douds (38) reported monoxenic culture of *G. intraradices* with Ri T-DNA transformed roots in two-compartment petri dishes as a very useful technique for physiological studies and the production of clean fungal tissues. Fortin *et al.* (45) developed successful propagation of some AM fungal strains on ROC allowed the cultivation of monoxenic strains that can be used either directly as inoculum or as starting inoculum for large-scale production. *In vitro* propagation on ROC consists of excised roots that proliferate under axenic conditions on a synthetic nutrient medium supplemented with vitamins, minerals, and carbohydrates. The excised roots are the Ri T-DNA transformed which has been used to obtain colonized root cultures. For the first time the use of Ri T-DNA transformed roots of *Daucus carota* as host by *Agrobacterium rhizogenes* has permitted an increase in spore production of *G. mosseae* (87).

Transformation is done by inserting *Agrobacterium rhizogenes* in transformed root copies of Ri T-DNA (transfer DNA), which are found in a large plasmid of *A. rhizogenes* (24). The transformed roots have an ability to grow profusely and vigorously (negative geotropism) and to provide opportunity for AM fungi hyphae to make contact (9). They also evaluated the ROCs for AM colonization in roots and identified rate limiting factors in the *G. margarita* colonized transformed roots of carrot. Continuous cultures of vigorous ROCs (Ri T-DNA transformed) have been obtained through transformation of roots by soil bacterium *A. rhizogenes* that provided the new way to obtain mass production of roots in a very short span of time. Several sub cultures (3-4) are necessary in this medium enriched with antibiotics such as carbenicillin or ampicillin to obtain free-living roots without bacteria (37) and for establishing dual culture. Monoxenic cultures is developed through extraradical spores and infected root fragments. Sporocarps of *G. mosseae* have also been attempted to establish *in vitro* cultures of AM fungi. In most cases, purified and surface sterilized spores (10) isolated from the field or from traps have been successful for establishing dual cultures under *in vitro* conditions. Spore sterilization can also be achieved with a solution containing chloramine T (oxidizing agent) and Tween 20 (surfactant) (45) and rinsed in a streptomycin-gentamycin antibiotic solution (10). While the success of spore germination is solely dependent on sterilization, the presence of root exudates and 2% $CO_2$ can stimulate germination and post germination hyphal growth (98).

The infected roots, which come from trap culture roots, can be used as a section of vesicles for *in vitro* establishment. The root sections can be disinfected in an ultrasonic processor under aseptic conditions and incubated in MSR (Modified Strullu- Romand) medium (31). Hyphal regrowth from root pieces is usually observed within 2-15 days (45). Using segments of vesicles have more advantages over spores as far as contamination is concerned but recovery of propagules is low compared to spores. Therefore, the vesicles are rarely used for routine inoculation for in vitro establishment (45). Regular subculturing is required to maintain higher infectivity. Continuous cultures can be obtained by transferring mycorrhizal roots to fresh medium either with or without spores. It is always preferable to use actively growing roots. Mass production of spores is a prerequisite and statistical models may be useful as descriptive and predictive tools for sporulation dynamics (32). Several forms of inoculum may be used and the substrate may influence the efficacy of the inoculum (20). Various inocula based on inert or sterilized substrata such as peat, expanded or calcined clays are used commercially and are less susceptible to contamination with pathogen (152). AMF are commercially available in various formulations and being sold throughout the world (Table 4).

**Table 4.** List of commercial products and sources of AM inoculants*

| Product/Company | Country |
|---|---|
| AgBio Inc., Westminster | Colorado, USA |
| Becker –Underwood, Ames | Iowa, USA |
| BioGrow TM | North America |
| Bio-organics | Medillin, Columbia |
| Biological Crop Protection Ltd | Kent, UK |
| Bio-Organics Supply, Camarillo | California, USA |
| Biorize | Dijon, France |
| BioScientific, Inc., Avondale | Arizona, USA |
| EcoLife Corporation, Moorpark | California, USA |
| Global Horticare | Lelystad, The Netherlands |
| Idemitsu Kosan Co., | Sodegaura, Chile |
| J.H. Biotech, Inc., | Ventura, California, USA |
| Cadila Pharmaceuticals | Ahmedabad, India |
| KCP Sugar and Industries Corporation Ltd | Andhra Pradesh, India |
| Majestic Agronomic (P) Ltd | Una, Himachal Pradesh, India |
| MicroBio, Ltd | Royston, Hertz, UK |
| Mikro-Tek Inc., | Ontario, Canada |

| | |
|---|---|
| MYCORISE® PRO (Premier Tech Ltd) | Premier Horticulture, Canada H9P 2W3 |
| MycorTM VAM Mini Plug TM | North America |
| N-Viron Sdn Bhd | Malaysia |
| Plant health Care, Inc., | Pennsylvania, USA |
| PlantWorks Ltd., Sittingbourne | UK |
| Roots Inc., Independence | Montana, USA |
| Soil moist | JRM Chemical, Inc. Ohio 44128 |
| T & J Enterprises, Spokane | Washington, USA |
| TIPCO, Inc., Knoxvile | Tennesse, USA |
| Tree of Life Nuresry, San Juan Capistrano | California, USA |
| Triton Umweltschutz GmbH | Bitterfeld, Germany |
| Vaminoc | MicroBio, Hemel Hempstead, UK |

*Modified from Tiwari *et al.,* 2003

## CONCLUSION

Due to complexity of the micro-soil-plant system and the decisive influence of prevailing environmental conditions, it is rather difficult to reach practical conclusions. We can certainly find the right combinations of different factors to exploit the prophylactic ability of AM fungi. It can be concluded from the available information that (1) AM associations can reduce the incidene and damage caused by soil borne nematodes, (2) the abilities of AM fungi to enhance resistance or tolerance in roots through symbioses can vary with the different AM fungi and (3) it is not possible to protect the plants effectively for all pathogens and it is modulated by soil and other environmental factors.Thus it can be expected that interactions between various AM fungi and nematodes will vary with the host plant and the system. Hence to achieve the high degree of control of target plant diseases in sustainable agrosystems particularly with nursery and horticultural crops in future, we must use indigenous and appropriate AM fungi alone or preferably in combinations with other pathogen-antagonistic microorganisms in the soil. The soil and crop management practices, which can sustain the level of effective AMF and other antagonistists to suppress disease, must be taken into condideration. Besides evaluation of the level of inoculum required to suppress a particular disease, the techniques of large-scale AMF production and application in the field must be standardized. The use of AM fungi in inducing the resistance against pathogen and parasitic nematodes is well elucidated and the efforts must be made for mass multiplication of these fungi for use in agriculture where soil borne parasites are a menace. The workers may also explore the possibilities to use AM fungal products or genetic transformations could be brought about expressing resistance against root parasitic nematodes and other microorganism by exuding toxic chemicals through roots.

## *Literature Cited*


1. Al-Raddad, A., Ahmad, M. 1995. Interaction of *Glomus mosseae* and *Paecilomyces lilacinus* on *Meloidogyne javanica* of tomato. *Mycorrhiza*, 5(3): 233-236.
2. Atkinson D., Berta G., Hooker J.E. 1994. Impact of mycorrhizal colonization on root architecture, root longevity and formation of growth regulators. In, "*Impact of arbuscular mycorrhizas on sustainable agriculture and natural ecosystems*", (eds. Gianinazzi, S. and Schüepp, H.) Birkhäuser, Basel, pp. 89-88.
3. Azcón-Aguilar, C., Barea, J.M. 1996. Arbuscular mycorrhizas and biological control of soil borne plant pathogens- an overview of the mechanism involved. *Mycorrhiza*, 6 : 457-464.
4. Azcón-Aguilar, C., Jaizme-Vega, M Calvet, C. 2002. The contribution of arbuscular mycorrhizal fungi to control of soilborne plant pathogens. *In Mycorrhizal technology in agriculture: from genes to bioproducts. Edited by* S. Gianinazzi, H. Schüepp, J.M. Barea, and K. Haselwandter. Birkhäuser Verlag, Basel. pp. 187–197.
5. Baghel, P.P.S., Bhatti, D.S., Jalali, B.L. 1990. Interaction of VA mycorrhizal fungus and *Tylenchulus semipenetrans* on citrus. In "*Proceedings of the National Conference on Mycorrhiza.* "Haryana Agricultural University, Hisar" (eds. Jalali, B.L and Chand, H)

February 14-16, 1990, New Delhi: TERI, pp. 210.

6. Bagyaraj, D.J., Manjunath, A., Reddy, D.D. 1979. Interaction of vesicular arbuscular mycorrhiza with root-knot nematodes in tomato. *Plant and Soil*, 51 : 397-403.

7. Baker, R. 1986. Biological control: an overview. *Canadian Journal of Plant Pathology*, 8: 218-221.

8. Barea, J.M., Jeffries, P. 1995. Arbuscular mycorrhizas in sustainable soil plant systems. In, "*Mycorrhiza structure, function, molecular biology and biotechnology*", (eds. Hock, B. and Varma, A) Springer, Heidelberg, pp. 521-559.

9. Becard, G., Fortin, J.A.1988. Early events of vesicular arbuscular mycorrhiza formation on Ri T-DNA transformed roots. *New Phytologist*, 108 : 211-218.

10. Becard, G., Piche, Y. 1992. Establishment of AM in root organ cultures: Review and proposed methodology. In: *Techniques for the Study of Mycorrhiza*. (Eds.) J. Norris, D. Read and A Verma, Academic Press, NY, USA, pp 89-108.

11. Bhagawati, B., Goswami, B. K., Singh, C.S. 2000. Management of disease complex of tomato caused by *Meloidogyne incognita* and *Fusarium oxysporum* f. sp. *lycopersici* through bioagent. *Indian J. of Nematol.*, 30, 16–22.

12. Bhat, M. S., Mahmood, I. 2000. Role of *Glomus mosseae* and *Paecilomyces lilacinus* in the management of root-knot nema-tode on tomato. *Archives of Phytopathology and Plant Protection*, 33, 131–140.

13. Bhowmik, S.N. Singh, C.S. 2004. Mass multiplication of AM inoculum: Effect of plant growth-promoting rhizobacteria and yeast in rapid culturing of *Glomus mosseae. Curr. Sci.* 86: 705-709.

14. Blee, K.A., Anderson, A.J. 1996. Defense-related transcript accumulation in *Phaseolus vulgaris* L. colonized by the arbuscular mycorrhizal fungus *Glomus intraradices* Schenck & Smith. *Plant Physiol* 110 : 675–688

15. Bonanomi, A., Oetiker, J.H., Guggenheim, R., Boller, T., Wiemken, A., Vögeli-Lange R. 2001. Arbuscular mycorrhizas in mini-mycorrhizotrons: first contact of *Medicago truncatula* roots with *Glomus intraradices* induces chalcone synthase. *New Phytologist* 150, 573–582.

16. Braunberger, P.G., Abbott, L.K., Robson, A.D. 1997. Early vesicular-arbuscular mycorrhizal colonisation in soil colle-

cted from an annual clover-based pasture in a Mediterranean environment: soil temperature and the timing of autumn rains. *Australian Jour. Agricultural Research,* 48(1) : 103-110.

17. Calvet, C.J. Pinochet, A. Hernandez-Dorrego, Estaun, V., Camprubi, A. 2001. Field micro-plot performance of the peach-almond hybrid GF-677 after inoculation with arbuscular mycorrhizal fungi in a soil infested with root-knot nematode. *Mycorrhiza* 10: 295-300

18. Cardoso, I.M., Kuyper, T.W. 2006. Mycorrhizas and tropical soil fertility Agriculture, *Ecosystems and Environment* 116: 72–84

19. Carling, D.E., Roncadori, R.W. and Hussey, R.S. 1995. Interactions of arbuscular mycorrhizae, *Meloidogyne are-naria* and phosphorus fertilization on peanut. *Mycorrhiza* 6: 9-13.

20. Caron, M., Fortin, J.A., Richard, C. 1985. Influence of substrate on the interaction of *Glomus intraradices* and *Fusarium oxysporum* f. sp. *radicislycopersici* on tomatoes. *Plant Soil* 87: 233–239.

21. Caron, M., Fortin, J. A. and Richar, C. 1986. Effect of phosphorus concentration and *Glomus intraradices* on Fusarium crown and root rot of tomatoes. *Phytopathology* 76: 942-946.

22. Cason, T.K.M., Hussey, R.S., Roncadori, R.W. 1983. Interaction of vesicular arbuscular mycorrhizal fungi and phosphorus with *Meloidogyne incognita* on tomato. *J. Nematol.* 15: 410-417

23. Chandrashekar, D.S., Shetty, H.S., Datta, R.K. 1995. Effect of VAM fungus, *Glomus fasciculatum* and *Acaulospora laevis* against root-knot nematode of mulberry. In *Mycorrhiza news*. Vol.7 No.2 pp.7-8

24. Chilton, M..D., Tefer, D.A., Petit, A., David, C., Casse-Delbart, F. Tempe, J. 1982. *Agrobacterium rhizogenes* inserts T-DNA into the genomes of the host plant root cells. *Nature* 295: 432-434.

25. Cooper, K., Grandison, G.S. 1986. Interaction of vesicular-arbuscular mycorrhizal fungi and root knot nematode as cultivars of tomato and white clover susceptible to *Meloidogyne hapla*. *Annals of Applied biology* 108: 555-565

26. Cooper, K.M., Grandison, G.S. 1987. Effects of vesicular-arbuscular mycorrhizal fungi on infection of tamarillo (*Cyphomandra betacea*) by *Meloidogyne incognita* in fumigated soil. *Plant Dis.* 71: 1101-1106

27. Cordier, C., Gianinazzi, S., Gianinazzi-Pearson,V. 1996. Colonization patterns of root tissues by *Phytophthora nicotianae* var. *parasitica* related to reduced disease in mycorrhizal tomato. *Plant Soil*, 185: 223–232.

28. Cranenbrouck, S., Voets,L., Bivort, C. 2005. Methodologies for *in vitro* cultivation of AM fungi with root organs. *In Vitro Culture of Mycorrhizas.* Declerck S, Strullu DG and Fortin A (Eds.) Springer Verlag Berlin Heidelberg pp 342-375.

29. Dalpé, Y., Monreal, M. 2004. Arbuscular mycorrhiza inoculum to support sustainable cropping systems. Online. *Crop Management doi*: 10. 1094/CM-2004-0301-09-RV.

30. Davis, R.M., Menge, J.A. 1980. Influence of *Glomus fasciculatus* and soil phosphorus on *Phytophthora* root rot of citrus. *Phytolopathology* 70 : 447-452.

31. Declerck, S., Strullu, D.G., Plenchette, C. 1998. Monoxenic culture of the intraradical forms of *Glomus* sp. isolated from a tropical ecosystem: A proposed methodology for germplasm collection. *Mycologia* 99: 579-585.

32. Declereck S, Cranenbrouck, S., Leboulenge, E. 2001. Modeling and sporulation dynamics of AM fungi in monoxenic culture. *Mycorrhiza* 11: 1178-1187.

33. De la Pena, E, Echeverria, S.R., Putten, W.H., Freitas, H., Moens, M. 2006. Mechanism of control of root-feeding nematodes by mycorrhizal fungi in the dune grass *Ammophila arenaria*. *New Phytol*. 169:829-840

34. Dehne, H.W. 1982. Interaction between vesicular-arbuscular mycorrhizal fungi and plant pathogens. *Phytopathology* 72 : 1114-1119.

35. Dehne, H.W. ,Schönbeck, F. 1979. The influence of endotrophic mycorrhiza on plant diseases. II. Phenolmetabolism and lignification). *Phytopathol. Z.* 95: 210-216.

36. Diedhiou, P. M., Hallmann, J., Oerke, E.C., Dehne, H.W. 2003. Effects of arbuscular mycorrhizal fungi and a nonpathogenic *Fusarium oxysporum* on *Meloidogyne incognita* infestation on tomato. *Mycorrhiza* 13, 199–204.

37. Diop, T.A. 2003. In vitro cultures of AM fungi: advances and future prospects: Mini review. *African J. Biotechnol.* 2: 692-697.

38. Douds, D.D. Jr. 2002. Increased spore production by *Glomus intraradices* in the split-plate monoxenic culture system by repeated harvest,

gel replacement, and resupply of glucose to the mycorrhiza. *Mycorrhiza* 12 : 163-167.

39. Douds, D.D. Jr., Nagahashi, G., Pfeffer PE, Reider, C., Kayser, W.M. 2006. On-farm production of AM fungus inoculum in mixtures of compost and vermiculite. *Biores. Technol.* 97: 809– 818.

40. Dumas-Gaudot, E., Furlan, V., Grenier, J., Asselin, A. 1992a. New acidic chitinase isoforms induced in tobacco roots by vesicular-arbuscular mycorrhizal fungi. *Mycorrhiza* 1 :133–136 463

41. Elsen, A., Declerck, S., De Waele, D. 2001. Effects of *Glomus intraradices* on the reproduction of the burrowing nematode (*Radopholus similis*) in dixenic culture, *Mycorrhiza* **11**: 49–51.

42. Elsen, A., Baimey, H., Swennen, R., De Waele, D. 2003a. Relative mycorrhizal dependency and mycorrhiza-nematode interaction in banana cultivars (*Musa* spp.) differing in nematode susceptibility, *Plant and Soil* **256** :. 303–313.

43. Elsen, A., Beeterens, R., Swennen, R., De Waele, D. 2003b. Effects of an arbuscular mycorrhizal fungus and two plant-parasitic nematodes on Musa genotypes differing in root morphology. *Biology and Fertility of Soils* 38, 367–376.

44. Fassuliotis, G.1970. Resistance in *Cucumis* spp. to root-knot nematode. *Meloidogyne incognita* acrita. *J. Nematol.* 2:174.

45. Fortin, J..A, Becard, G., Declerckm, S., Dalpe, Y., St Arnaud, M., Coughlan, A.P. Piche, Y. 2002. Arbuscular mycorrhiza on root-organ cultures. *Can. J. Bot.* 80: 1-20.

46. Francl, L.J., Dropkin, V.H. 1985. *Glomus fasciculatum*, a weak pathogen of *Heterodera glycines*, *Journal of Nematology* **17** :470–475.

47. Fusconi, A., Gnavi, E., Trotta, A., Berta, G. 1999. Apical meristems of tomato roots and their modifications induced by arbuscular mycorrhizal and soil-borne pathogenic fungi. *New Phytol.* 142: 505–516

48. Gaur, A., Adholeya, A. 2000b. Effects of the particle size of soil-less substrates upon AM fungus inoculum production. *Mycorrhiza* 10 : 43-48

49. Gianinazzi-Pearson, V., Dumas-Gaudot, E., Gollotte, A., Tahiri-Alaoui, A., Gianinazzi, S. 1996. Cellular and molecular defencerelated root responses to invasion by arbuscular mycorrhizal fungi. *New Phytol* 133 : 45–57

50. Gianinazzi-Pearson, V., Gollotte, A., Dumas-Gaudot, E., Franken, P., Gianinazzi, S. 1994. Gene expression and molecular modifications associated with plant responses to infection by arbuscular mycorrhizal fungi. In, "*Advances in molecular genetics of plant-microbe interactions*", (eds. Daniels, M., Downic, J.A. and Osbourn, A.E.) Kluwer, Dordrecht, pp. 179-186.

51. Graham, J.H., Leonard, R.T., Menge, J.A. 1981. Membrane mediated decreases in root exudation responsible for phosphorus inhibition of vesicular-arbuscular mycorrhiza formation. *Plant Physiology*, 68 : 548-552.

52. Grandison, G.S., Cooper, K.M. 1986. Interaction of vesicular-arbuscular mycorrhizae and cultivars *Meloidogyne hapla*. *Journal of Nematology* 18: 141-149.

53. Habte, M., Zhang, Y.C., Schmitt, D.P. 1999. Effectiveness of *Glomus* species in protecting white clover against nematode damage. *Can. J. Bot.* 77(1):135-139

54. Harley, J.L., Smith, S.E. 1983. *Mycorrhizal Symbiosis*. Academic Press, New York, 483 pp

55. Harrison, M.J., Dixon, R.A. 1993. Isoflavonoid accumulation and expression of defense gene transcripts during the establishment of vesicular-arbuscular mycorrhizal associations in roots of *Medicago truncatula*. *Mol. Plant Microbe Interact* 6: 643– 654

56. Hasan, A. 2001. Arbuscular mycorrhizal colonization and incidence of root-knot nematode in some rabe vegetables in eastern Uttar Pradesh: a preliminary survey report. *Mycorrhiza News* 12(4) : 15-17.

57. Heald, C.M., Bruton, B.D., Davis, R.M. 1989. Influence of *Glomus intraradices* and soil phosphorus on *Meloidogyne incognita* infecting Cucumis melo. *Journal of Nematology* 21(1) : 69-73

58. Heinzemann, J., Weritz, J. 1990. Rockwool: a new carrier system for mass multiplication of vesicular-arbuscular mycorrhizal fungi. *Angew Bot* 64 : 271-274

59. Hepper, C.M., Azcon-Aguilar, C., Rosendahl, S., Sen, R. 1988. Competition between three species of *Glomus* used as spatially separated, introduced & indigenous mycorrhizal inocula for leek (*Allium porrum* L.). *New Phytol* 93 : 401–413

60. Hepper, C.M. 1984. Isolation and culture of VA mycorrhizal (VAM) fungi. In, "*VA Mycorrhiza*" (ed. Powell C.L. and

Bagyaraj D.J.) CRC Press Inc., Boca Raton, Florida, pp. 95-112.

61. Hoagland, D.R., Arnon, D.I. 1938. The water-culture method for growing plants without soil. University of California, College of Agriculture, Agriculture Experiment Station Circular 347, Berkeley, CA, USA.

62. Hol, W.H.G., Cook, R.2005 An overview of arbuscular mycorrhizal fungi–nematode interactions. *Basic and Applied Ecology* 6: 489-503

63. Hooker, J.E., Gianinazzi, S., Vestberg, M., Barea, J.M., Atkinson, D. 1994. The application of arbuscular mycorrhizal fungi to micropropagation systems: *an opportunity to reduce chemical inputs.* Agricultural Science in Finland, 3 : 227-232.

64. Hussey, R.S., Roncadori, R.W. 1978. Interaction of *Pratylenchus brachyurus* and *Gigaspora margarita* on cotton. *Journal of Nematology* 10 : 18-20.

65. Hussey, R.S., Roncadori, R.W. 1982. Vesicular-arbuscular mycorrhizae may limit nematode activity and improve plant growth. *Plant Disease* 66 : 9-14.

66. Jain, R.K., Sethi, C.L. 1987. Pathogenicity of *Heterodera cajani* on cowpea as influenced by the presence of VAM fungi *Glomus fasciculatum* or *G. epigaeus. Indian J.of Nematol.* 17(2) : 165-170.

67. Jaizme-Vega, M. C., Tenoury, P., Pinochet, J., Jaumot, M. 1997. Interactions between the root-knot nematode *Meloidogyne incognita* and *Glomus mosseae* in banana. *Plant and Soil*, 196, 27–35.

68. Jalali, B.L., Jalali, I. 1991. Mycorrhiza in plant disease control. In, *Handbook of Applied Mycology, Soil and Plants* Vol. 1, (eds. Arora, D.K, *et al.*), Marcel Dekker, New York, pp. 131-154.

69. Jarstfer, A.G., Sylvia, D.M. 1992. Inoculum production and inoculation strategies for vesicular-arbuscular mycorrhizal fungi. In: B. Meting (Ed.), *Soil Microbial Ecology; Application in Agriculture and Environmental Management.* Marcel Dekker, New York, pp. 349-377.

70. Jarstfer, A.G., Sylvia, D.M. 1995. Aeroponic culture of VAM fungi. In, "*Mycorrhiza Structure, Function, Molecular Biology and Biotechnology*" (Varma A and Hock B (Eds.) Springer-Verlag, Heidelberg, pp. 521-559.

71. Kassab, A.S., Taha, A.H.Y. 1990. Interaction between plant-parasitic nematode,

vesicular –arbuscular mycorrhiza, rhizobia and Egyptian clover. *Ann. Agric. Sci. (Cairo)* 35: 509-520.

72. Krishna Prasad, K.S. 1991. Influence of a vesicular arbuscular mycorrhiza on the development and reproduction of root-knot nematode affecting flue cured tobacco. *Afro-Asian Journal of Nematology* 1(2) : 130-134.

73. Kuszala, C., Gianinazzi, S., Gianinazzi-Pearson, V. 2001. Storage conditions for the long -term survival of AM fungal propagules in wet sieved soil fractions. *Symbiosis* 30 : 287-299.

74. Lambais, M.R.,Mehdy, M.C. 1993. Suppression of endochitinase, b-1,3-endoglucanase and chalcone isomerase expression in bean vesicular-arbuscular mycorrhizal roots under differ-ent soil phosphate conditions. *Mol. Plant Microbe Interact* 6 : 75– 83

75. Lambais, M.R. 2000. Regulation of plant defence-related genes in arbuscular mycorrhizae. In: Podila GK, Douds DD, eds. *Current advances in mycorrhizae research.* Minnesota, USA: The American Phytopathological Society, 45–59.

76. Lee, Y.J. George, E. 2005. Development of a nutrient film technique culture system for arbuscular mycorrhizal plants. *Hort Sci.* 40: 378-380.

77. Linderman, R.G. 2000. Effects of mycorrhizas on plant tolerance to diseases. In: Kapulnik Y, Douds Jr DD, eds. *Arbuscular mycorrhizas: physiology and function.* Dordrecht, The Netherlands: Kluwer Academic Publishers, 345–365

78. Linderman, R.G. 1994. Role of VAM fungi in biocontrol. In, "*Mycorrhizae and plant health*", (Eds. Pfleger, F.L. and Linderman, R.G), APS Press, St. Paul, MN, pp. 1-26.

79. MacGuidwin, A.E., Bird, G.W., Safir, G.R. 1985. Influence of *Glomus fasciculatum* on *Meloidogyne hapla* infecting *Allium cepa*. *Journal of Nematology* 17 : 389-395.

80. Mader, P., Edenhofer, S., Boller, T., Wiemken, A., Niggli, U. 2000. Arbuscular mycorrhizae in a longterm field trial comparing low-input (organic, biological) and high-input (conventional) farming systems in a crop rotation. *Biology and Fertility of Soils* 31, 150–156.

81. Mallesha, B.C., Bagyaraj, D.J., Pai, G. 1992. Perlite-soilrite mix as a carrier for mycorrhiza and rhizobia to inoculate *Leucaena leucocephala*. *Leucaena Res Rep*. 13 : 32-33.

82. Menge, J.A., Steirle, D., Bagyaraj, D.J., Johnson, E.L.V., Leonard, R.T. 1978. Phosphorus concentrations in plants responsible for inhibition of mycorrhizal infection. *New Phytol.* 80 : 575-578.

83. Meyer, J.R., Linderman, R.G. 1986. Selective influence on populations of rhizosphere or rhizoplane bacteria and actinomycetes by mycorrhizas formed by *Glomus fasciculatum. Soil Biology and Biochemistry* 18 : 191-196.

84. Mishra, A., Shukla, B.N. 1995. Studies on management of root-knot (*Meloidogyne incognita*) of tomato by *Glomus fasciculatum* and some pesticides. In *Mycorrhiza news.*Vol.7 No.4 pp.8-11.

85. Mohammad, A., Khan, A.G., Kuek, C. 2000. Improved aeroponic culture of inocula of arbuscular mycorrhizal fungi. *Mycorrhiza* 9 : 337-339.

86. Mosse, B., Thompson, J.P. 1984. Vesicular-arbuscular endomycorrhizal inoculum production. I. Exploratory experiments with beans (*Phaseolus vulgaris*) in nutrient flow culture. *Canadian Journal of Botany* 62 : 1523-1530.

87. Mugnier, J., Mosse, B. 1987. AM infection in transformed root-inducing T-DNA roots grown axenically. *Phytopathol.* 77: 1045-1050.

88. Nagesh, M., Reddy, P.P. 1997. Management of *Meloidogyne incognita* on *Crossandra undularefolia* using vesicular arbuscular mycorrhiza, *Glomus mossae* and oil cakes. In *Mycorrhiza news.* Vol.9 No.1 pp.12-14

89. Nagesh, M., Reddy, P.P., Kumar, M.V.V. Nagaraju, B.M. 1999. Studies on correlation between *Glomus fasiculatum* spore density, root colonization and *Meloidogyne incognita infection* on *lycopersicon esculantum.* Zeitschrift Fur Pflanzenk-rankheiten Und Pflanzensc-hutz- *Journal of Plant Disease and Protection.* 106(1): 82-87.

90. Nehra, S., Pandey, S., Trivedi, P.C. 2003. Interaction of arbuscular mycorrhizal fungi and different level of root-knot nematode on ginger, *Indian Phytopathology* **56** : 297–299.

91. Niemira, B.A., Hammerschmidt, R., Safir, G.R. 1996. Postharvest suppression of potato dry rot (*Fusarium sambucinum*) in prenuclear minitubers by arbuscular mycorrhizal fungal inoculum. *Am. Potato J.* 75: 509–515.

92. Pandey, R., Bhargava, S. 2004. Phytochemicals in the management of phytoparasitic

nematodes. *Ann. Rev. of Plant Pathology* 3: 309-329.

93. Pandey, R., Gupta, M.L., Singh, H.B., Kumar, S. 1999. The influence of vesicular-arbuscular mycorrhizal fungi alone or in combination with *Meloidogyne incognita* on *Hyoscyamus niger* L, *Bioresource Technology* **69** : 275–278.

94. Pandey, R., Kalra, A., Singh, H. N., Gupta, M. L., Singh, H. B., Kumar, S. 2000. Suppressive response of bioagent, organic material and pesticide on population development of *Meloidogyne incognita* on *Matricaria chamomilla. Jour. Medicinal and Aromatic Plant Sciences* 22: 649-651.

95. Pandey, R., Singh, H.B., Gupta, M.L. 1997. Antagonistic impact of Vesicular- arbuscular (VAM) on *Meloido-gyne incognita* population development in Japanese mint. *International Journal of Tropical Plant Disease* 15:237-245

96. Pinochet, J., Ferna´ndez, C., de Carmen Jaizme, M., Tenoury, P. 1997. Micropropagated banana infected with *Meloidogyne javanica* responds to Glomus intraradices and phosphorus. *Hort. Science* 32, 101–103.

97. Plenchette, C., Declerck, C., Diop, T.A., Strullu, D.G. 1996. Infectivity of monoaxenic subcultures of the arbuscular mycorrhizal fungus Glomus versiforme associated with Ri-T-DNA-transformed carrot root. *Applied Microbiology and Biotechnology* 46(5/6) : 545-548

98. Poulin, M.J., Bel-Rhlid, R., Piche, Y., Chenevert, R. 1993. Flavonoids released by carrot (*Daucus carota*) seedlings stimulate hyphal development of VAM in the presence of optimal $CO_2$ enrichment. *J. Chem. Ecol.* 19:2317-2327.

99. Pozo, M.J., Azcón-Aguilar, C., Dumas-Gaudot, E., Barea, J.M.. 1999. ß-1,3-glucanase activities in tomato roots inoculated with arbuscular mycorrhizal fungi and/or *Phytophthora parasitica* and their possible involvement in bioprotection. *Plant Science* 141, 149–157.

100. Pozo, M.J., Cordier, C., Dumas-Gaudot, E., 2002. Localized versus systemic effect of arbuscular mycorrhizal fungi on defence responses to Phytophthora infection in tomato plants. *J. Exp. Bot.* 53(368): 525-534.

101. Pozo, M.J., *et al.* 1996. Detection of new chitinase isoforms in arbuscular mycorrhizal tomato roots: possible implications in protection against *Phytoph-*

*thora nicotianae* var. parasitica. *Agronomie* 16, 689–697.

102. Pozo, M.J., Cordier, C., Dumas-Gaudot, E., Gianinazzi, S., Barea, J.M., Azcón-Aguilar, C. 2002*a*. Localized versus systemic effect of arbuscular mycorrhizal fungi on defence responses to *Phytophthora* infection in tomato plants. *J. Exp. Bot.* 53: 525–534.

103. Price, N.S., Roncadori, R.W., Hussey, R.S. 1995. The growth of nematode 'tolerant' and 'intolerant' soyabeans as affected by phosphorus, *Glomus intraradices* and light. *Plant Pathol.* 44 : 597–603

104. Priestel, G. 1980. Wechseibeziehung zwischen der endotropics Mycorrhiza und dem Wurzelg-allennematoden *Meloidogyne incognita*. (Kofoid and White, 1919) Chitwood, 1949 an Gurke. *Dissertation,* Hannover, West Germany, 103 pp.

105. Rao, M. S., Kerry, B.R., Gowen, S.R., Bourne, J.M., Reddy, P.P. 1997. Management of *Meloidogyne incognita* in tomato nurseries by integration of *Glomus deserticola* with *Verticillium chlamydosporium, Zeitschrift für Pflanzenkrankheiten und Pflanzenschutz* **104** : 419–422.

106. Rao, M. S., Naik, D., Shylaja, M., Reddy, P.P. 2003. Management of *Meloidogyne incognita* on eggplant by integrating endomycorrhiza, *Glomus fasciculatum* with bio-agent *Verticillium chlamydosporium* under field conditions. *Indian J. of Nematol.* 33, 29–32.

107. Rao, M.S., Reddy, P.P., Mohandas, M.S. 1998. Biointensive management of *Meloidogyne incognita* on egg plant by integrating *Paecilomyces lilacinus* & *Glomus mosseae. Nematologia Mediterranea* 26, 213–216.

108. Ratti, N., Khaliq, A., Shukla, P.K., Haseeb, A., Janardhanan, K. K. 2000. Effect of *Glomus mosseae* (Nicol. & Gerd.) Gerdemann and Trappe on rootknot disease of menthol mint (*Mentha arvensis* sub sp. *haplocalyx* Briquet) caused by *Meloidogyne incognita* (Kofoid and White) Chitwood. *Journal of Spices and Aromatic Crops* 9, 129–132.

109. Redecker, D., Thierfelder, H., Werner, D. 1995. A new cultivation system for arbus-cular-mycorrhizal fungi on glass beads. *Angewandte Botanic* 69 : 189-191

110. Ryan, N.A., Deliopoulos, T., Jones, P., Haydock, P.P.J. 2003. Effects of a mixed-isolate mycorrhizal inoculum on the

potato–potato cyst nematode interaction. *Annals of Applied Biology* 143, 111–119.

111. Ryan, N. A., Duffy, E. M., Cassells, A.C., Jones, P. W. 2000. The effect of mycorrhizal fungi on the hatch of potato cyst nematodes. *Applied Soil Ecology* 15, 233–240.

112. Saleh, H.M., Sikora, R.A. 1989. Effect of quintozen, benomyl and carbendazim on the interaction between the endomycorrhizal fungus *Glomus fasciculatum* and the root knot nematode, *Meloidogyne incognita* on cotton. *Nematologica* 34(4) : 432-442.

113. Sankaranarayanan, C., Sundarababu, R. 1997b. Effect of oil cakes and nematicides on the growth of blackgram (*Vigna mungo*) inoculated with VAM fungus (*Glomus fasciculatum*) and root knot nematode (*Meloidogyne incognita*). *Ind. J. of Nematol*. 27, 128–130.

114. Schönbeck, F. 1979. Endomycorrhiza in relation to plant diseases. In, "*Soil-borne Plant Pathogens*" (eds. Schipper, B. and Gams, W.) Academic Press, New York, U.S.A., pp. 271-280

115. Schubler, A., Schwarzott, D., Walker, C. 2001. A new fungal phylum, the Glomeromycota, phylogeny and evolution. *Mycol. Res.* 105: 1413–1421.

116. Schübler, A. 2002. Molecular phylogeny, taxonomy, and evolution of Geosiphon pyriformis and arbuscular mycorrhizal fungi. *Plant and soil 244*, 75-83.

117. Secilia, J., Bagyaraj, D.J. 1987. Bacteria and actinomycetes associated with pot cultures of vesicular-arbuscular mycorrhizas. *Canadian Journal of Microbiology* 33 : 1069-1073.

118. Sharma, M. P., Bhargava, S. 1993. Potential of VAM fungus *G. fasciculatum* against the root–knot nematode *Meloidogyne incognita* on tomato. *Mycorrhiza News* 4(4) : 4-5.

119. Sharma, M.P., Bhargava, S., Verma, M.K., Adholeya, A. 1994. Interaction between the endomycorrhizal fungus *Glomus fasciculatum* and the root-knot nematode, *Meloidogyne incognita* on tomato. *Ind. J. Nematol*. 24(2) : 34-39.

120. Sharma, MP, Gaur, A., Tanu, Sharma, O.P. 2004. Prospects of arbuscular mycorrhiza in sustainable management of root and soil borne diseases of vegetable crops. In "*Disease management of fruits and vegetables*" vol. 1: Fruit and vegetable diseases editor K.G Mukerji. Kluwer Academic Publishers, The Netherlands 501-539.

121. Sharma, M.P. Sharma, S.K. 2006. Arbuscular Mycorrhizal Fungi: *An emerging Bioinoculant for Production of Soybean.* SOPA Digest, Volume III, Issue IX pages10-16.

122. Sharma, W., Trivedi, P.C. 1997. Concomitant effect of *Paecilomyces lilacinus* and vesicular arbuscular mycorrhizal fungi on root-knot nematode infested okra. *Annals of Plant Protection Science* 5, 70–74.

123. Shivaprasad, P., Jacob, A., Nair, S.K., George, B. 1990. Influence of VA mycorrhizal colonization on root-knot nematode infestation in Piper nigrum L. In, "*Current Trends in Mycorrhizal Research: Proceedings of the National Conference on Mycorrhiza*", (eds. Jalali B L. and Chand, H.) Haryana Agricultural University, Hisar, India, 14-16 February, 1990, pp. 210

124. Siddiqui, Z. A., Mahmood, I. 1995. Some observations on the management of the wilt disease complex of pigeonpea by treatment with a vesicular arbuscular fungus and biocontrol agents for nematodes. *Bioresource Technology* 54, 227–230

125. Siddiqui, Z.A., Mahmood, I. 1998. Effect of a plant growth promoting bacterium, an AM fungus and soil types on the morphometrics and reproduction of *Meloidogyne javanica* on tomato. *Applied Soil Ecology* 8, 77–84

126. Siddiqui, Z. A., Mahmood, I., Hayat, S. 1998. Biological control of *Heterodera cajani* and *Fusarium udum* on pigeonpea using *Glomus mosseae, Paecilomyces lilacinus* and *Pseudomonas fluorescens. Thai Journal of Agricultural Science* 31, 310–321.

127. Sikora, R.A. 1992. Management of the antagonistic poten-tial in agricultural ecosystem for the biological control of plant parasitic nematode. *Ann. Rev. Phytopathol.* 32:245-270.

128. Simmons, C.R. 1994. The physiology and molecular biology of plant 1,3-ß-D-glucanases and 1,3;1,4-ß-D-glucanases. *Critical Reviews in Plant Sciences* 13, 325–387

129. Singh, C.S. 1992. Mass inoculum production of vesicular-arbuscular (VA) mycorrhizae. 2. Impact of N2-fixing and Psolubilizing bacterial inoculation on VA-mycorrhiza. *Zentralblatt Four Mikrobiologie* 147, 503–508.

130. Singh, Y.P., Singh, R.S., Sitaramaiah, K. 1990. Mechanism of resistance of mycorrhizal tomato against root-knot

nematode. 96-97. In, "*Proceedings of the National Conference on Mycorrhiza*", Haryana Agricultural University, Hisar, India (eds. Jalali, B.L. and Chand, H.) 14-16 February 1990, pp. 210

131. Sitaramaiah, K, Sikora, R. A. 1982. Effect of mycorrhizal fungus *Glomus fasciculatus* on the host-parasitic relationship of *Rotylenchus reniformis* in tomato. *Nematologica* 28 : 412-419.

132. Smith, G.S., Hussey, R. S., Roncadori, R.W. 1986. Penetration and post-infection development of *Meloidogyne incognita* as affected by *Glomus intraradices* and phosphorus. *Journal of Nematology* 18 : 429-435.

133. Smith, S.E., Gianinazzi-Pearson, V. 1988. Physiological interactions between symbionts in vesicular-arbuscular mycorrhizal plants. *Annual Review of Plant Physiology and Molecular Biology,* 39 : 221-244.

134. Smith, S.E., Read, D.J.1997. *Mycorrhizal symbiosis*, 2nd Edition. Academic Press, London.

135. Spanu, P., Bonfante-Fasolo, P. 1988. Cell wall-bound peroxidase activity in roots of mycorrhizal Allium porrum. *New Phytologist* 109, 119–124.

136. Sreenivasa, M.N., Bagyaraj, D.J. 1988. Selection of a suitable host for mass multiplication of *Glomus fasciculatum. Plant and Soil* 106: 289-290.

137. Sundarababu, R., Mani, M. P., Arulraj, P. 2001. Management of *Meloidogyne incognita* in Chilli nursery with *Glomus mosseae. Annals of Plant Protection Science* 9, 117–170.

138. Suresh, C.K., Bagyaraj, D. J. 1984. Interaction between a vesicular arbuscular mycorrhiza and a root knot nematode arbuscular mycorrhiza and a root knot nematode and its effect on growth and chemical composition of tomato. *Nematolgia Mediterranea* 12: 31-39.

139. Suresh, C.K., Bagyaraj, D.J., Reddy, D.D.R. 1985. Effect of vesicular-arbuscular mycorrhiza for survival, penetration and development of root-knot nematode in tomato. *Plant and Soil* 87 : 305-308.

140. Sylvia, D.M., Hubbell, D.H. 1986. Growth and sporulation of vesicular-arbuscular mycorrhizal fungi in aeroponic and membrane systems. *Symbiosis* 1 : 259-267.

141. Talavera, M., Itou, K., Mizukubo, T. 2001. Reduction of nematode damage by root colonization with arbuscular mycorrhiza (*Glomus* spp) in tomato - *Meloidogyne incog-*

*nita* and carrot-*Pratylenchus penetrans* pathosystems. *Appl. Entomol. Zool.* 36:387-392.

142. Talavera, M., Itou, M., Mizukubo, T. 2002. Combined application of *Glomus* sp. and *Pasturia penetrans* for reducing *Meloidogyne incognita* populations and improving tomato growth. *Appl. Entomol Zool.* 37 :61-67.

143. Tiwari, P., Adholeya, A., Prakash, A. 2003. Commercialization of arbuscular mycorrhizal biofertilizer. In: *Fungal Biotechnology in Agricultural, Food, and Environmental Applications,* DK Arora (ed.), Marcel Dekker Inc., NY, USA, 195-201.

144. Todd, T.C., Winkler, H.E., Wilson, G.W.T. 2001. Interaction of *Heterodera glycines* and *Glomus mosseae* on soybean. *Supplement to the Journal of Nematology* 33, 306–310.

145. Torelli, A., Trotta, A., Acerbi, L., Arcidiacono, G., Berta, G., Branca, C. 2000. IAA and ZR content in leek (*Allium porrum* L.), as influenced by P nutrition and arbuscular mycorrhizae, in relation to plant development. *Plant Soil* 226: 29–35.

146. Tylka, G.L., Hussey, R.S., Roncadori, R.W. 1991. Interactions of vesicular-arbuscular mycorrhizal fungi, phosphorus, and *Heterodera glycines* on soybean. *Journal of Nematology* 23, 122–133.

147. Vaast, Ph., Caswell-Chen, E. P., Zasoski, R.J. 1998. Influence of a root-lesion nematode, Pratylenchus coffeae, and two arbuscular mycorrhizal fungi, *Acaulospora mellea* and *Glomus clarum* on coffee (*Coffea arabica* L.). *Biology and Fertility of Soils* 26, 130–135.

148. Villenave, C., Leye, K., Chotte, J.L., Duponnois, R. 2003. Nematofauna associated with exotic and native leguminous plant species in West Africa: effect of Glomus intraradices arbuscular mycorrhizal symbiosis. *Biol. Fertil. Soils* 38, 161–169.

149. Waceke, J. W., Waudo, S.W., Sikora, R. 2001. Suppression of *Meloidogyne hapla* by arbuscular mycorrhizal fungi (AMF) on pyrethrum in Kenya, *International Journal of Pest Management* **47**: 135–140.

150. Waceke, J. W., Waudo, S. W., Sikora, R. 2002. Effects of organic phosphatic fertilizers on the efficacy of an arbuscular mycorrhizal fungus against root-knot nematode on pyrethrum. *International Journal of Pest Management* 48, 307–313.

151. Wallace, H.R. 1983. Interactions between nematodes and other factors on plants. *J. Nematol.* 15: 221-227.

152. Whipps, J.M. 2004. Prospects and limitations for mycorrhizas in biocontrol of root pathogens. *Canadian J. Bot.* 82:1198-1227

153. Wu, C.G., Liu, Y.S., Hung, L.L. 1995. Spore development of *Entrophospora kentinensis* in an aeroponic system. *Mycologia* 87: 582-587.

154. Wyss, P., Boller, T., Wiemken, A. 1991. Phytoalexin response is elicited by a pathogen (*Rhizoctonia solani*) but not by a mycorrhizal fungus (*Glomus mosseae*) in soybean roots. *Experientia* 47:395–399.

155. Yao, M.K., Tweddell, R.J., Desilets, H. 2002. Effect of two vesicular-arbuscular mycorrhizal fungi on the growth of micropropagated potato plantlets and on the extent of disease caused by *Rhizoctonia solani*. *Mycorrhiza* 12(5): 235-242.

# हिन्द पॉकेट बुक्स

## एक रुका हुआ फैसला

मूलत: गोपालगंज(बिहार) के रहने वाले प्रभाकर कुमार मिश्र एक टीवी पत्रकार हैं और करीब पंद्रह सालों से सुप्रीम कोर्ट की ख़बरों की रिपोर्टिंग करते आ रहे हैं। इस दौरान वे अयोध्या विवाद पर आए फैसले समेत सुप्रीम कोर्ट के कई अन्य महत्वपूर्ण फैसलों के गवाह रहे हैं। वे नियमित टीवी रिपोर्टिंग के अलावा अलग-अलग अख़बारों और पत्रिकाओं में कानूनी और संवैधानिक मुद्दों पर लिखते रहे हैं। वे दिल्ली विश्वविद्यालय, दक्षिण परिसर के पत्रकारिता विभाग में अतिथि अध्यापक के तौर पर भी जुड़े हुए हैं।

प्रभाकर कुमार मिश्र मगध विश्वविद्यालय से कानून में स्नातक हैं और इलाहाबाद विश्वविद्यालय से राजनीति शास्त्र में परास्नातक हैं। उन्होंने दिल्ली विश्वविद्यालय से पत्रकारिता की पढ़ाई की है। संप्रति वे न्यूज़ *24* में असिस्टेंट एडिटर के पद पर कार्यरत हैं।